MODERN CEREAL SCIENCE AND TECHNOLOGY

MODERN CEREAL SCIENCE AND TECHNOLOGY

Y. Pomeranz

VCH
Publishers

Y. Pomeranz
Department of Food Science
   and Human Nutrition
Washington State University
Pullman, WA 99164-2032

Library of Congress Cataloging-in-Publication Data

Pomeranz, Y. (Yeshajahu), 1922-
   Modern cereal science and technology.

   Includes index.
   1. Grain. 2. Cereal products. 3. Bread.
I. Title.
SB189.P65   1987   633.1   87-13364
ISBN 0-89573-326-9

Printed in the United States of America.

✓ISBN 0-89573-326-9 VCH Publishers
ISBN 3-527-26512-0 VCH Verlagsgesellschaft

Distributed in North America by:

VCH Publishers, Inc.
220 East 23rd Street, Suite 909
New York, New York 10010-4606

Distributed Worldwide by:

VCH Verlagsgesellschaft mbH
P.O. Box 1260/1280
D-6940 Weinheim
Federal Republic of Germany

Preface

Writing a book about modern cereal science, when one has been professionally active in the field for over 40 years, should be a fairly simple task of summarizing what one has contributed to the field and what one has learned about it from others. Other books, reviews, scientific papers, notes, and references on the subject are all available. All that seemingly needs to be done is to select the most important information, put it in the correct order, and discuss it so that it is in the proper perspective.

Unfortunately the process is not quite that easy. Writing this book was an arduous task in decision making. It involved deciding how much background and detailed information to include or exclude, and how to organize it so that the book would be useful to students, researchers, novices, and experts in the areas of grain production, processing, marketing, agronomy, food sciences, nutrition, cereal chemistry, and technology. Decisions had to be made as to how big and costly to make the book, and how many diagrams, figures, tables, and references to include. Last but not least, it had to be determined how much of my own work (with which I am obviously most familiar and know first hand about its significance, potential, and limitations) and the work of others (including those who may have disagreed with my own work) to include in the volume.

I have written, rewritten, revised, and edited this book many times in order to do justice to all of the above-mentioned challenges and requirements. It is my hope that most of the decisions were correct. It is my further hope that the reader will let me know, after reading and perusing the book, to what extent the need to produce a book for the senior undergraduate and graduate student, worker, and researcher has been met and where deletions, changes, and additions are needed in future editions.

Pullman, WA   Y. Pomeranz

Contents

Chapter 1

Introduction

INTRODUCTION

Cereals supply the most calories per acre, can be stored safely for a long time, and can be processed into many products acceptable throughout the world. They are adaptable in a variety of soil and climatic conditions, can be cultivated both on a large-scale mechanically or on a small garden scale. They have a high food value in a multitude of products in which they are excellent sources of energy with relatively good sources of inexpensive protein, minerals, and vitamins. More than two-thirds of the world's cultivated area is planted with grain crops. Most developing countries rely on cereals as their major food sources, providing more than half the calories consumed for human energy and well over two-thirds of their total food. Cereals are major sources of the proteins which contribute to the production of animal feeds.

GENERAL CLASSIFICATION OF CEREAL GRAINS

The term cereal encompasses members of the grass family, Gramineae, that are grown for their edible grains and supply the basic nutritional needs of much of mankind. Members of the family Gramineae comprise the species listed in Table 1-1, the most important being wheat, rice, maize (corn), barley, oats, grain sorghum, and rye. The cereals were grown by primitive peoples before the recorded history of man. The wild forms from which they evolved, therefore, are unknown.

WORLD CEREAL PRODUCTION AND TRADE

Average world cereal production in the years 1979–1981 is summarized in Table 1-2. World production and grain production by country in 1983/84

Table 1-1. Botanical Classification of Common Grains (Family Gramineae)

Tribe	Genus	Species	Subspecies	Common name
Andropogoneae	*Andropogon*	*sorghum*	*vulgare*	sorghum grain
	Andropogon	*sorghum*	*bicolor*	sorghum grain
Aveneae	*Avena*	*sativa* L.		white and yellow oats
	Avena byzantina			red oats
Hordeae	*Hordeum distichon* L.			two-rowed barley
	Hordeum vulgare L.			six-rowed barley
	Secale cereale L.			rye
	Triticum sativum		*compactum* Host.	club wheat
	Triticum sativum		*vulgare* Vill.	common wheat
	Triticum sativum		*durum* Desf.	durum wheat
	Triticosecale Witt.			triticale
Maydeae	*Zea mays*		*indentata* Sturt.	dent corn
	Zea mays		*indurata* Sturt.	flint corn
	Zea mays		*everta* Sturt.	pop corn
	Zea mays		*saccharata*	sweet corn
Oryzeae	*Oryza sativa* L.		*indica*	rice
	Oryza sativa L.		*japonica*	rice
	Oryza sativa L.		*javanica*	rice
	Oryza glabberrima			African rice
Paniceae	*Panicum miliaceum*			common millet
	Echinocloa spp.			Japanese millet
Zinanieae	*Zinania aquatica*			wild rice

are shown in Figures 1-1 and 1-2. Figure 1-3 shows export and imports of wheat and coarse grains between 1970/71 and 1984/85. Whole grain weights and measures and conversion factors are listed in Table 1-3.

Cereals are the most important staple food of the world. Only cereals have the potential to support the world human population with its needed amounts of calories (Table 1-4). Cereals also provide proteins, even though of low biological value. An improvement in the biological value of cereal proteins through conventional plant breeding or genetic engineering could make them a single satisfactory source of proteins from a nutritional standpoint. It is estimated that almost 10% (about 500 million) of our population is hungry and that an even greater percentage is undernourished. This is a result, however, of socioeconomic and political conditions combined with skewed resources in productive land versus birth rate, rather than of global shortage. Whereas annual per capita grain consumption (direct as food and/or indirectly as feed) averages about 300 kg in some countries in the Far East and Africa, the consumption is up to 1,500 kg in Western Europe and in North America. On a worldwide basis, however, world cereal production, mainly because of increased yields, has been able to more than keep up with the world population. During the past two

Table 1-2. World Grain: Area, Yield, Total Production, and Protein Production (Average of 1979–1981)[a]

Grain	Area (million ha)	Area (% of total)	Yield (tons/ha)	Production (million ton)	Production (% of total)	Main producer (Country)	Main producer (million ton)	Protein[b] (million ton)	Protein[b] (% of total)
Wheat	233.3	34.5	1.88	438.2	28.3	USSR	92.1	54	33
Corn	129.0	19.1	3.26	420.8	27.2	USA	193.0	40	25
Rice	143.9	21.3	2.76	394.2	25.5	China	142.4	18	18
Barley	82.6	12.2	1.95	161.3	10.4	USSR	44.4	12	12
Sorghum[c]	45.0	6.6	1.45	65.3	4.2	USA	18.0	9	6
Oats	26.9	4.0	1.68	45.1	2.9	USSR	15.3	6	4
Rye	14.3	2.1	1.66	23.8	1.5	USSR	9.3	3	2
Total[d]	675.0	100.0		1,548.7	100.0		300.8	162	100.0

[a]USDA Agricultural Statistics, Washington, D.C., 1982.
[b]Estimate; calculated data, average protein content.
[c]Does not include millet, mixed grains, and triticale.
[d]Includes millet.

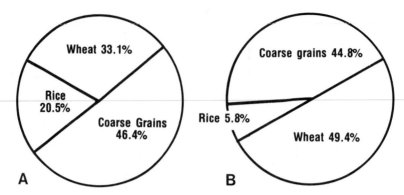

Figure 1-1. (A) World production of grains, 1983–84. (B) World trade by crop, 1983–84. (Source: Foreign Agricultural Service (FAS), USDA, Washington, D.C.)

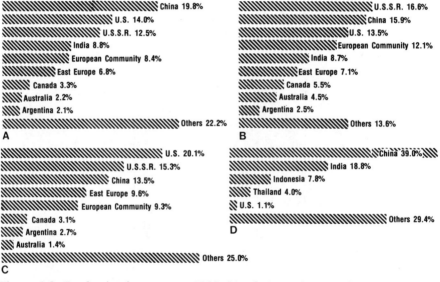

Figure 1-2. Production by country, 1983–84, of (A) grain, (B) wheat, (C) coarse grain, and (D) rice. (Source: FAS-USDA.)

decades total production of cereal grains and production of the leading cereals has increased by about one-third (31% to 43%). At the same time, however, the gap between the surplus and deficit regions has widened (Table 1-5).

According to the International Wheat Council, the 1985 World Wheat Statistics show the following for 1984 (in millions of tons): PRODUCTION. China 118.7; European Community 102.9; USSR 101.0; USA 95.2; India 61.0; Canada 28.6; Australia 25.1; Turkey 23.2; and Argentina 17.8. EXPORTS. USA 48.9; Canada 27.7; Australia 15.5; European Economic Community 15.0; Argentina 12.9. IMPORTS. USSR 27.4; China 13.1; Japan 8.0; Egypt 6.4; Brazil 5.8.

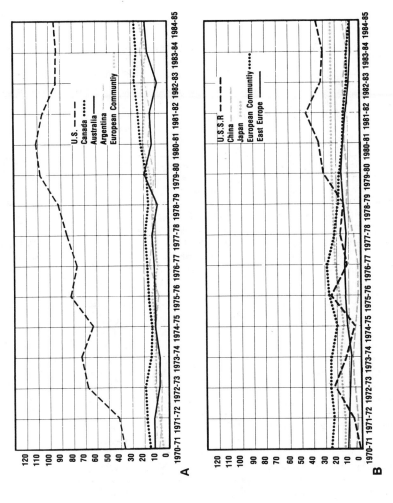

Figure 1-3. (A) Exports and (B) imports of wheat and coarse grains, by country (millions of tons: Source: FAS-USDA).

Table 1-3. International Weights and Measures

1 hectare (ha)	= 2.47109 acres
1 acre	= 0.40469 ha
1 metric ton (t)	= 1,000 kg
	= 0.98421 large ton
	= 1.10231 short ton
	= 2,204.62 lb
1 long ton	= 2,240 lb
	= 1.01605 t
1 short ton	= 2,000 lb
	= 0.90718 t
1 hundredweight (ctw)	= 112 lbs = 50.802 kg (UK)
1 ctw	= 100 lbs = 45.359 kg (USA, Canada)
1 hectoliter (hl)	= 100 liters
1 Winchester bushel (bu)	= 35.23812 liters (USA)
	= 36.34866 liters (UK, Canada)
1 quarter (qr)	= 8 bushels (bu)

1 bushel =	grain	1 metric ton =
60 lb = 27.216 kg	wheat, soybean	36.743 bu
56 lb = 25.402 kg	rye, corn, milo	39.367 bu
50 lb = 22.680 kg	barley (UK, Australia)	44.092 bu
48 lb = 21.773 kg	barley (USA, Canada)	45.929 bu
40 lb = 18.144 kg	oats (UK, Australia)	55.115 bu
34 lb = 15.422 kg	oats (Canada)	64.841 bu
32 lb = 14.515 kg	oats (USA)	68.894 bu

Conversion from bushels/acre to tons/ha

Wheat, soybeans	No. bu × 0.0672
Corn, milo, rye	No. bu × 0.0627
Barley (UK, Australia)	No. bu × 0.0560
Barley (USA, Canada)	No. bu × 0.0538
Oats (UK, Australia)	No. bu × 0.0448
Oats (Canada)	No. bu × 0.0381
Oats (USA)	No. bu × 0.0359

Conversion from lb/bu to kg/hl

Winchester bushel (USA)	No. bu × 1.2872
Imperial bushel (UK, Canada, Australia)	No. bu × 1.2479

Some general approximate figures

100 kg wheat	= 78 kg flour
	= 22 kg feed products
	= 104 kg baked goods
100 kg soybean	= 18 kg oil
	= 80 kg meal
100 kg malting barley	= 100 kg green malt
	= 80 kg kilned malt
	= 96 kg spent grains (75–80% water)
	= 4 hl beer

Table 1-4. World Human Population in Relation to Its Cereal Production

Requirement per capita:	
Energy; (2,500 kcal/day)	900,000 kcal/yr
Protein (70 g/day)	25 kg/yr
Requirement of human world population (4,300,000,000):	
Energy	3,870,000,000,000,000 kcal/yr
Protein	108,000,000 ton/yr
(Cereals contain approx 3,000 kcal and 100 g protein/kg)	
Energy in weight of cereals	1,290,000,000 ton/yr
Protein in weight of cereals	1,080,000,000 ton/yr
World production of cereals	1,500,000,000 ton/yr
Other harvest products suitable for food	1,800,000,000 ton/yr

Source: MacKey, 1981.

Table 1-5. The Changing Pattern of World Grain Trade (in Millions of Metric Tons)

Region	1934–38	1948–52	1960	1970	1978
North America	+ 5	+23	+39	+56	+104
Australia and New Zealand	+ 3	+ 3	+ 6	+12	+ 14
Latin America	+ 9	+ 1	± 0	+ 4	± 0
Western Europe	−24	−22	−25	−30	− 21
Africa	+ 1	± 0	− 2	− 5	− 12
Eastern Europe and USSR	+ 5	± 0	± 0	± 0	− 27
Asia	+ 2	− 6	−17	−37	− 53

Plus sign indicates net exports; minus sign, net imports.
Source: Brown, 1980.

ORIGIN AND TYPES OF CEREAL GRAINS

The common ancestor of all wheats is believed by most researchers to be a species called wild einkorn, which crossbred with an unknown grass to produce the seven-chromosome wheat einkorn (single seed). Kernels of wild and cultivated einkorn were found in ruins of a 6700 BC village Jermo, situated in the upper Tigris-Euphrates basin in southwest Asia, the Fertile Crescent. Some authorities classify 14 species of wheat according to the number of chromosomes in which are found genes that govern plant characteristics. The species *Triticum aegilopoides* (wild einkorn) and *T. monococcum* have seven chromosomes; the species *T. dicoccoides* (wild emmer), *T. dicoccum* (emmer), *T. durum* (durum wheat), *T. persicum* (Persian wheat), *T. turgidum* (rivet wheat), *T. polonicum* (Polish wheat), and *T. timophevi* have 14 chromosomes, and *T. aestivum* (common wheat), *T. sphaerococcum* (shot wheat), *T. compactum* (club wheat), *T. spelta* (spelt), and *T. macha* (macha wheat) have 21 chromosomes. Bread wheats spread over Europe from southern Russia about 2000 BC and were brought to North America by early explorers. Of some 600 genera of the family Gramineae, the genus *Triticum* has about 5,000 species. Only three species (*T. aestivum*,

T. durum, and *T. compactum*) are grown to a significant extent today; of these *T. aestivum* is over 90% of the cultivated wheat.

Much research has been and continues to be conducted to improve new wheat varieties that have better resistance to insects and diseases, shorter straw to reduce damage to wind and heavy rains and hail, a shorter growing season to reduce frost and sprouting damage and fit into a multicropping system, better milling, baking, or other end-use properties, and, foremost, higher yields. High-yielding dwarf or semidwarf wheats have resulted from crosses with Japanese wheats, judicious selections and crosses have increased protein content of the grain without adversely affecting the biological value of the proteins and the total yield, and much promising research is under way to increase the yield through production of hybrid wheat (similar to hybrid corn and sorghum). New approaches are being explored to improve resistance to drought and diseases and overall quality through genetic engineering—ie, recombinant DNA.

Rice probably originated in southern Asia. Rice production in China dates back to about 2000 BC and in India to 1000 BC. It was cultivated in the Euphrates valley in 400 BC and was spread throughout southern Europe in medieval times by the Saracens. Early colonists experimented with rice in Virginia as early as 1609. Commercial U.S. rice production began in the late 17th century, and the grain is now a major crop.

Corn, or maize, originated in the western hemisphere. To the best of our knowledge, it was the only cereal cultivated by the American Indians, although some other grains were harvested in the wild state.

Among cultivated crops, barley occupies a less prominent role than wheat, rice, or corn. Barley is a highly adaptable cereal grain; it grows well at high and low altitudes, in semiarid and moist, and in hot and cold climates. It shows relatively high resistance to soil salinity, and in many parts of the world is grown because of its relatively short growing period, making it attractive for inclusion in multicrop cycles or in areas where there is danger of early frost. Ancient records show that cultivated barley was used by Neolithic cultures in Egypt between 5000 and 6000 BC. Major gene centers, where cultivated barley may have developed, include Ethiopia and the highlands of Sikkim and southern Tibet. Most cultivated barleys are covered. However, in some Far Eastern countries and in some mountainous areas in Africa and Asia where barley is used for human food, naked (hull-less) barleys are found and grown.

Oats is not considered as old a crop as wheat or barley. Oats apparently was not known or cultivated by the ancient Chinese, Hebrews, or Hindus. In the Franchtin cave in Greece, however, deposits from 10,500 BC contain two grains of *Avena* species. The common oat (*A. sativa*) originated in Asia Minor. Oats were cultivated in China as early as 400–500 AD. The grain was brought to the western Hemisphere in the 16th century. Among the many oat species, the hexaploids ($2n = 6\times = 42$) are the most important. *A. sativa* is the most important cultivated oat and is particularly adapted to moderate climates. *A. byzantina* is grown as a winter oat in the southern United States. *A. nuda* is a naked form in which the hulls (lemma and palea) are loosely attached to the kernel and are removed during threshing. *A. fatua* is a wild species that causes as a weed billions of dollars in crop losses.

Sorghum and millet are major food crops in Asia and Africa, where they are grown in semiarid areas, usually as dry land crops. They are considered a poor man's crop. Relatively little millet is grown in the western hemisphere. Sorghum culture probably originated in Ethiopia or the Sudan about 6,000 years ago. The earliest record of sorghum is in a carving in the plains of Sennacherib at Ninveh, Assyria, dated back about 700 BC. The grain reached Europe and India at the beginning of the Christian era and southern Asia by the 13th century. Benjamin Franklin is believed to have introduced it to North America in the 18th century.

In many countries, rye is cultivated for the production of rye or mixed wheat-rye breads. Some rye is used to produce specialty breads. Cultivated rye seems to have originated from rye grown as a weed in crops of barley and wheat. Alternatively, cultivated rye could have originated from perennial wild ryes, forerunners of annual wild ryes. Rye entered Europe from Transcaucasia and/or Turkestan/Afghanistan. Rye was introduced to the United States during colonial times by English and Dutch settlers. Triticale is a man-made cereal species synthesized by combining the genomes of wheat (*Triticum* sp.) and rye (*Secale* sp.). The common name "triticale" was coined as a contraction of the two generic names. The objectives in synthesizing triticale were to combine the highly desirable winter hardiness and vigor of rye with the agronomic and quality characteristics of wheat. The first report and description of a wheat-rye hybrid dates back to 1875, and serious efforts to develop triticale as a commercial crop were started about 50 years ago. Early programs involved development of octoploid triticales ($2n = 8\times = 56$) from hexaploid ($2n = 6\times = 42$) wheats and diploid ($2n = 2\times = 14$) ryes. Subsequent studies have demonstrated the agronomic advantages of hexaploid triticales by combining tetraploid wheat (containing A and B genomes) with rye (containing R genome) and lacking the D genome contributed by hexaploid wheat (genomes ABD).

BIBLIOGRAPHY

I. *General*

Adrian, W. "And Thus Bread Was Made from Stock and Blood." Ceres-Verlag; Bielefeld, Germany; in German; **1959**.

Anon. "From Wheat to Flour." Wheat Flour Institute; Chicago; **1965**.

Anon. "Grain Sorghum Research in Texas—1970." Texas Agr. Exp. Sta. Consolidated Progr. Reps.; **1971**; pp. 2938–2949.

Araullo, E. V.; De Padua, D. B.; Graham, M. (Eds.). "Rice—Post Harvest Technology." International Development Centre; Ottawa, publication No. IDRC-053e; **1976**.

Akroyd, W. R.; Doughty, J. "Wheat in Human Nutrition." FAO Nutritional Studies No. 23; FAO; Rome; **1970**.

Bailey, C. H. "The Constituents of Wheat and Wheat Products." Reinhold; New York; **1944**.

Barber, S.; Mitsuda, H.; Desikachar, H. S. R. (Eds.). "Rice Report." Instituto de Agroquimica y Tecnologia de Alimentos: Valencia, Spain; **1975**.

Bennion, E. B. "Breadmaking: Its Principles and Practice," 4th ed. Oxford University Press; London; **1967**.

Bohn, R. M. "Biscuit and Cracker Production." American Trade Publishing Co.; New York; **1957**.

Brown, L. R.; Eckholm, E. P. "By Bread Alone." Pergamon Press; Oxford; **1975**.

Bushuk, W. (Ed.). "Rye: Production, Chemistry, and Technology." Am. Assoc. Cereal Chem.: St. Paul, MN; **1976**.

Canadian International Grains Institute. "Grains and Oilseeds, Handling, Marketing, Processing," 2nd ed. Can. Int. Grains Inst.: Winnipeg, Canada; **1975**.

Christensen, C. M. (Ed.). "Storage of Cereal Grains and Their Products." Am. Assoc. Cereal Chem.: St. Paul, MN; **1974**.

Christensen, C. M.; Kaufmann, H. H. "Grain Storage—The Role of Storage in Quality Loss." University of Minnesota Press; Minneapolis; **1969**.

Cook, A. H. "Barley and Malt, Biology, Biochemistry, and Technology." Academic Press; New York; **1962**.

Daniels, R. "Modern Breakfast Cereal Processes." Noyes Data Corp.: Park Ridge, NJ; **1970**.

Daniels, R. "Breakfast Cereal Technology." Noyes Data Corp.; Park Ridge, NJ; **1974**.

Dunlap, F. L. "White Versus Brown Flour." Wallace and Tiernan Co.; Newark, NJ; **1945**.

Fance, W. J. Breadmaking and Flour Confectionary, 3rd ed. Avi Publishing Co.; Westport, CT; **1969**.

Feistritzer, W. P. (Ed.). Cereal Seed Technology. FAO: Rome; **1975**.

Findley, W. P. K. (Ed.). "Modern Brewing Technology." Macmillan: London; **1971**.

De Renzo, D. J. "Bakery Products—Yeast Leavened." Food Technology Review No. 200; Noyes Data Corp.; Park Ridge, NJ; **1975**.

De Renzo, D. J. "Doughs and Baked Goods; Chemical, Air and Nonleavened." Noyes Data Corp.; Park Ridge, NJ; **1975**.

Doggett, H. "Sorghum." Longmans, Green and Co.; London; **1970**.

Horder, T. J.; Dodds, C.; Moran, T. "Bread: The Chemistry and Nutrition of Flour and Bread." Constable; London; **1954**.

Hough, J. S.; Briggs, D. E.; Stevens, R. "Malting and Brewing Science." Chapman & Hall; London; **1971**.

Houston, D. F. (Ed.). "Rice: Chemistry and Technology." Am. Assoc. Cereal Chem.; St. Paul, MN; **1972**.

Houston, D. F.; Kohler, G. O. "Nutritional Properties of Rice." Food and Nutrition Board, National Research Council; Washington, D.C.; **1970**.

Huelsen, W. A. "Sweet Corn." Interscience Publishers; New York; **1954**.

Hummel, C. "Macaroni Products—Manufacture, Processing, and Packaging," 2nd ed. Food Trade Press; London; **1966**.

Inglett, G. E. (Ed.). "Corn: Culture, Processing, Products." Avi Publishing Co.; Westport, CT; **1970**.

Inglett, G. E. (Ed.). "Wheat: Production and Utilization." Avi Publishing Co.; Westport, CT; **1974**.

Jacob, H. E. "Six Thousand Years of Bread, Its Holy and Unholy History." Doubleday Co.; New York; **1944**.

Jugenheimer, R. W. "Corn: Improvement, Seed Production, and Uses." John Wiley & Sons; New York; **1976**.

Kent, N. L. "Technology of Cereals with Special Reference to Wheat." Pergamon Press; Oxford; **1966**.

Kent-Jones, D. W.; Amos, A. J. "Modern Cereal Chemistry," 6th ed. Food Trade Press; London; **1967**.

Kent-Jones, D. W.; Mitchell, S. "The Practice and Science of Breadmaking," 3rd ed. Northern Publ. Co.; Liverpool; **1962**.

Kretovich, V. L. (Ed.). "Biochemistry of Grain and of Breadmaking." (Transl. from Russian.) Israel Program for Scientific Translations; Jerusalem; **1965**.

Kuprits, Y. N. "Technology of Grain Processing and Provender Milling." Translated from Russian (1967). U.S. Department of Commerce (TT 67-51273); Springfield, VA; **1965**.

Lockwood, J. F. "Flour Milling," 4th ed. Northern Publishing Co.; Liverpool; **1960**.

Luh, B. S. (Ed.). "Rice: Production and Utilization." Avi Publishing Co.; Westport, CT; **1980**.

MacIntyre, R.; Campbell, M. "Triticale." International Development Research Center; Ottawa; **1974**.

Matz, S. A. (Ed.). "The Chemistry and Technology of Cereals as Food and Feed." Avi Publishing Co.; Westport, CT; **1959**.

Matz, S. A. "Cookie and Cracker Technology." Avi Publishing Co.; Westport, CT; **1968**.

Matz, S. A. "Cereal Science." Avi Publishing Co.; Westport, CT; **1969**.

Matz, S. A. "Cereal Technology." Avi Publishing Co.; Westport, CT; **1970**.

Matz, S. A. "Bakery Technology and Engineering," 2nd ed. Avi Publishing Co.; Westport, CT; **1972**.

Matz, S. A. "Snack Food Technology." Avi Publishing Co.; Westport, CT; **1976**.

McCance, R. A.; Widdowson, E. M. "Breads, White and Brown; Their Place in Thought and Social History." J.B. Lippincott Co.; Philadelphia; **1956**.

Mertz, E. T. (Ed.). "High-Quality Protein Maize." Dowden, Hutchinson & Ross, Inc.; Stroudsburg, PA; **1975**.

Neuman, M. P.; Pelshenke, P. F. "Bread Grain and Bread." Verlag, Paul Parey; Berlin; in German; **1954**.

Pomeranz, Y. *Adv. Food Res.*, **1968**, *16*, 335–455.

Pomeranz, Y. *CRC Crit. Rev. Food Technol.*, **1970**, *1*, 453–478.

Pomeranz, Y. *CRC Crit. Rev. Food Technol.*, **1971**, *2*, 45–80.

Pomeranz, Y. (Ed.). "Wheat Chemistry and Technology," Monogr. 3, 2nd ed. Am. Assoc. Chem.: St. Paul, MN; **1971**.

Pomeranz, Y. *CRC Crit. Rev. Food Technol.*, **1973**, *4*, 377–395.

Pomeranz, Y. (Ed.). "Industrial Uses of Cereals." Am. Assoc. Cereal Chem.; St. Paul, MN; **1973**.

Pomeranz, Y. (Ed.). "Advances in Cereal Science and Technology," Vols. I–VIII. Am. Assoc. Cereal Chem.; St. Paul, MN; **1976–1986**.

Pomeranz, Y. (Ed.). "Cereals '78: Better Nutrition for the World's Millions." Am. Assoc. Cereal Chem.; St. Paul, MN; **1978**.

Pomeranz, Y.; MacMasters, M. M. In "Kirk-Othmer's Encyclopedia of Chemical Technology," Vol. 21, Standen, A., Ed.; John Wiley & Sons; New York; **1971**.

Pomeranz, Y.; Munck, L. "Cereals: A Renewable Resource, Theory and Practice." Am. Assoc. Cereal Chem.; St. Paul, MN; **1981**.

Pomeranz, Y.; Shellenberger, J. A. "Bread Science and Technology." Avi Publishing Co.; Westport, CT; **1971**.

Pyler, E. J. "Baking Science and Technology," 2 Vols. Siebel Publishing Co.; Chicago; **1976**.

Quisenberry, K. S.; Reitz, L. P. (Eds.). "Wheat and Wheat Improvement." American Society Agronomy, Inc.; Madison, WI; **1967**.

Rohrlich, M.; Bruckner, G. "Cereals," Vol. I. "Cereals and Their Processing." Verlag Paul Parey; Berlin; in German; **1966**.

Scott, J. H. "Flour Milling Processes," 2nd ed. Chapman and Hall; London; **1951**.

Sherman, H. C.; Pearson, C. S. "Modern Bread from the Viewpoint of Nutrition." Macmillan Co.; New York; **1942**.

Smith, L. "Flour Milling Technology," 3rd ed. Northern Publishing Co.; Liverpool, England; **1944**.

Smith, W. H. "Biscuits, Crackers, and Cookies," 2 Vols. Applied Science Publ.; Barking, Essex, England; **1972**.

Spicer, A. (Ed.). "Bread: Social, Nutritional and Agriculture Aspects of Wheat Bread." Applied Science Publ. Ltd.; London; **1975**.

Storck, J.; Teague, W. D. "A History of Milling. Flour for Man's Bread." University of Minnesota Press; Minneapolis; **1952**.

Sultan, W. J. "Practical Baking." Avi Publishing Co.; Westport, CT; **1965**.

Swanson, C. O. "Wheat and Flour Quality." Burgess Publishing Co.; Minneapolis; **1938**.

Tsen, C. C. (Ed.). "Triticale; First Man-Made Cereal." Am. Assoc. Cereal Chem., St. Paul, MN; **1974**.

Wall, J. S.; Ross, W. M. (Eds.). "Sorghum Production and Utilization." Avi Publishing Co.; Westport, CT; **1970**.

Wiebe, G. A. (Ed.). "Barley: Origin, Botany, Culture, Winterhardiness, Genetics, Utilization, Pests." Agr. Handbook No. 338, ARS-U.S.D.A., Washington, D.C.; **1968**.

Yamazaki, W. T.; Greenwood, C. T. (Eds .). "Soft Wheat: Production, Breeding, Milling, and Uses." Am. Assoc. Cereal Chem., St. Paul, MN; **1981**.

II. Selected Topics—Recent Books

Allen, J. C.; Hamilton, R. J. (Eds.). "Rancidity in Foods." Elsevier Publishing Co.; New York; **1984**.

Barnes, P. J. (Ed.). "Lipids in Cereal Technology." Academic Press, New York; **1983**.

Barrows, A. B. "Bakery Specialties." Elsevier Applied Science Publ.; London, New York; **1984**.

Bodwell, C. E.; Petit, L. (Eds.). "Plant Proteins for Human Food." Martinus Nijhoff, The Hague, The Netherlands; **1983**.

Bowmans, G. "Grain Handling and Storage." Elsevier Science Publ., Amsterdam, The Netherlands; **1985**.

Cherry, J. P. (Ed.). "Food Protein Deterioration: Mechanism and Functionality." Am. Chem. Society, Washington, DC.; **1982**.

Edwardson, W.; MacCormack, C. W. (Eds.) "Improving Small-Scale Food Industries in Developing Countries." IDRC, Ottawa, Canada; **1984**.

Faridi, H. (Ed.). "Rheology of Wheat Products." Am. Assoc. Cereal Chem., St. Paul, MN; **1986**.

Finley, J. W.; Hopkins, D. T. (Eds.) "Digestibility and Amino Acid Availability in Cereals and Oilseeds." Am. Assoc. Cereal Chem., St. Paul, MN; **1985**.

Fjell, K. M. (Ed.). "Analyses as Practical Tools in the Cereal Field—An ICC Symposium." Norvegian Grain Corp., Oslo; **1985**.

Forsberg, R. A. (Ed.). "Triticale." Am. Soc. Agronomy, Madison, WI; **1985**.

Gallagher, E. J. "Cereal Production." Butterworth Publ., Stoneham, MA; **1984**.

Godon, B.; Loisel, W. (Eds.). "Practical Guide for Analysis in the Cereal Industries." Technique et Documentation Lavoisier, Paris, France (in French); **1984**.

Graveland, A.; Moonen, J. H. E. (Eds.). "Gluten Proteins." TNO Inst. Cereals, Flour, Bread. Wageningen, The Netherlands; **1983**.

Grenby, T. H.; Parker, K. J.; Lindley, M. G. (Eds.). "Developments in Sweeteners." Applied Science Publ., Inc., New York; **1983**.

Gunstone, F. D.; Norris, F. A. (Eds.). "Lipids in Foods: Chemistry, Biochemistry and Technology." Pergamon Press, Inc., Elmsford, N. Y.; **1983**.

Gustafson, J. P. (Ed.). "Manipulation in Plant Improvement." Plenum Press, New York; **1984**.

Hanson, H.; Borlaug, N. E., Anderson, R. G. "Wheat in the Third World." Westview Press, Inc., Boulder, CO; **1982**.

Holas, J.; Kratochvil, J. (Eds.). "Progress in Cereal Chemistry and Technology." Vol. 1 and 2. Elsevier Science Publ., Amsterdam, The Netherlands; **1983**.

Hudson, B. J. F. (Ed.). "Developments in Food Proteins." Vol. 2. Applied Science Publ., London; **1983**.

Hulse, J. H.; Laing, E. M. "Nutritive Value of Triticale Protein." IDRC, Ottawa, Canada; **1974**.

International Atomic Energy Agency. "Cereal Grain Protein Improvement. Proc. Final Res. Coordination Mtg., FAO/IAEA/GSF/SIDA Coordinated Research Programme." UNIPUB, New York; **1984**.

Jowitt, R. (Ed.). "Extrusion Cooking Technology." Elsevier Applied Science Publ., London; **1984**.

Juliano, B. O. (Ed.). "Rice; Chemistry and Technology." 2nd. ed., Am. Assoc. Cereal Chem., St. Paul, MN; **1986**.

Matz, S. A. "Snack-Food Technology." 2nd. ed. Avi Publ. Co., Westport, CT; **1984**.

McKenna, B. M. (Ed.). "Engineering and Food. Vol. I. Engineering Sciences in the Food Industry. Vol. II. Processing Applications." Elsevier Applied Science Publ., London; **1984**.

Meuser, F.; Kulikowski, W. (Eds.). "Carbohydrates, Proteins, Lipids: Basic Views and New Approaches in Food Technology." Techn. Univ. Berlin; **1984**. (Mostly in German).

Mohsenin, N. N. "Electromagnetic Radiation Properties of Foods and Agricultural Products." Gordon and Breach Science Publ., New York; **1984**.

Osborne, B. G. "A Bibliography of Near-Infrared Reflectance Spectroscopy in Food Analysis." Flour Milling and Baking Research Assoc., Chorleywood, Herts. England; **1983**.

Peleg, M.; Bagley, E. B. (Eds.). "Physical Properties of Foods." Avi Publ. Co., Westport, CT; **1983**.

Pomeranz, Y. "Functional Properties of Food Components." Academic Press, Orlando, FL; **1985**.

Prentice, J. H. "Measurements in the Rheology of Foodstuffs." Elsevier Applied Science Publ., London; **1984**.

Rexen, F.; Munck, L. "Cereal Crops for Industrial Use in Europe." Eur. 9617 EN, The Commission of European Communities. Brussels and Copenhagen; **1984**.

Ripp, B. E.; Banks, H. S.; Bond, E. J.; Calverly, D. J.; Jay, E. G.; Navarro, S. (Eds.). "Controlled Atmosphere and Fumigation in Grain Storages." Elsevier Science Publ. Co., New York; **1984**.

Rooney, L. W.; Murty, D. S.; Mertin, J. V. (Eds.). "International Symposium on Sorghum Grain Quality. Proceedings." INCRASAT Patancheru, A. P. India; **1982**.

Scientific Advisory Committee of the Wheat Industry Council. Wheat Foods: Nutritional Implications in Health and Disease. Proc. Symp. Am. J. Clin. Nutr., 41, 1069–1176, **1985**.

Shurtleff, W.; Aoyagi, A. (Eds.). "Soymilk. Industry and Market." The Soyfoods Center. Lafayette, CA; **1984**.

Suderman, D. R.; Cunningham, F. F. 1983. "Batter and Breading Technology." Avi Publ. Co., Westport, CT; **1983**.

Waller, G. R.; Feather, M. S. (Eds.). "The Maillard Reaction in Foods and Nutrition." Am. Chem. Soc., Washington, DC; **1983**.

Whistler, R. L.; BeMiller, J. N.; Paschall, E. F. (Eds.). "Starch Chemistry and Technology." 2nd ed. Academic Press, Orlando, FL; **1984**.

Williams, A. (Ed.). "Breadmaking—The Modern Revolution." Hutchinson Benham. London; **1975**.

Young, B. (Ed.). "Prospects for Sovjet Grain Production." Vestview Press, Boulder, CO; **1983**.

Zeuthen, P.; Cheftel, J. C.; Eriksson, C.; Jul, M.; Leniger, H.; Linko, P.; Varela, G.; Vos, G. (Eds.). "Thermal Processing and Quality of Foods." Elsevier Applied Science Publ., New York; **1984**.

III.

The following publications are excellent sources of information on grain production and trade:

World Crop Production (12 issues annually);

Grains: World Grain Situation and Outlook (16);

Grains: Export Markets for U.S. Grain and Products (12).

The circulars are published by the Foreign Agricultural Service of the U.S. Department of Agriculture, Washington, DC, and are approved by the World Agricultural Outlook Board of the U.S.D.A.

Chapter 2

Cereal Crops—General

GRAIN CROPS—GENERAL

Wheat grows in almost every kind of arable land, from sea level to elevations of 3,000 meters, wherever water is sufficient, in regions that are relatively arid, and in well-drained loams and clays. It has become increasingly a basic food throughout the world, even though higher incomes and increased meat, milk, and egg production and consumption have decreased the per capita wheat consumption in some affluent, highly developed countries. In developing countries, wheat consumption is increasing and accompanies rising living standards. The increase in wheat consumption is at the expense of the more expensive rice or the less expensive barley and sorghum. In some temperate areas with high annual rain, high-yielding feed wheat is grown. Wheat is the most highly cultivated grain crop in temperate areas, where it is the highest yielding crop and commands the highest price. Winter wheat may be multicropped prior to soybeans or grown as spring wheat in rotation with barley and oats to reduce loss of soil fertility. In areas with limited rain and a short growing period, barley may replace wheat. Rye is better adapted than wheat to poor soil and is more winter-hardy. Oats thrive better in areas with cold summers. Corn and sorghum are particularly suited for the tropics and warmer temperate climates. Grain sorghum, like maize and rice, is a tropical cereal but is better suited than corn to cultivation under semiarid conditions.

Corn is now the most important grain used for animal feed in temperate regions. In many tropical areas it is a basic human food. There are five major types of corn: flint, dent, floury, pop, and sweet. Dent varieties, grown most widely in the United States, are high yielding and are used to a large extent as feed grain. Flint corn has a somewhat higher food value, is common in Europe and South America, and is valued for its physical properties and as a raw material in the production of certain Central American foods (tortilla and arepa). High-yielding hybrid dent corn is replacing flint corn in many countries. There are white and yellow varieties of dent and flint corn; yellow corn varieties are rich in carotenoid pigments. Pop, sweet, and floury corn are relatively small food crops; the friable floury corn is hand-milled in some parts of South America.

The starch in waxy corn is almost entirely amylopectin, in amylomaize up to 70–80% amylose, and in regular corn about 25% amylose and 75% amylopectin (see Chapter 17). The starches vary widely in their physical properties (gelatinization characteristics, water-binding capacity, gel properties, etc) and usefulness in various foods and industrial applications. High-lysine and high-lysine-tryptophan genetic mutants have improved the nutritional properties of corn and are now available as dent and flint corn. The latter is likely to be acceptable in the production of maize-based foods.

Barley is a winter-hardy and drought-resistant grain. It matures more rapidly than wheat, oats, or rye. It is used mainly as feed for livestock and in malting and distilling industries. The two main types of barley, depending on the arrangement of grains in the ear, are two-rowed and six-rowed. The former predominate in Europe and parts of Australia; the latter type is more resistant to extreme temperatures and is grown in North America, India, and the Middle East.

Oats are grown most successfully in cool, humid climates and on neutral to slightly acid soils. The bulk of the crop is consumed as animal feed (mainly for horses). Heavy oats (with high groat to hull ratios) are processed into rolled oats, breakfast foods, and oatmeal.

WHEAT

Wheat provides about one-fifth of all calories consumed by humans. It accounts for nearly 30% of the world's grain production and for over 50% of the world grain trade. World wheat production more than doubled from 1950 to 1977, increasing from 172 to 382 million metric tons. During the same period world trade increased over 2.5-fold (from 28 to 73 million metric tons). World wheat production in 1981 increased to 450 million metric tons, of which about one-fifth entered the world grain trade.

Wheat is harvested somewhere in the world in nearly every month of the year (Figure 2-1). The United States, Canada, Australia, Argentina, and France are the main wheat exporters. In the period 1950–1977, yield increases have been more responsible for increases in world wheat production than increases in harvested area. The average yield was 1.0 tons/ha in 1950 and 1.67 tons/ha in 1977 (a 67% increase). The acreage increased from 173 to 228 million ha (a 32% increase). In 1981 the acreage increased only to 236 million ha (a 3.5% increase), but the average yield increased to 1.91 tons/ha (a 14% increase). Wheat production in North America

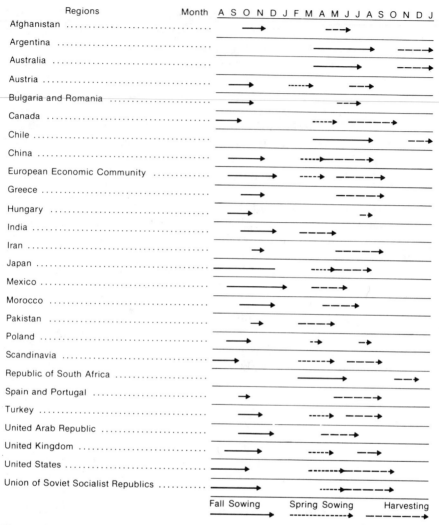

Figure 2-1. Worldwide sowing and harvesting seasons. (Source: Menze et al., 1974.)

and Western Europe has declined relative to the rest of the world since 1950. Those two areas accounted for 41% of the world production in 1950, 33% in 1977, and 36% in 1981. Eastern Europe and Asia accounted for 48% of the total world production in 1950, 60% in 1977, and 56% in 1981. The USSR leads all countries in wheat production: 31 million tons in 1950, 92 million tons in 1977, and 88 million tons in 1981. The United States is the second largest wheat-producing country. In 1981, the top five countries accounted for 62.0% and the top 12 for 78.6% of the total world production. The situation is changing, however, see pages 3-5.

Hexaploid wheats are the most widely grown around the world. They can be classified according to many criteria:

1. Growth habitat: Spring wheats are sown in the spring and harvested in late summer, and winter wheats are sown in the fall and harvested early the next summer. Winter wheats are preferred because of their higher (generally) yield.
2. Bran color: white, amber, red, or dark.
3. Kernel hardness (hard or soft) and vitreousness (glassy-vitreous or meally-starchy-floury). Whereas in many European countries all bread wheats (hard and soft) are called soft and the term hard is reserved for the very hard durum wheats, in the United States the term hard is used for bread wheats; soft for wheats for production of cookies, cakes, etc; and durum for wheats for the the production of alimentary pastes.
4. Geographic region of growth (Pool, 1948).

   a. The USSR region.
   b. The European region (the plains of Hungary, the Danube basin, Western Europe, the Mediterranean area).
   c. The Northern India–Pakistan region.
   d. The north-central China region.
   e. The southeastern Australia region.
   f. The South American region (the pampas of Argentina, Southern Chile).
   g. The North American region (the southern Great Plains of the U.S., the northern Great Plains of Canada and the U.S., the Columbia Plateau area).
5. End-use properties: protein range, gluten content and quality, physical dough properteis, bread-making quality, strength.
6. Composite designations: U.S. dark northern spring, U.S. hard red winter, Canadian hard red spring, Argentina plate, Queensland prime hard.
7. Variety.

The major hard red winter wheat producing areas are the United States, southern European Russia, the Ukraine, Australia, China, Europe, central India, and northern Africa. Australia and Argentina grow a medium hard wheat, 9–13% protein. Hard red spring wheat is produced in the three prairie provinces of Canada (Saskatchewan, Alberta, and Manitoba—in decreasing amounts and under the name of Manitoba Wheat); most of it is exported. Other major spring wheat-producing countries are the Soviet Union, China, and some of the northern states of the United States. The

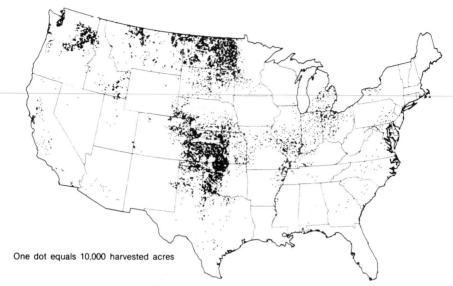

One dot equals 10,000 harvested acres

Figure 2-2. Distribution of U.S. wheat acreage. (Source: USDA, 1978.)

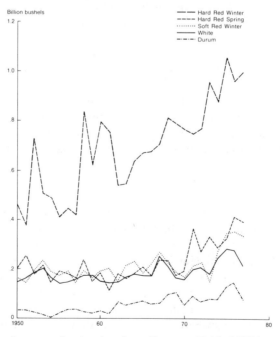

Figure 2-3. U.S. wheat production by class. (Source: Heid, 1979.)

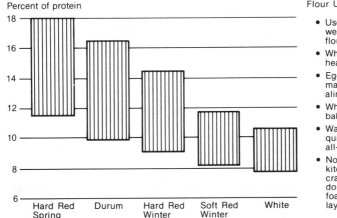

Percent of protein   Flour Uses

- Used to blend with
  weaker wheats for bread
  flour

- Whole wheat bread,
  hearth breads

- Egg noodles (U.S.),
  macaroni, and other
  alimentary pastes

- White bakers' bread,
  bakers' rolls

- Waffles, muffins,
  quick yeast breads,
  all-purpose flour

- Noodles (Oriental)
  kitchen cakes and
  crackers, pie crust,
  doughnuts, and cookies,
  foam cakes, very rich
  layer cakes

Hard Red   Durum   Hard Red   Soft Red   White
Spring   Winter   Winter

Figure 2-4. Protein range and flour use of major wheat classes. (Source: Menze et al., 1974.)

main soft wheat–producing countries are the European Economic Community, the United States, and Australia. Major durum wheat–producing countries are Canada, the United States, Argentina, the Soviet Union, and North Africa. The United States produces about 90 million tons of wheat on about 30 million ha in 42 states (Figures 2-2, 2-3). Kansas and North Dakota combined produce about one-third of the total. Hard red springwheat (one-fifth of the total) is produced in the northern semiarid portion of the Great Plains (North Dakota, Minnesota, South Dakota, Montana). The hard red winter region (about one-half of the total) of the Great Plains is south of the hard red spring wheat area with the highest acreages in Kansas, Nebraska, Texas, and Oklahoma. Soft red winter wheat (up to 15% of the total) is grown east of the Great Plains, in the Ohio River valley; the leading producers are Ohio and Indiana. White wheats (about 15% of the total) are grown principally in the Pacific northwest states (Washington, Oregon, and Idaho) and in New York and Michigan. Durum wheats (less than 10% of the total) are adapted to regions of limited rainfall, and most of the crop is produced in North Dakota, with small amounts in Minnesota and South Dakota. Protein range and flour uses of major wheat classes are given in Figure 2-4.

Wheat is produced in the northern and central states of Mexico; the northern state of Sonora produces under irrigation about 50% of the 2.5-million-ton wheat crop. Europe produces up to 90 million tons of wheat on 25 million ha; the average yield in western Europe is well above 4 tons/ha, the highest in the world. Europe is an exporter of low-protein soft wheats and an importer of high-protein hard wheats. The USSR produces more wheat than any other nation; most of it is produced north of the 45th parallel, about equivalent to the U.S.-Canada border. Wheat of spring habitat is preferred; in the southern region, mainly wheat of winter habitat is grown. In the eastern Mediterranean, diverse types of wheat are grown (durum, common, club). The main wheat-producing areas in Asia are China, India, Turkey, and Pakistan. India, occasionally, and Turkey, frequently, are wheat exporters. Wheat is second only to rice as a source of human food in China; soft, white winter wheat is preferred in the south, and hard, red spring wheat in the north. Japan is a large wheat importer. Argentina and Australia are major wheat exporters.

RICE

Rice is the staple food of about half of the human race. It provides over one-fifth of the total food calories consumed by the people of the world, and most of it is produced in the Far East. Rice is consumed primarily within the borders of producing countries. Asia, which has nearly 60% of the world population, produces and consumes about 90% of the world's rice. The United States produces less than 2% of the world crop but accounts for about 30% of the world rice trade. World production of 168 million tons in 1950 more than doubled to 373 million metric tons in 1978. Production growth was on the average about 7 million tons per year and relatively steady except for years of abnormally favorable or unfavorable weather. The increase in production was both from increase in harvested area (from 103.4 to 145.1 million ha) but primarily from increased yield (from 1.63 to 2.58 tons/ha), largely because of the development of new, high-yielding varieties. Assuming that milled rice is 67.5% of rough rice, annual per capita consumption of rice increased from 45.6 to 60.4 kg. Rice is a major crop and industry in Arkansas, California, Louisiana, Mississippi, Missouri, and Texas. The five southern states account for over 80% of the national production. Arkansas ranks first with 35–40% of the national total. Long grain accounts for about 50%, medium grain for 40%, and short grain for about 10% of U.S. rice production. California is, practically, the exclusive producer of short-grain rice. California and Louisiana lead in medium-grain rice production. Growers in Arkansas, Texas, and Mississippi produce mostly long-grain rice.

CORN (Maize)

U.S. corn production accounts for over half the total world production, and the United States accounts for about 80% of the annual world corn exports. U.S. grain production expanded rapidly during the 1970s as corn became the nation's largest crop and represented well over half the total grain output. Corn is harvested in 41 states and accounts for about 30% of the acreage planted to principal crops. Corn production is concentrated in the "Corn Belt," most of which extends from eastern Ohio to the western parts of Iowa and Missouri and accounts for at least 80% of the corn produced in the United States. The most important corn-producing states are Illinois and Iowa, which generally grow about 40% of the total. In the United States, yields climbed from less than 2.50 tons/ha in 1952 to 3.35 in 1959 and to above 5.00 in the late 1970s. These dramatic yield increases resulted primarily from the development of the new high-yielding hybrids, improved cultural practices, and high levels of fertilizer application. In the United States, most corn is used as feed, alcoholic beverages, seed, and livestock feed. Only about 10% is used as food (production of starch, corn syrups, breakfast cereals, and various other foods). Livestock feed accounts for almost 85% of the total domestic use. Use of corn expanded during the 1970s for food (mainly corn sweeteners) and industrial (gasohol-alcohol fuel) purposes. Exports of corn from the United States increased from about 2.1 million tons in 1950/51 to over 60 million tons in 1979/80. The export accounted for 5% of the total use (disappearance) in the 1950s, for 13% between 1962 and 1972, and for 27% since 1972. The main importers of U.S. corn are western Europe and Japan, primarily for livestock feed; they import about half the total U.S. exports.

Some dramatic changes have taken place in harvesting and handling of corn in the United States, most notably an increase in the average farm size, use of larger and more efficient machinery, sophisticated management, and increased on-farm storage capacity. This change in harvesting method requires field shelling of corn with a high moisture (around 20%) and artificial drying to a safe moisture (about 14.5%). About 80% of the corn produced in the major corn-producing states is dried artificially on the farm. Practically all of the acreage seeded to corn is planted with yellow hybrids, which are highly productive and responsive to high plant populations, high levels of fertilization, and improved cropping practices. Most of the field corn (well over 90%) harvested for grain is dent corn. The name is derived from the indentation in the crown caused by shrinkage of the kernel during drying. Some early-maturing flint corn is grown in Pennsylvania, New York, the New England states, California, and the western and northern edges of the Corn Belt. Sweet corn, adapted to cooler climates, is grown in the North Atlantic and central states for canning purposes and fresh consumption (corn on the cob). Popcorn is grown mainly in the Corn Belt. Some high-lysine corn, limited amounts of waxy corn, and a very small amount of high-amylose corn are grown under contract with wet corn millers.

BARLEY

Although barley is grown throughout the world, production is concentrated in the northern latitudes. Since 1960, world production of barley tripled, with most of the increase in Europe. World production of barley in the late 1980s was about 157 to 170 million tons per year, two-thirds of which were in Europe and the Soviet Union (the largest barley producer). Average yields of barley are comparable to those of wheat (slightly below 2 tons/ha). Barley is adaptable to a variety of conditions and is produced commercially in 36 U.S. states. In many areas, however, barley must compete for land with more profitable corn, sorghum, and soybeans. Consequently, its production has become concentrated in the northern plains and Pacific coast states, where the other grains cannot adapt well to the climate. Six-rowed barleys in the central states are grown more extensively than two-rowed barleys in the western states. North Dakota, California, and Montana account for 50% of the total production; Idaho and Minnesota produce about 20%. Practically all the barley produced in North Dakota and Minnesota is for malting; almost all the California barley is seeded to feed varieties. In the United States, the use of barley as a livestock feed is declining, and the use for processing into malt is increasing. During the years 1960–1975, malting barley increased from 25% to 40% of the total. In the United States and Canada, a dual pricing system is in effect for feed and malting barleys.

OATS

Slightly over 25 million ha is seeded to oats. The yield per hectare is about 1.7 tons. World production of oats (45 million metric tons) is about one-tenth of the world production of wheat and slightly over one-fourth that of barley. The chief oat-producing states in the United States are in the north central Corn Belt and further north. Oat production in 1980 in the United States was close to 10 million metric tons on over 5 million ha—about half of that in 1965/66. About 90% of the oat is left on the farm as a feed; only a small proportion is processed as food or used for industrial products.

SORGHUM

Sorghum is the fifth most widely grown crop in the world: about 60–70 million tons on 45–60 million ha. Sorghum gained significance as a grain as a result of the development of high-yielding hybrids. Still, sorghum production is less than 5% of the total grain production. There has been only a small increase in the area harvested to sorghum but a substantial increase in sorghum production due to a rise in yields since 1960 (over 50%). The main sorghum-producing countries are the United States, China, India, Argentina, Nigeria, and Mexico. Those countries produce more than three-fourths, with the United States producing one-third, of the world total. More than half the world's production of sorghum is for human consumption. In the United States, sorghum is grown almost entirely as a feed grain for local use or export. Sorghum is produced under a wide range of climatic conditions. It withstands limited moisture conditions and adapts to high temperatures. In the United States, about 80% of the sorghum grain is produced in Texas, Kansas, and Nebraska. The largest exporters of sorghum are the United States, Argentina, and Australia (combined 90% of the total). The largest importers are Japan, Western Europe, and Central America.

Grain sorghum, one of the four general classes (sorgo, or sweet sorghum, broom corn, grass sorghum, grain sorghum) thrives relatively well under drought conditions and is grown primarily in the southern sections of the Great Plains and in parts of the southwestern United States. Several groups of grain sorghum (milo, kafir, feterita, durra, kaoliang, and shallu) have been grown in the United States, but in recent years crosses of milo with kafir are the leading varieties. In some varieties of sorghum, the nucellar tissue contains pigments that complicate production of acceptable white sorghum starch. Varieties with no nucellar tissue persisting to maturity could be used in starch production, but the availability of corn has discouraged such attempts. Waxy (starch) sorghums are potentially promising for special food uses. Some sorghums are rich in pigments, notably tannins, which reduce palatability and feed value.

RYE

Rye is characterized by good resistance to cold, pests, and diseases, but it cannot compete with wheat or barley on good soil and under improved cultivation practices. Total world production of rye is about 24 million tons on 14 million ha. About 90% of the rye is produced in Europe (mainly Poland and Germany) and the Soviet Union. Rye is considered a bread grain in Europe, but its proportion in mixed wheat-rye bread is decreasing. Much of the rye is used as a feed grain, and a small proportion is used in the distilling industry.

BIBLIOGRAPHY

Heid, W. G. "U.S. Barley Industry." Agr. Econ. Report, No. 395. U.S. Dept. Agric. Economics, Statistics and Cooperatives Service: Washington, D.C., **1978**.

Heid, W. G. "U.S. Wheat Industry." Agr. Econ. Report No. 432. U.S. Dept. Agric. Economics, Statistics and Cooperative Service: Washington, D.C., **1979**.

Leath, M. N.; Meyer, L. H.; Hill, L. D. "U.S. Corn Industry." Report No. 479. U.S. Dept. Agric., Economic Research Service: Washington, D.C., **1982**.

Menze, R. E.; Heid, W. G.; Wirak, O. S. "Factors Determining the Price of White Wheat in the Pacific Northwest." Washington State University Press: Pullman; **1974**.

Pool, R. J. "Marching with the Grasses." University of Nebraska Press: Lincoln; **1948**.

U.S. Dept. Agric. "1974 Census of Agriculture. Graphic Summary," Vol. IV, Part I. Bureau of Census: Washington, D.C.; **1978**.

Chapter 3

Physical Properties and Structure

Some physical properties of cereal grains are listed in Table 3-1. Table 3-2 summarizes approximate grain size and proportions of the principal parts comprising the mature kernels of different cereals.

Table 3-1. Some Physical Properties of Cereal Grains

Name	Length (mm)	Width (mm)	Grain wt (mg)	Bulk density (kg/m³)	Unit density (kg/m³)
Rye	4.5–10	1.5– 3.5	21	695	—
Sorghum	3 – 5	2.5– 4.5	23	1,360	—
Paddy rice	5 –10	1.5– 5	27	575–600	1,370–1,400
Oats	6 –13	1 – 4.5	32	356–520	1,360–1,390
Wheat	5 – 8	2.5– 4.5	37	790–825	1,400–1,435
Barley	8 –14	1 – 4.5	37	580–660	1,390–1,400
Maize	8 –17	5 –15	285	745	1,310
Bullrush millet	3 – 4	2 – 3	11	760	1,322
Wild rice	8 –20	0.5– 2	22	388–775	—
T'ef	1 – 1.5	0.5– 1	0.3	880	—
Findi	1	1 – 1.5	0.4	790	—
Finger millet	1.5	1.5	—	—	—

Source: Muller, 1973.

Optimum utilization of cereal grains requires knowledge of their structure and composition. The practical implications of kernel structure are numerous. They relate to the various stages of grain production, harvest, storage, marketing, and use. Some of the implications are listed in Table 3-3.

KERNEL STRUCTURE—GENERAL

The cereal grain is a one-seeded fruit called a caryopsis, in which the fruit coat is adherent to the seed. As the fruit ripens, the pericarp (fruit wall)

Table 3-2. Approximate Grain Size and Proportions of the Principal Parts Comprising the Mature Kernels of Different Cereals

Cereal	Grain weight (mg)	Embryo (%)	Scutellum (%)	Pericarp (%)	Aleurone (%)	Endosperm (%)
Barley	36– 45(41)[a]	1.85	1.53	18.3		79.0
Bread wheat	30– 45(40)	1.2	1.54	7.9	6.7–7.0	81–84
Durum wheat	34– 46(41)		1.6	12.0		86.4
Maize	150–600(350)	1.15	7.25		5.5	82
Oats	15– 23(18)	1.6	2.13		28.7–41.4	55.8–68.3
Rice	23– 27(26)	2–3	1.5	1.5	4–6	89–94
Rye	15– 40(30)	1.8	1.73	12.0		85.1
Sorghum	8– 50(30)		7.8–12.1		7.3–9.3	80–85
Triticale	38– 53(48)		3.7	14.4		81.9

Source: Simmonds, 1978.
[a]Average weight in parentheses.

becomes firmly attached to the wall of the seed proper. The pericarp, seed coats, nucellus, and aleurone cells form the bran. The embryo occupies only a small part of the seed. The bulk of the seed is taken up by the endosperm, which constitutes a food reservoir.

The floral envelopes (modified leaves known as lemma and palea), or chaffy parts, within which the caryopsis develops, persist to maturity in the grass family. If the chaffy structures envelope the caryopsis so closely that they remain attached to it when the grain is threshed (as in rice and most varieties of oats and barley), the grain is considered to be covered. However, if the caryopsis readily separates from the floral envelopes, on threshing, as with common wheats, rye, hull-less barleys, and the common varieties of corn, these grains are considered to be naked.

WHEAT

The structure of the wheat kernel is shown in Figure 3-1. The dorsal side of the wheat grain is rounded, and the ventral side has a deep groove or crease along the entire longitudinal axis. At the apex or small end (stigmatic end) of the grain is a cluster of short, fine hairs known as brush hairs. The pericarp, or dry fruit coat, consists of four layers: epidermis, hypodermis, cross cells, and tube cells. The remaining tissues of the grain are the inner bran (seed coat and nucellar tissue), endosperm, and embryo (germ). The aleurone layer consists of large, rectangular, heavy-walled, starch-free cells. Botanically, the aleurone is the outer layer of the endosperm, but as it tends to remain attached to the outer coats during wheat milling, it is shown in the diagram as the innermost bran layer.

The embryo (germ) consists of the plumule and radicle, which are connected by the mesocotyl. The scutellum serves as an organ for food

Table 3-3. Some Implications of Kernel Structure

Significance	Parameter	Effect	Commodity
Threshing	Germ damage or skinning	Reduced germinability, impaired storability	All cereal grains
Drying	Cracks, fissures, and breakage; hardening	Reduced commercial value; lowered grade, impaired storability, dust formation, reduced starch yield	Mainly corn and rice
	Discoloration	Reduced commercial value, lowered grade	Mainly rice
Marketing	Breakage	Reduced commercial value in food processing	Mainly corn and rice
General use	High husk; caryopsis ratio or high pericarp: endosperm ratio	Reduced nutritional value as food or feed	All cereal grains
General use	Kernel shape and dimensions; proportions of tissues in the kernel; distribution of nutrients in the tissues	Yield of food products; nutritional value of cereal (or cereal products) as food or feed	All cereal grains
Malting	Germ damage, skinning, or inadequate husk adherence	Reduced germinability, uneven malting	Mainly barley
Milling	Uneven surface, deep crease, or uneven aleurone	Reduced milling yield	Mainly wheat and rice
Milling	Steely texture	Increased power requirements, starch damage, high water absorption, difficulty in air classification	Wheat and malt milling
Germination-malting	Starch granule size	Uneven degradation	All cereal grains
Consumption-nutrition	Distribution and composition of proteins	Change in nutritional value	All cereal grains

storage. The outer layer of the scutellum, the epithelium, may function as either a secretory or an absorption organ. In a well-filled wheat kernel, the germ comprises 2–3% of the kernel, the bran 13–17%, and the endosperm the remainder. The inner bran layers (the aleurone) are high in protein, whereas the outer bran (pericarp, seed coats, and nucellus) is high in cellulose, hemicelluloses, and minerals; biologically, the outer bran functions as a protective coating and remains practically intact when the seed germinates. The germ is high in proteins, lipids, sugars, and minerals; the

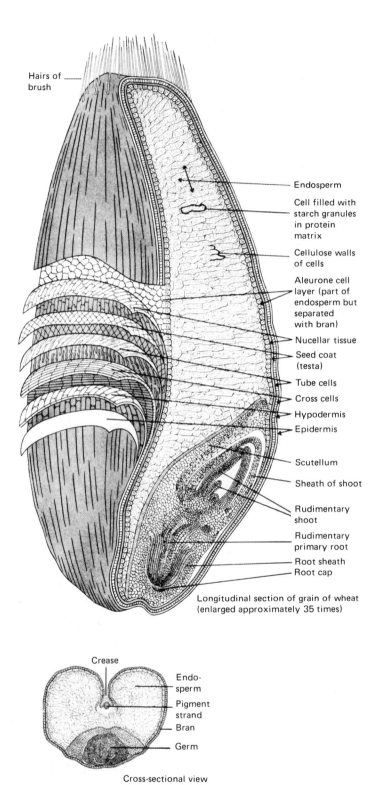

Figure 3-1. Longitudinal and cross section through a wheat kernel. (Source: Wheat Flour Institute, Chicago.)

endosperm consists largely of starch granules imbedded in a protein matrix.

Grains of rice, barley, oats, rye, and triticale are similar in structure to wheat.

RICE

Rice is a covered cereal. In the threshed grain (or rough rice), the kernel is enclosed in a tough siliceous hull, which renders it unsuitable for human consumption. When this hull is removed during milling, the kernel (or caryopsis), composed of the pericarp (outer bran) and the seed proper (inner bran, endosperm, and germ), is known as brown rice or sometimes as unpolished rice. Brown rice is in little demand as a food. Unless stored under favorable conditions, it tends to become rancid and is more subject to insect infestation than are the various forms of milled rice. When brown rice is subjected to further milling processes, the bran and germ are removed and the purified endosperms are marketed as white rice or polished rice. Milled rice is classified according to size as head rice (whole endosperm) and various classes of broken rice, known as second head, screenings, and brewer's rice, in order of decreasing size.

CORN

The corn grain is the largest of all cereals (Figure 3-2). The kernel is flattened, wedge-shaped, and relatively broad at the apex of its attachment to the cob. The aleurone cells contain much protein and oil and also contain the pigments that make certain varieties appear blue, black, or purple. Two types of starchy endosperms—horny and floury—are found beneath the aleurone layer. The horny endosperm is harder and contains a higher level of protein. In dent corn varieties, the horny endosperm is found on the sides and back of the kernel and bulges in toward the center at the sides. The floury endosperm fills the crown (upper part) of the kernel, extends downward to surround the germ, and shrinks as dent corn matures, causing an indentation at the top of the kernel. In a typical dent corn, the pericarp comprises approximately 6%, the germ 11%, and the endosperm 83% of the kernel. Flint corn varieties contain more horny than floury endosperm.

BARLEY

In barley, the husks are cemented to the kernel and remain attached after threshing. The husks protect the kernel from mechanical injury during

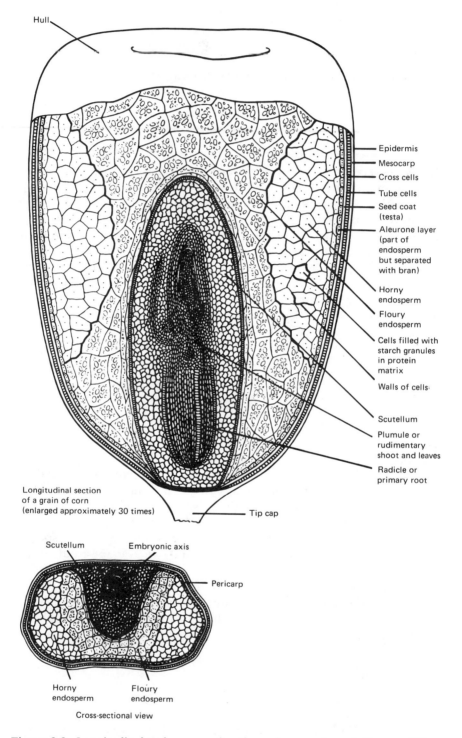

Hull

Epidermis
Mesocarp
Cross cells
Tube cells
Seed coat (testa)
Aleurone layer (part of endosperm but separated with bran)
Horny endosperm
Floury endosperm
Cells filled with starch granules in protein matrix
Walls of cells
Scutellum
Plumule or rudimentary shoot and leaves
Radicle or primary root

Longitudinal section of a grain of corn (enlarged approximately 30 times)

Tip cap

Scutellum   Embryonic axis

Pericarp

Horny endosperm   Floury endosperm

Cross-sectional view

Figure 3-2. Longitudinal and cross section through a corn kernel. (Source: Wheat Flour Institute, Chicago.)

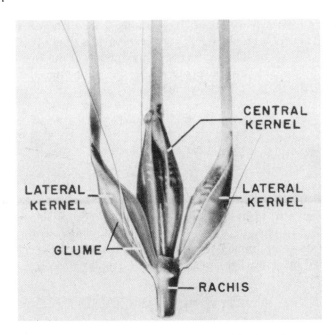

Figure 3-3. Arrangement of kernels in six-rowed barley. (Source: Malting Barley Improvement Assoc., Milwaukee.)

commercial malting, strengthen the texture of barley steeped to a moisture of about 42%, and contribute to a more uniform germination of the kernels. The husks are also important as a filtration bed in the separation of extract components during mashing and contribute to the flavor and astringency of beer. The main types of cultivated covered barleys, depending on the arrangement of grains in the ear, are two-rowed and six-rowed. The axis of the barley ear has nodes throughout its length; the nodes alternate from side to side. In the six-rowed types of barleys, three kernels develop on each node—one central kernel and two lateral kernels (Figure 3-3). In the two-rowed barleys, the lateral kernels are sterile and only the central kernel develops.

The kernel of covered barley consists of the caryopsis and the flowering glumes (or husks). The husks consist of two membraneous sheaths that completely enclose the caryopsis. One of the husks, the lemma, is drawn out into a long awn. The color of the grain in covered barleys depends on the color of the caryopsis and the second husk (the palea). Color in the caryopsis is due to anthocyanin pigments or to a black melaninlike compound. Anthocyanin, when present, is red in the pericarp and blue in the aleurone layer. During the development of the growing barley, a cementing substance causing adherence is secreted by the caryopsis within the first 2 weeks after pollination. The husk in cultivated malting barleys amounts to about 8–15% of the grain. The proportion varies according to type, variety, grain size, and climatic conditions. Large kernels have less husk than small kernels. The husk in two-rowed barleys is, generally, lower than in six-rowed barleys. The caryopsis, as in wheat, is a one-seeded fruit in which the outer pericarp layers enclose the aleurone, the starchy endosperm, and the germ. The aleurone layer in barley is at least two layers thick; in other cereal grains, except rice, it is one layer thick.

OATS

The common varieties of oats have the fruit (caryopsis) enveloped by a hull composed of certain floral envelopes. Naked or hull-less oat varieties are not extensively grown. In light, thin oats, hulls may comprise as much as 45% of the grain; in very heavy or plump oats, they may represent only 20%. The hull normally makes up approximately 30% of the grain. Oat kernels, obtained by removing hulls, are called groats.

SORGHUM

Sorghum kernels are generally spherical, have a kernel weight of 20–30 mg, and may be white, yellow, brown, or red. Sorghum grains contain in the endosperm, but mainly in the outer kernel layers, polyphenolic compounds that are colored and impart various colors to fractions of foods from sorghums. Tannins, present in some sorghum grains, are polyphenols that interact with and precipitate proteins. Tannins that are not hydrolyzed by enzymes impart some resistance of sorghum to attack by birds, but they also reduce nutritional value and germinability. Pearl millet is the largest-seeded and most widely grown millet. It is known as *Pennisetum americanum*, L. Leeke, or *P. typhoides* and in older literature as *P. glaucum*. The common name in Africa is bullrush millet, and in India it is bajra. Pearl millet kernels are generally tear-shaped and weigh about 9 mg, or one-third the weight of sorghum grain. Other millets include proso millet (*Panicum miliaceum*) grown as food in East Asia and as a birdseed in the USA and South America. Finger millet (*Eleusine coracana*), known as ragi, is an important food crop in parts of India and Africa. Foxtail millet (*Setaria italica*) is known in China and India as a food grain or as a forage.

THE HULL AND BRAN LAYERS

The hull silica can slow the attack of storage insects on rice and barley. The palea and lemma in barley are held together by two hooklike structures. In rice the ability of these structures to hold the palea and lemma together without gaps probably depends on the variety. Varieties of rice that have many gaps and separations have greater insect infestations than varieties with tight husks.

The outer pericarp layers of wheat (epidermis and hypodermis) have no intercellular spaces and are composed of closely adhering, thick-walled cells. The inner layers of the pericarp, on the other hand, consist of thinner-walled cells and often contain intercellular spaces, through which water can move rapidly and in which molds are commonly found. Molds can also enter through the large intercellular spaces at the base of the kernel where the grain was detached from the plant at harvest and where there is no protective epidermis. The structure of the pericarp, seed coats, and nucellus also explains how the kernel reacts to water. Following initial rapid water absorption, the rate decreases significantly. The seed coat offers more water resistance than the nucellus. The ability of the germ to absorb and hold considerable amounts of water probably accounts in part for the susceptibility of the germ to attack by molds.

An intact grain stores much better than damaged or ground grain. Deteriorative changes (ie, rancidity, off-flavors, etc) occur slowly in the whole grain but rapidly after the grain has been ground. The hull, apparently, prevents the grain from becoming rancid by protecting the bran layers from mechanical damage during harvesting and subsequent handling.

THE GERM

The site of the germ in the kernel and whether it protrudes or is well protected by adjacent layers govern the extent to which the germ is retained intact during threshing and determines the grain's usefulness for seeding or malting. The germ is a separate structure that generally can be easily separated from the rest of the cereal grain. However, the scutellar epithelium (located next to the endosperm) has fingerlike cells. The free ends protrude toward the adjacent starch endosperm cells and form an amorphous cementing layer between the germ and endosperm. If part of the layer projects into the spaces between the fingerlike cells of the scutellar epithelium and into the folds of the scutellar structure, it may be difficult to separate the germ from the endosperm unless the cementing layer is softened. The softening may be accomplished by steeping in corn wet milling or by conditioning in wheat milling. In rice, a layer of crushed cells separating the scutellar epithelium from the starchy endosperm provides a line of easy fracture.

Germ separation is also enhanced by the fact that the germ takes up water faster and swells more readily than the endosperm. The strains resulting from differential swelling contribute to easy separation in milling.

GRINDING AND MILLING

The starchy endosperm is the portion of the caryopsis that is used primarily as a food source. In the milling of wheat and rice, the bran layers and the germ are separated, thereby removing much of the fiber, ash, lipid, and protein.

Endosperm structure, composition, and associated texture govern both power requirements in milling and particle size of the products. In wheat flour milling, the size and shape of the crease affect yield and composition of flour, because the bran in the crease area is difficult to separate from the starchy endosperm (Figure 3-4). The thickness and irregularity of the aleurone around the starchy endosperm and in the crease make it impossible to increase flour yields in wheat milling by processes that successively "peel" outer layers. The thickness and tenacity with which the pericarp is bound to the aleurone layer may be responsible for difficulties in the milling of some European wheat cultivars that yield well in the field but create problems in processing.

The irregular thickness of the aleurone layer in wheat makes it difficult to separate all the starchy endosperm from the aleurone. The cells of the

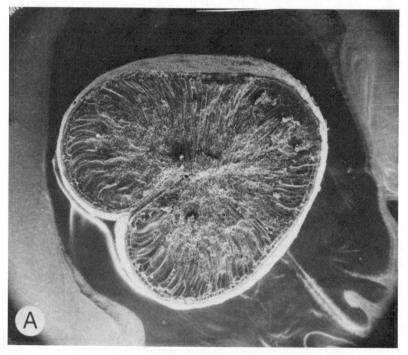

Figure 3-4. Scanning electron micrographs of (A) cross section through a wheat kernel, (B) section through crease, (C) central starchy endosperm, and (D) aleurone layer, and (E) high magnification of aleurone grains.

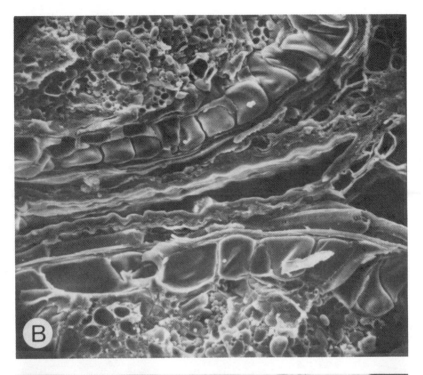

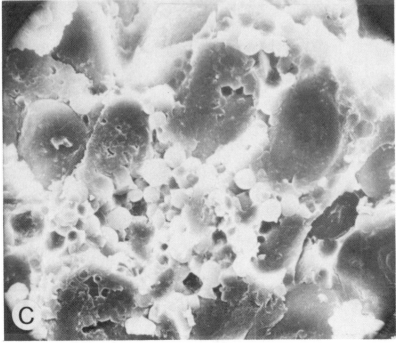

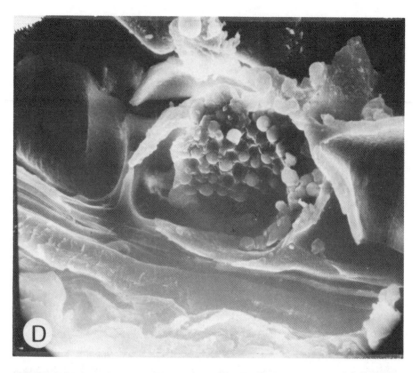

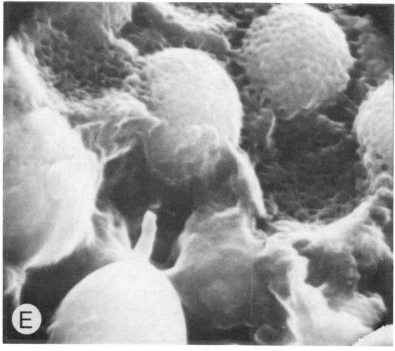

starchy endosperm are large and irregularly shaped; in milling the cells are broken. Cell walls in the starchy inner endosperm can be broken with greater ease than cell walls in the protein-rich outer endosperm. The outer cells are smaller, and their walls are thicker than the inner cells.

Wheat varieties differ in vitreosity and hardness of their mature endosperms. The importance of grain hardness lies in its effect on the milling properties and, in particular, on the amounts of semolina and farina and mechanically damaged starch produced during milling.

Mechanically damaged starch absorbs more water than undamaged starch. That is of particular importance to bakers in countries in which the amount of water added to a dough is essentially limited only by the amount required to obtain a dough of adequate consistency for handling and for producing a well-developed bread of adequate shelf life. Damaged starch granules in flour from hard wheat are more susceptible to enzymatic degradation by amylases than are those from soft wheats; that is important in producing bread in sugar-free formulas in which fermentable sugars are primarily those produced by the action of malt on damaged starch rather than by added sucrose.

GRAIN DRYING

Changes in texture and structure during the drying of corn and rice are important in minimizing breakage during handling. Excessive cracks reduce the value of corn in processing, and especially for production of breakfast cereals. Harsh heat treatment during grain drying may reduce starch yields, impair quality of the starch, and create difficulties in corn wet milling. Overheated corn may also cause difficulties in brewing beer and distilling alcoholic beverages. The starch granules are imbedded in a proteinaceous matrix that hardens during overheating. The hardened matrix prevents the starch from being attacked by enzymes and converted to alcohol. In rice milling, harsh drying and accompanying structural cracks substantially reduce yields of head rice and increase the amounts of "brokens," causing economic losses to the miller. The method of rice drying may also affect the texture and color of the milled rice and result in off-color, especially objectionable browning.

NUTRITIONAL IMPLICATIONS

The various components are not uniformly distributed in the kernel. The hulls and bran are high in cellulose, pentosans, and ash; the germ is high in lipid and rich in proteins, sugars, and, generally, ash. The endosperm contains the starch, has a lower protein content than the germ and the bran (in some cereals), and is low in fat and ash.

Breeding efforts to improve the nutritional value of cereal grains have concentrated on increasing protein content without decreasing protein quality (mainly retaining lysine concentration in the protein). The significance of protein distribution in the endosperm depends on the type of product that is likely to be consumed. In the production of highly refined products, in which some of the subaleurone layer is removed, a high concentration of protein in the subaleurone layer would not be desirable. However, if the whole kernel is to be consumed, distribution of protein in the kernel is of limited nutritional consequence. In all cereal grains, the storage proteins form a matrix, which surrounds the starch granules. The concentration of protein increases from the inner to the outer starch endosperm. The increase may be relatively gradual, as in some soft wheats, or quite steep, as in some high-protein wheat types in which some of the outer subaleurone cells contain few, if any, starch granules.

IMPROVING NUTRITIONAL QUALITY BY MODIFYING GRAIN MORPHOLOGY

Many studies concern the improvement of the nutritional value of cereal grains. Simple changes in grain morphology could be the basis of improvement. The embryo of cereal seeds is rich in protein (up to 38%), and the protein may contain about 7% lysine. Selection for larger embryos is important if the whole seed (rather than starchy endosperm) is to be consumed. Variations in the number of aleurone cells of the endosperm exist in corn, rice, and barley. The aleurone layer is rich in protein having a good amino acid balance. Selection for a high aleurone cell number could be useful, provided the high number is associated with improved nutritional value. Increasing the relative surface area of the seed could lead to an increase in the aleurone layer. In early triticale (crosses between wheat and rye) and in *Avena sterilis* (a wild hexaploid oat), such an increase results from the undesirable development of long and thin or shrunken kernels. Both in wheat and in rice, much of the protein is concentrated in the aleurone and outermost subaleurone. Those tissues are diverted to feed during the milling and polishing of rice or during the milling of highly refined wheat flour. "Restructuring" cereal grains to obtain a more even distribution of protein throughout the whole endosperm would increase the protein content of milled products.

The hulls are rich in fibrous materials and low in protein. Many hull-less varieties of barley and oats have substantially more protein than the hulled varieties do. However, the low yield of the available hull-less varieties discourage their cultivation.

BIBLIOGRAPHY

Muller, H. G. "Industrial Uses of Cereals." Pomeranz, Y. Ed. Am. Assoc. Cereal Chem.; St. Paul, MN; **1973**.

Pomeranz, Y.; Bechtel, D. B. In "Cereals '78. Better Nutrition for the World's Millions." Pomeranz, Y., Ed. Am. Assoc. Cereal Chem.; St. Paul, MN; **1978**.

Simmonds, D. H. In "Cereals '78: Better Nutrition for the World's Millions." Pomeranz, Y., Ed. Am. Assoc. Cereal Chem.; St. Paul, MN; **1978**.

Chapter 4

Composition

Like that of other foods of plant origin, the chemical composition of the dry matter of different cereal grains varies widely. Proximate analysis of important cereal grains is summarized in Table 4-1. Variations are encountered in the relative amounts of proteins, lipids, carbohydrates, pigments, vitamins, and ash; mineral elements present also vary widely. As a food group, cereals are characterized by relatively low protein and high carbohydrate contents; the carbohydrates consist essentially of starch (90% or more), dextrins, pentosans, and sugars.

The various components are not uniformly distributed in the different kernel structures. Table 4-2 compares the weights and compositions of the main anatomical parts of the wheat kernel with the composition of flours, which vary in milling extraction rate. The hulls and pericarp are high in cellulose, pentosans, and ash; the germ is high in lipid content and rich in

Table 4-1. Proximate Analysis of Important Cereal Grains (% Dry Weight)

Cereal		Nitrogen	Protein[a]	Fat	Fiber	Ash	NFE[b]
Barley	Grain	1.2–2.2	11	2.1	6.0	3.1	—
	Kernel	1.2–2.5	9	2.1	2.1	2.3	78.8
Maize		1.4–1.9	10	4.7	2.4	1.5	72.2
Millet		1.7–2.0	11	3.3	8.1	3.4	72.9
Oats	Grain	1.5–2.5	14	5.5	11.8	3.7	—
	Kernel	1.7–3.9	16	7.7	1.6	2.0	68.2
Rice	Brown	1.4–1.7	8	2.4	1.8	1.5	77.4
	Milled			0.8	0.4	0.8	—
Rye		1.2–2.4	10	1.8	2.6	2.1	73.4
Sorghum		1.5–2.3	10	3.6	2.2	1.6	73.0
Triticale		2.0–2.8	14	1.5	3.1	2.0	71.0
Wheat (Bread)		1.4–2.6	12	1.9	2.5	1.4	71.7
Wheat (Durum)		2.1–2.4	13			1.5	70.0
Wild Rice		2.3–2.5	14	0.7	1.5	1.2	74.4

[a]Typical or average figure.
[b]Nitrogen-free extract. This is an approximate measure of the starch content.
Source: Simmonds, 1978.

Table 4-2. Weight, Ash, Protein, Lipid, and Crude Fiber Contents of Main Anatomical Parts of the Wheat Kernel and of Flours of Different Milling Extraction Rates

Parameter	Wheat kernel fractions				Milling extraction (%)		
	Peri-carp	Aleurone layer	Starchy endosperm	Germ			
Weight (%)	9	8	80	3	75	85	100
Ash (%)	3	16	0.5	5	0.5	1	1.5
Protein (%)	5	18	10	26	11	12	12
Lipid (%)	1	9	1	10	1	1.5	2
Crude fiber (%)	21	7	> 0.5	3	> 0.5	0.5	2

Source: Pomeranz and MacMasters, 1968.

proteins, sugars, and ash constituents. The endosperm contains the starch and is lower in protein content than the germ and, in some cereals, bran; it is also low in crude fat and ash constituents. As a group, cereals are low in nutritionally important calcium, and the concentration of calcium and other ash constituents is greatly reduced by the milling processes used to prepare refined foods. In these processes, hulls, germ, and bran, which are the structures rich in minerals and vitamins, are more or less completely removed. Mineral contents of cereal grains and cereal products are given in Table 4-3.

All cereal grains contain vitamins of the B group, but all are completely lacking in vitamin C (unless the grain is sprouted) and vitamin D. Yellow corn differs from white corn and the other cereal grains in containing carotenoid pigments (principally cryptoxanthin, with smaller quantities of carotenes), which are convertible in the body to vitamin A. Wheat also contains yellow pigments, but they are almost entirely xanthophylls, which are not precursors of vitamin A. The oils of the embryos of cereal grains are rich sources of vitamin E. The relative distribution of vitamins in kernel structures is not uniform, although the endosperm invariably contains the least (Table 4-4).

Protein contents of wheat and barley are important indexes of their quality for manufacture of various foods. The bread-making potentialities of bread wheat are largely associated with the quantity and quality of its protein.

The cereal grains contain water-soluble proteins (albumins), salt-soluble proteins (globulins), alcohol-soluble proteins (prolamins), and acid- and alkali-soluble proteins (glutelins). The prolamins are characteristics of the grass family and, together with the glutelins, comprise the bulk of the proteins of cereal grains. The following are names given to prolamins in proteins of the cereal grains: gliadin in wheat, hordein in barley, zein in maize, avenin in oats, kafferin in grain sorghum, and secalin in rye. Distribution of cereal endosperm proteins according to their solubility is summarized in Table 4-5.

Table 4-3. Mineral Content of Cereal Grains and Cereal Products (mg/100 g)

	Ca	Fe	Mg	P	K	Na	Cu	Mn	Zn
Barley	80	10	120	420	560	3	0.76	1.63	1.53
Buckwheat	110	4	390	330	450	—	0.95	3.37	0.87
Maize (grain)	30	2	120	270	280	1	0.21	0.51	1.69
Maize (bran)	30	—	260	190	730	—	—	1.61	—
Maize (germ)	90	90	280	560	130	—	1.10	0.90	—
Millett (proso)	50	10	160	280	430	—	2.16	2.91	1.39
Oats	100	10	170	350	370	2	0.59	3.82	3.4
Rice (grain)	40	3	60	230	150	9	0.33	1.76	1.8
Rice (white)	30	1	20	120	130	5	0.29	1.09	1.3
Rye	60	10	120	340	460	1	0.78	6.69	3.05
Sorghum	40	4	170	310	340	—	0.96	1.45	1.37
Wheat (grain)	50	10	160	360	520	3	0.72	4.88	3.4
Wheat (bran)	140	70	550	1170	1240	9	1.23	11.57	9.8
Triticale	20	4	—	—	385	—	0.52	4.26	0.02

From Lockhart and Nesheim, 1978.

The various proteins are not distributed uniformly in the kernel. Thus, the proteins fractionated from the inner endosperm of wheat consist chiefly of a prolamin (gliadin) and glutelin (glutelin), apparently in approximately equal amounts. The embryo proteins consist of nucleoproteins, an albumin (leucosin), a globulin, and proteoses, whereas in wheat bran a prolamin predominates with smaller quantities of albumins and globulins. When water is added, the wheat endosperm proteins, gliadin and glutenin, form a tenacious colloidal complex, known as gluten (see Figure 4-1). Gluten is responsible for the superiority of wheat over the other cereals for the manufacture of leavened products, since it makes possible the formation of a dough that retains the carbon dioxide produced

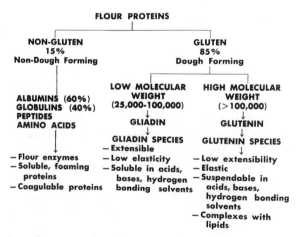

Figure 4-1. Wheat flour proteins. (Source: Holme, 1966.)

Table 4-4. Vitamin Contents of Cereal Grains and Cereal Products

	Thiamin (mg/100 g)	Riboflavin (mg/100 g)	Niacin (mg/100 g)	Vitamin B_6 (mg/100 g)	Folic acid (µg/100 g)	Pantothenic acid (mg/100 g)	Biotin (µg/100 g)	Vitamin E (IU/100 g)
Barley (pearled)	0.23	0.13	4.5	0.26	67		6	1
Buckwheat	0.60		4.4					
Maize	0.37	0.12	2.2	0.47	26	1	21	2
Oats	0.67	0.11	0.8	0.21	104	1	13	3
Millet	0.73	0.38	2.3					1
Rice (brown)	0.34	0.05	4.7	0.62	20	2	12	2
Rice (polished)	0.07	0.03	1.6	0.04	16	1	5	1
Rye	0.44	0.18	1.5	0.33	34	1		2
Sorghum	0.38	0.15	3.9					
Wheat	0.57	0.12	7.4	0.35	78	1	6	1
Wheat germ	2.01	0.68	4.2	0.92	328	2		
Wheat bran	0.72	0.35	21.0	1.38	258	3	14	
Wheat flour (patent)	0.13	0.04	2.1	0.05	25	1	1	

From Lockhart and Nesheim, 1978.

Table 4-5. Distribution of Cereal Endosperm Proteins According to Their Solubility: Expressed as a Percentage of Total Protein

Cereal	Protein range (% dwb)	Albumins (water-soluble)	Globulins (salt-soluble)	Prolamin (alcohol-soluble)	Residue and glutelin (alkali-soluble)
Barley	10–16	3–4	10–20	35–45	35–45
Maize	7–13	2–10	10–20	50–55	30–45
Oats	8–20	5–10	50–60	10–15	5
Rice	8–10	2–5	2–8	1–5	85–90
Rye	9–14	20–30	5–10	20–30	30–40
Sorghum	9–13	Trace	Trace	60–70	30–40
Triticale	12–18	20–30	5–10	20–30	30–40
Wheat (HRS)	10–15	5–10	5–10	40–50	30–40
Wheat (Durum)	12–16	10–15	5–10	40–50	30–40

Source: Simmonds, 1978.

by yeast or chemical leavening agents (see Chapter 9). The gluten proteins collectively contain about 17.55% nitrogen; hence, in estimating the crude protein content of wheat and wheat products from the determination of total nitrogen, the factor 5.7 is normally employed rather than the customary value of 6.25, which is based on the assumption that, on the average, proteins contain 16% nitrogen.

As a class, cereal proteins are not so high in biological value as those of certain legumes, nuts, or animal products (Tables 4-6, 4-7). Zein, the prolamin of corn, lacks lysine and is low in tryptophan. The limiting amino acid in wheat endosperm proteins is lysine. While biological values of the proteins of entire cereal grains are greater than those of the refined mill products, which consist chiefly of the endosperm, the American and North European diets normally include various cereals, as well as animal products. Under those conditions, different proteins tend to supplement each other, and the cereals are important and valuable sources of amino acids for the synthesis of body proteins. The distribution of kernel parts and proteins in those parts are shown in Table 4-8 for wheat and corn; in Table 4-9 for rice, covered barley, and sorghum; and in Table 4-10 for oats.

In most cereal grains, as total protein contents increases to about 14% the concentration of the albumins plus globulins (and consequently of lysine) in the protein decreases. Results of a study for barley are shown in Figure 4-2. At the same time, since the reports on a biochemical mutant for lysine in opaque-2 corn by Mertz et al (1964) and in floury-2 corn by Nelson et al (1965), there were reports on the possibility of increasing by genetic means the lysine concentration in the proteins of barley (Munck et al, 1970) and in sorghum (Singh and Axtell, 1973).

The carbohydrate contents, including crude fiber, of various grains, are shown in Table 4-11. The main form of carbohydrate is starch, which is the

Table 4-6. Protein Quality of Cereal Grains (PER[a])

	Actual	Estimate[b]
Barley	—	1.6
Buckwheat	—	1.8
Foxtail millet	—	1.0
Job's tears	—	1.0
Maize (normal)	1.2	1.2
Maize (opaque-2)	2.3	1.9
Pearl millet	1.8	1.6
Proso millet	—	1.4
Rogi (finger) millet	0.8	—
Rice (brown)	1.9	1.8
Rice (polished)	1.7	1.7
Rye	1.6	1.6
Oats	1.9	1.7
Sorghum	1.8	0.9
Wheat	1.5	1.3
Wheat germ	2.5	2.5
Wheat gluten	—	0.7
Wheat flour (80–90% extraction)	—	1.1
Wheat flour (70–80% extraction)	—	1.0
Wheat flour (60–70% extraction)	—	0.8
Bulgur wheat	—	1.2
Triticale	1.6	1.4

From Lockhart and Nesheim, 1978.
[a]Protein Efficiency Ratio
[b]Estimated from amino acid content assuming availability of amino acids the same as the amino acids in casein.

main source of calories provided by the grains. The major portion of the carbohydrates is in the starchy endosperm. For a discussion of starches, see Chapter 17 on corn.

Fatty acids in cereals occur in three main types—neutral lipids, glycolipids, and phospholipids. Total lipids and individual fatty acids in cereal grains and some of their products are shown in Table 4-12. The lipids in cereals are relatively rich in the essential fatty acid, linoleic acid. Saturated fatty acids (mainly palmitic) represent less than 25% of the total fatty acids for most grains. For a discussion of lipids, see Chapter 9 on wheat flour components in bread making. A summary of the chemical composition and nutritive value of cereal grains is presented in Table 4-13.

In summary, cereal grains are a diversified and primary source of nutrients. Their high starch contents make them major contributors of calories; they also contribute to our needs for proteins, lipids, vitamins, and minerals. Vitamins and minerals lost during milling into refined food products (wheat flour or white rice) can be (and in many countries are) replaced by nutrient fortification. The composition of cereal grains and their milled products make them uniquely suited in the production of wholesome, nutritional, and consumer-acceptable foods.

Table 4-7. Amino Acid Composition of Cereals (% by Weight)

Amino acid	Rice (brown)	Wheat (HRS)	Maize (field)	Sorghum	Pearl millet	Barley	Oats	Rye	Triticale	WHO requirement (1973)[a]
Tryptophan	1.08	1.24	0.61	1.12	2.18	1.25	1.29	1.13	1.08	1.0
Threonine	3.92	2.88	3.98	3.58	4.00	3.38	3.31	3.70	3.11	4.0
Isoleucine	4.69	4.34	4.62	5.44	5.57	4.26	5.16	4.26	3.71	4.0
Leucine	8.61	6.71	12.96	16.06	15.32	6.95	7.50	6.72	6.87	7.0
Lysine	3.95	2.82	2.88	2.72	3.36	3.38	3.67	4.08	2.77	5.5
Methionine	1.80	1.29	1.86	1.73	2.37	1.44	1.47	1.58	1.44	3.5
Cystine	1.36	2.19	1.30	1.66	1.33	2.01	2.18	1.99	1.55	
Phenylalanine	5.03	4.94	4.54	4.97	4.44	5.16	5.34	4.72	5.26	6.0
Tyrosine	4.57	3.74	6.11	2.75	—	3.64	3.69	3.22	2.14	
Valine	6.99	4.63	5.10	5.71	5.98	5.02	5.95	5.21	4.39	
Arginine	5.76	4.79	3.52	3.79	4.60	5.15	6.58	4.88	4.99	
Histidine	1.68	2.04	2.06	1.92	2.11	1.87	1.84	2.28	2.48	
Alanine	3.56	3.50	9.95	—	—	4.60	6.11	5.13	3.53	
Aspartic acid	4.72	5.46	12.42	—	—	5.56	4.13	7.16	5.00	
Glutamic acid	13.69	31.25	17.65	21.92	—	22.35	20.14	21.26	31.80	
Glycine	6.84	6.11	3.39	—	—	4.55	4.55	4.79	4.05	
Proline	4.84	10.44	8.35	—	—	9.02	5.70	5.20	12.06	
Serine	5.08	4.61	5.65	5.05	—	4.65	4.00	4.13	4.70	
Total protein (%)	7.5	14.0	10.0	11.0	11.4	12.8	14.2		17.3	

Source: Simmonds, 1978.
[a]Nutritionally essential amino acids.

Table 4-8. Distribution of Kernel Parts and of Proteins in Wheat and Corn[a]

Part of grain	Distribution of kernel parts (% kernel weight)		Protein content[b] ($N \times 6.25$, % part weight, db)		Protein distribution (% kernel protein)	
	Wheat	Corn	Wheat	Corn	Wheat	Corn
Pericarp	8.0	6.5	5.1	3.5	4.0	2.2
Aleurone	7.0	2.2	22.9	22.3	15.5	4.7
Germ[c]	(2.5)	(11.7)	(34.1)	(19.8)	(8.0)	(22.1)
Embryo	1.0	1.1	38.7	30.8	3.5	3.2
Scutellum	1.5	10.6	31.0	18.6	4.5	18.9
Starchy endosperm[c]	(82.5)	(79.6)	—	—	(72.5)	(71.0)
Outer	12.5	3.9	15.9	32.2	19.4	11.9
Middle	12.5	58.1	10.2	8.7	12.4	48.2
Inner	57.5	17.6	7.2	6.5	40.7	10.9
Whole grain	100	100	10.1	10.5	100	100

[a]From Hinton, 1953.
[b]Expressed on a dry-weight basis.
[c]Calculated values.

Table 4-9. Distribution of Kernel Parts and Proteins in Rice, Covered Barley, and Sorghum

Grain	Part	Weight distribution (% grain)	Protein content (% tissue)	Protein distribution[a] (% grain proteins)
Rice[b]	Hull	26.2	1.5	5
	Brown rice	73.8	7.8	95
Brown rice		100	7.8	100
	Pericarp	2.2	16.0	4
	Aleurone	3.4	15.8	7
	Embryo	2.2	25.3	7
	Milled rice	92.2	7.2	82
Covered barley[c]	Whole	100	12.4	100
	Hull	10.4	1.9	2
	Germ	3.7	35.0	11
	Dehulled-degermed	85.9	12.3	87
Sorghum[d]	Whole	100	—	100
	Pericarp	5.4	5.1	3
	Germ	8.0	16.3	12
	Endosperm	86.5	10.2	85

[a]Calculated values by using data from b–d.
[b]From Bechtel and Pomeranz, 1980. Source: IRRI Annual Report, 1974, p. 102.
[c]From Robbins and Pomeranz, 1971.
[d]From Haikerwal and Mathieson, 1971.

Table 4-10. Groat Weight; Weight Distribution of Fractions in Groats; Protein Contents of Oat Hulls, Groats, Groat Fractions; and Protein Distribution in Groats of *A. sativa* and *A. sterilis*

Analyses	A. sativa[a]		A. sterilis[b]	
	Mean	(Range)	Mean	(Range)
Groat weight (mg/kernel)	19.4	(14.4–22.7)	12.9	(8.0–17.8)
Weight distribution of fractions in groats (% groat weight)				
Groats:	100		100	
Embryonic axis	1.2	(1.1–1.4)	1.3	(0.8–2.0)
Scutellum	1.9	(1.7–2.6)	1.6	(1.0–2.4)
Bran	34.6	(28.7–41.4)	39.1	(33.0–48.1)
Starchy endosperm	62.3	(55.8–68.3)	58.0	(49.5–64.7)
Protein content ($N \times 6.25$, %, db)				
Hulls	1.6	(1.4–1.9)	3.4	(1.4–5.3)
Groats:	16.9	(13.8–22.5)	23.2	(19.3–25.5)
Embryonic axis	37.8	(26.3–44.3)	35.9	(27.8–45.7)
Scutellum	28.7	(24.2–32.4)	31.1	(20.2–43.2)
Bran	22.8	(18.5–32.5)	31.9	(22.7–38.9)
Starchy endosperm	11.7	(9.6–17.0)	17.5	(14.7–21.2)
Protein distribution in groats (% total protein in groat)[c]				
Groats:	100		100	
Embryonic axis	2.8	(1.9–4.0)	2.0	(1.1–3.5)
Scutellum	3.4	(2.3–4.5)	2.1	(1.2–4.7)
Bran	48.8	(41.3–57.3)	52.9	(44.5–62.8)
Starchy endosperm	45.0	(37.2–50.2)	43.0	(35.3–52.7)

[a]The average of seven varieties and their ranges reported by Youngs (1972).
[b]The average of 25 selections and their ranges reported by Youngs and Peterson (1973).
[c]Calculated values.

Table 4-11. Carbohydrate Contents of Cereal Grains and Their Products (g/100 g)

	Carbohydrate	
	Total	Fiber
Barley, pearled	78.8	0.5
Buckwheat, whole grain	72.9	9.9
Buckwheat flour, dark	72.0	1.6
Buckwheat flour, light	79.5	0.5
Bulgur, club wheat	79.5	1.7
Bulgur, hard red winter wheat	75.7	1.7
Bulgur, white wheat	78.1	1.3
Maize (field corn)	72.2	2.0
Maize (sweet corn, raw)	22.1	0.7
Maize flour (corn flour)	76.8	0.7
Malt, dry	77.4	5.7
Malt extract, dried	89.2	trace
Millet, proso	72.9	3.2
Oatmeal, dry	68.2	1.2
Popcorn, unpopped	72.1	2.1
Popcorn, popped, plain	76.7	2.2
Rice, brown	77.4	0.9
Rice, white	80.4	0.3
Rice bran	50.8	11.5
Rice polish	57.7	2.4
Rye	73.4	2.0
Rye flour, light	77.9	0.4
Rye flour, medium	74.8	1.0
Rye flour, dark	68.1	2.4
Sorghum grain	73.0	1.7
Wheat, hard red spring	69.1	2.3
Wheat, hard red winter	71.1	2.3
Wheat, soft red winter	72.1	2.3
Wheat, white	75.4	1.9
Wheat, durum	70.1	1.8
Wheat flour, 80% extraction	74.1	0.5
Wheat flour, straight, hard wheat	74.5	0.4
Wheat flour, straight, soft wheat	76.9	0.4
Wheat flour, patent, all purpose	76.1	0.3
Wheat bran	61.9	9.1
Wheat germ	46.7	2.5

From Lockhart and Nesheim, 1978.
Source: Watt and Merrill, 1964.

Table 4-12. Fatty Acid Composition of Cereals and Related Products (g/100 g food, edible portion)

Food	Water	Total lipid	Saturated					Unsaturated				
			Sum	14:0	16:0	18:0	20:0	Sum	16:1	18:1	18:2	18:3
Barley (*Hordeum vulgare*), whole grain	14	2.8	0.48	0.01	0.45	0.02	0	1.52	0.01	0.24	1.14	0.13
Buckwheat, domestic (*Fagopyrum* spp.), whole grain	11	2.4	0.46	0.01	0.36	0.05	0.04	1.66	0	0.80	0.74	0.12
Corn (*Zea mays*)												
Whole grain, raw	13.8	4.1	0.47	0	0.40	0.06	0.01	3.07	0.01	0.91	2.12	0.03
Flour	12	2.6	0.30	0	0.25	0.04	0.01	2.00	—[b]	0.64	1.34	0.02
Germ	0	30.8	3.93	0	3.30	0.54	0.09	25.57	0.04	7.58	17.7	0.25
Grits, degermed, enriched or unenriched dry form	12	0.8	0.09	0	0.08	0.01	—	0.60	—	0.18	0.41	0.01
Corn oil (commercial)	0	100	12.73	0	10.7	1.74	0.29	82.96	0.14	24.6	57.4	0.82
Cornmeal, white or yellow												
Bolted (nearly whole grain)	12	3.4	0.39	0	0.33	0.05	0.01	2.54	—	0.75	1.76	0.03
Degermed, enriched or unenriched dry form	12	1.2	0.14	0	0.12	0.02	—	0.90	—	0.27	0.62	0.01
Farina, enriched, regular dry form	10.3	1.5	0.24	—	0.22	0.01	0.01	0.83	—	0.11	0.69	0.03
Millet, pearl millet (*Pennisetum glaucum*) whole grain	11.8	4.1	0.86	0	0.68	0.16	0.02	2.67	0.02	0.83	1.69	0.13
Oatmeal or rolled oats (*Avena sativa*) dry form	8.3	7.4	1.37	0.02	1.21	0.10	0.04	5.65	0.02	2.60	2.87	0.16
Rice (*Oryza sativa*)												
Brown, dry form	12	2.3	0.62	0.03	0.54	0.04	0.01	1.36	0.01	0.54	0.78	0.03
White, fully milled or polished, enriched dry form	12	0.8	0.21	0.01	0.19	—	—	0.47	—	0.19	0.27	0.01
Bran oil	0	100	19.50	0.69	16.7	1.58	0.53	74.56	0.26	39.2	33.5	1.60

Fatty acid[a]

Rye (*Secale cereale*), whole grain	12.1	2.2	0.27	—	0.25	0.02	0	1.30	0.01	0.22	0.95	0.12
Sorghum (*Sorghum vulgare*) whole grain	11	3.3	0.48	0.01	0.44	0.03	0	2.74	0.04	1.15	1.46	0.09
Triticale (*Triticum × Secale*), whole grain flour	14	3.4	0.49	0.02	0.45	0.02	0	1.91	0	0.28	1.48	0.15
Wheat (*Triticum aestivum*)												
Whole grain												
Hard red spring	14	2.7	0.37	—	0.36	0.01	—	1.56	0.01	0.25	1.20	0.10
Hard red winter	14	2.5	0.35	—	0.33	0.02	—	1.47	0.01	0.28	1.08	0.10
Soft red winter	14	2.4	0.35	—	0.33	0.02	0	1.40	0.01	0.25	1.07	0.07
White	14	2.0	0.30	—	0.28	0.02	0	1.14	0.01	0.18	0.88	0.07
Flours												
Hard red spring	14	1.5	0.23	—	0.21	0.01	0.01	0.84	—	0.12	0.69	0.03
Hard red winter	14	1.5	0.20	—	0.19	0.01	—	0.74	—	0.10	0.64	0.03
Soft red winter	14	1.4	0.22	—	0.20	0.01	0.01	0.76	—	0.08	0.65	0.03
All-purpose	14	1.4	0.23	—	0.22	0.01	—	0.72	—	0.11	0.58	0.03
Bran	14	4.6	0.74	0.01	0.69	0.04	—	3.09	0.02	0.71	2.20	0.16
Germ	14	10.9	1.88	—	0.01	1.81	0.06	8.18	0.04	1.54	5.86	0.74
Wheat, durum (*Triticum durum*)												
Whole grain	14	3.3	0.54	—	0.51	0.03	0	1.88	0.01	0.40	1.36	0.11
Semolina	14	1.8	0.33	—	0.31	0.02	0	0.90	0.01	0.17	0.68	0.04

From Lockhart and Nesheim, 1978.

[a] 14:0 = myristic; 16:0 = palmitic; 18:0 = stearic; 20:0 = arachidic; 16:1 = palmitoleic; 18:1 = oleic; 18:2 = linoleic; 18:3 = linolenic.

[b] Dashes denote <0.005 g.

Table 4-13. Chemical Composition (on Dry-Matter Basis) and Nutritional Value of Cereal Grains

Chemical composition and nutritional value	Wheat[a]	Corn		Rice			Barley[a]	Triticale[a]	Millet[a]	Sorghum		
		Whole[b]	Degermed[b]	Brown[c]	Brown[d]	Milled[d]				Whole[c]	Whole[c]	Flour[e] (64%)
Protein (N × 6.25) (g/100 g)	12.3	9.9	8.7	8.5	9.9	9.2	15.8	15.5	13.4	9.6	15.6	16.7
Fat (g/100 g)	2.2	5.2	1.4	2.6	3.0	1.2	3.9	2.6	5.5	4.5	4.2	1.7
Available CHO (g/100 g)[f]	81.1	76.0	89.2	74.8	85.8	92.5	64.9	67.4	73.7	67.4	72.9	81.5
Crude fiber (g/100 g)	1.2	2.1	0.5	0.9	0.9	0.2	4.3	2.7	1.8	4.8	2.2	0.8
Ash (g/100 g)	1.6	1.4	0.4	1.6	1.3	0.6	2.2	2.0	1.8	3.0	2.0	0.9
Tannin (g/100 g)	0.5	—[g]	—[g]	0.1	0.4	0.3	0.8	0.7	0.7	1.9	0.5	0.2
Energy (kcal/100 g)	436	449	437	447	437	432	454	442	459	447	454	447
TD (%) = Total digestible	96.0	95.6	101.1	99.7	95.0	99.3	88.0	93.0	93.0	84.8	92.9	96.1
BV (%) = biological value	55.0	61.7	54.1	74.0	66.1	64.4	70.0	66.0	66.0	59.2	58.9	54.4
NPU (%) = net protein utilization	53.0	58.9	54.7	73.8	62.8	63.9	62.0	61.0	56.0	50.0	54.7	52.3
Utilizable protein (g/100 g)	6.5	5.8	4.8	6.3	6.2	5.9	7.9	9.5	7.5	4.8	8.5	8.7
Digestible energy (%)	86.4	89.9	95.8	96.3	96.0	98.2	81.0	—[g]	87.2	79.0	87.0	90.5

[a] From Khan and Eggum, 1978.
[b] From Pedersen and Eggum, 1983b.
[c] From Eggum, 1979.
[d] From Pedersen and Eggum, 1983a.
[e] From Pedersen and Eggum, 1983c.
[f] CHO = Carbohydrate.
[g] Not reported.

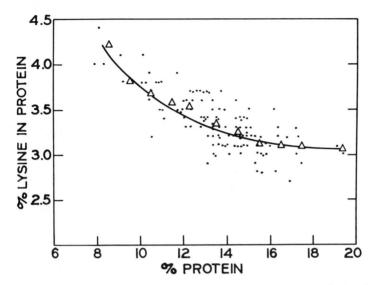

Figure 4-2. Plot of relation between percent protein and percent lysine in protein of 113 barleys from the USDA World Collection. Dots represent individual samples; triangles are averages of percent lysine in protein for 1% intervals in protein content. (From Pomeranz et al., 1976.)

BIBLIOGRAPHY

Bechtel, D. B.; Pomeranz, Y. *Adv. Cereal Sci. Technol.* **1980**, *3*, 73–113.

Eggum, B. O. In "Chemical Aspects of Rice Grain Quality," Proc. of Workshop. Int. Rice Res. Inst.: Los Banos, Philippines, **1979**, pp. 91–111.

FAO. "Amino Acid Content of Foods and Biological Data on Proteins," FAO Nutr. Study 24. FAO: Rome, **1970**.

Haikerwal, M.; Mathieson, A. R. *Cereal Chem.* **1971**, *48*, 690–699.

Holme, J. *Baker's Dig.* **1966**, *40*(5), 38–42.

Hinton, J. J. C. *Cereal Chem.* **1953**, *30*, 441–445.

Khan, M. A.; Eggum, B. O. *J. Sci. Food Agric.* **1978**, *29*, 1069–1075.

Lockhart, H. B.; Nesheim, R. O. In "Cereals '78: Better Nutrition for the World's Millions," Y. Pomeranz, Ed. Am. Assoc. Cereal Chem.; St. Paul, MN, **1978**.

Mertz, E. T.; Bates, L. S.; Nelson, O. E. *Science* **1964**, *145*, 279–280.

Munck, L.; Karlson, K. E.; Hagberg, A.; Eggum, B. O. *Science* **1970**, *168*, 985–987.

Nelson, O. E.; Mertz, E. T.; Bates, L. S. *Science* **1965**, *150*, 1469–1470.

Pedersen, B.; Eggum, B. O. *Qual. Plant. Plant Food Hum. Nutr.* **1983a**, *33*, 267–278.

Pedersen, B.; Eggum, B. O. *Qual. Plant. Plant Food Hum. Nutr.* **1983b**, *33*, 299–311.

Pedersen, B.; Eggum, B. O. *Qual. Plant. Plant Food Hum. Nutr.* **1983c**, *33*, 313–326.

Pomeranz, Y.; MacMasters, M. M. *Baker's Dig.* **1968**, *42*(4), 24–28.

Pomeranz, Y.; Robbins, G. S.; Smith, R. T.; Craddock, J. C.; Gilbertson, J. T.; Moseman, J. G. *Cereal Chem.* **1976**, *53*, 497–504.

Robbins, G. S.; Pomeranz, Y. *Am. Soc. Brewing Chem.* **1971**, 15–21.

Simmonds, D. H. In "Cereals '78: Better Nutrition for the World's Millions," Y. Pomeranz, Ed. Am. Assoc. Cereal Chem.: St. Paul, MN, **1978**, p. 107.

Singh, R.; Axtell, J. D. *Crop Sci.* **1973**, *13*, 535–539.

USDA. "Agricultural Statistics." U.S. Government Printing Office: Washington, D.C.; **1982**.

Watt, B. K.; Merrill, A. L. "Composition of Foods, Raw, Processed, Prepared." Agriculture Handbook No. 8, U.S. Dept. Agric.: Washington, D.C.; **1963**.

Youngs, V. L. *Cereal Chem.* **1972**, *49*, 407–411.

Youngs, V. L.; Peterson, D. M. *Crop Sci.* **1973**, *13*, 365–367.

Chapter 5

Classification and Standards

At the turn of the century, many organizations in the United States and in several other grain-producing countries tried to develop uniform grain standards. The demand for uniform grades and application of the standards resulted in the introduction in the years 1903–1916 of 26 bills in the U.S. Congress. The bills provided either for federal supervision of grain grading or for actual federal grain inspection.

Those bills and the extensive hearings culminated in the United States Grain Standards Act that was passed August 11, 1916. The Act was amended in 1940 to include soybeans. The Act provides in part for (1) the establishment of official grain standards, (2) the federal licensing and supervision of the work of grain inspectors, and (3) the mechanism of filing appeals from grades assigned by licensed inspectors. The official U.S. standards for grain cover wheat, corn, barley, oats, rye, sorghum, flaxseed, soybeans, mixed grain, and triticale.

Tables 5-1 through 5-4 summarize wheat standards used in the People's Republic of China, the Soviet Union, and Canada. Table 5-1 shows that the requirements in China are rather minimal; the wet gluten requirement there of 30% in bread wheats and 26% in other wheats is not always met. Requirements listed in Tables 5-2 and 5-3 govern the market price of grain by stipulating the premium for high-quality grain in the Soviet Union. In addition, the gluten in classes 1 and 2 should be of type I quality and in class 3 of type II quality. This requires determination of rheological

Table 5-1. Chinese Standards for Wheat

| Class | kg/hl (min) | | Dockage (max) |
	Shensi	Shanghai	
1	75.0	79.0	0.4
2	73.0	77.0	0.8
3	71.0	75.0	1.2
4	69.0	73.0	1.6

This grain should be sound, show no live insects, and be free of scab.

Table 5-2. USSR Guidelines for Premiums on Strong Hard Wheat Cultivars

Class	kg/hl (min)	Vitreous-ness (min) (%)	Sprouted (%) (max)	Mixtures of kernels difficult to separate (ie, tartar, buckwheat) (max)	Gluten of first quality (%)	Other wheat types (max)	Premium above soft wheat (%)
1	750	60	1.0	2.0	Over 32	10	50
2	750	60	1.0	2.0	28–31	10	30
3					At least 25 second qualtiy		10

Moisture, waste, damaged (shrunken, heat-damaged, underdeveloped, immature, etc), and insect-damaged grain must conform to requirements for limiting factors. Strong hard wheat should be neither discolored nor bleached and have a normal taste and smell.

Table 5-3. USSR Guidelines for Premiums on Durum Wheat Cultivars

Class	kg/hl (min)	Other wheat classes (max) (%)	Gluten (min) (%)	Gluten quality (min)	Barley, rye, and sprouted seeds (max) (%)	Sprouted seeds (max) (%)	Premium above soft wheat (%)
1	770	10	28	Class 2	2.0	0.5	65
2	745	15	25	Class 2	2.0	0.5	40
3	745	15	22	Class 2	4.0	3.0	20

Durum wheat grain not answering the quality of class 3 is paid 10% premium above soft wheat. Durum wheat (winter and spring habitat) must have normal taste, color, and smell. Moisture, waste, damaged (shrunken, heat-damaged, underdeveloped, immature, etc), and insect-damaged grain must conform to requirements for limiting factors.

properties of the washed-out gluten. The seven classes of Canadian wheat offered for marketing are:

1. Canada Western Red Spring, 13.5% protein, a high-quality milling and bread making hard wheat;
2. Canada Prairie Spring Wheat (a new class introduced August 1, 1985), 11.5–12.5% protein, medium to soft kernel characteristics, for production of French-type hearth breads and flat breads;
3. Canada Western Red Winter Wheat, protein 11.5%, a hard wheat for French-type hearth breads, flat breads, steamed breads, and certain types of noodles;
4. Canada Utility Wheat, 12.5% protein, hard wheat for production of pan and hearth breads;
5. Canada Western Soft White Spring Wheat, 10.5% protein; for production of cookies, pastry, flat breads, noodles, steamed breads, and chappattis;

Table 5-4. Official Standards for Canada Western Red Spring Wheat

| Grades | | Standard of quality | | | Maximum limits of | | | |
| | | | | | Foreign material other than wheat | | Wheats of other classes or varieties | |
Grade name	Minimum weight per measured bushel in pounds	Variety	Minimum percentage by weight of hard vitreous kernels	Degree of soundness	Matter other than cereal grains	Total including cereal grains other than wheat	Durum	Total including durum
No. 1 Canada western red spring	59	Marquis or any variety equal to Marquis	65	Reasonabaly well matured, reasonably free from damaged kernels	Practically free	About 0.75%	About 1%	About 3%
No. 2 Canada western red spring	57	Marquis or any variety equal to Marquis	35	Fairly well matured, may be moderately bleached, or frost damaged, but reasonably free from severely weather damaged kernels	Reasonably free	About 1.5%	About 3%	About 6%
No. 3 Canada western red spring	54	Any variety of fair milling quality	—	Excluded from higher grades on account of frosted, immature, or otherwise damaged kernels	Reasonably free	About 3.5%	About 5%	10%

Maximum tolerances—grading factors

Grade	Sprouted	Severely bin-burned	Heated, incl. severely bin-burned	Fire-burned	Dark immature	Inseparable seeds	Stones	Ergot	Sclerotina
No. 1 CW	0.5%	2K	0.1%	Nil	1%	20K	3K	3K	3K
No. 2 CW	1.5%	5K	0.75%	Nil	2.5%	50K	3K	6K	6K
No. 3 CW	5.0%	10K	2%	Nil	10%	100K	5K	24K	24K

Note: The letter "K" refers to kernels or kernel size pieces in 500 g.
Red Spring will only be graded "rejected" on account of "stones" or "dried." Samples containing any other factors affecting quality will be degraded to a lower grade.
Stones: Samples containing stones in excess of grade tolerances, up to 2.5%, grade "rejected" account stones; over 2.5% or when containing any conspicuous material normally found in salvaged grain, grade wheat, sample salvage.
Dried: Wheat that has been damaged in artificial drying will be graded rejected "grade" account dried.

Maximum tolerances—grading factors (cont'd)

Grade	Artificial stain, no residue	Natural stain	Smudge	Total smudge and blackpoint	Degermed kernels	Insect damage[a]			Pink kernels
						Sawfly or midge	Grasshopper or army-worm	Grass green kernels[b]	
No. 1 CW	Nil	0.5%	30K	10%	4%	2%	1%	0.75%	1.5%
No. 2 CW	5K	2%	1%	15%	7%	8%	3%	2%	5%
No. 3 CW	10K	5%	5%	35%	13%	25%	8%	10%	10%

Note: The letter "K" refers to kernels or kernel size pieces in 500 g.
[a] Insect damage: Sawfly or midge damage refers to kernels that are shriveled or distorted; grasshopper or armyworm damage refers to kernels that are chewed, usually on the sides. The above tolerances are not absolute maximums. Consideration is given to the degree of damage in conjunction with the overall quality of the sample.
[b] Grass green kernels tolerances are given as a general guide and may be increased or reduced after consideration of the general quality of a sample.
From Grain Grading Handbook for Western Canada, Canadian Grain Commission, Winnipeg, Manitoba, August 1, 1977, Agricultural, Canada.

Table 5-5. Official U.S. Standards for Wheat

| Grade | Minimum test weight per bushel (lbs) | | Percent maximum limits of | | | | | Wheat of other classes[c] | |
	Hard red spring wheat or white club wheat	All other classes and sub-classes	Heat-damaged kernels	Damaged kernels (total)[a]	Foreign material	Shrunken and broken kernels	Defects (total)[b]	Contrasting classes	Wheat of other classes (total)[d]
U.S. No. 1	58.0	60.0	0.2	2.0	0.5	3.0	3.0	1.0	3.0
U.S. No. 2	57.0	58.0	0.2	4.0	1.0	5.0	5.0	2.0	5.0
U.S. No. 3	55.0	56.0	0.5	7.0	2.0	8.0	8.0	3.0	10.0
U.S. No. 4	53.0	54.0	1.0	10.0	3.0	12.0	12.0	10.0	10.0
U.S. No. 5	50.0	51.0	3.0	15.0	5.0	20.0	20.0	10.0	10.0

U.S. sample grade shall be wheat which: (1) does not meet the requirements for the grades U.S. Nos. 1, 2, 3, 4, or 5 or (2) contains a quantity of smut so great that one or more of the grade requirements cannot be determined accurately; or (3) contains eight or more stones, two or more pieces of glass, three or more crotalaria seed (*Crotalaria* spp.), three or more castor beans (*Ricinus communis*), four or more particles of an unknown foreign substance(s) or a commonly recognized harmful or toxic substance(s) or two or more rodent pellets, bird droppings, or an equivalent quantity of other animal filth per 1,000 g of wheat; or (4) has a musty, sour, or commercially objectionable foreign odor (except smut or garlic odor); or (5) is heating or otherwise of distinctly low quality.

[a] Includes heat-damaged kernels.
[b] Defects (total) include damaged kernels (total), foreign material, and shrunken and broken kernels. The sum of these three factors may not exceed the limit of defects.
[c] Unclassed wheat of any grade may contain not more than 10% of wheat of other classes.
[d] Includes contrasting classes.

6. Canada Western Amber Durum Wheat, 13.5% protein, very hard wheat, for production of semolina for alimentary pastes (ie. spaghetti);
7. Canada Eastern White Winter Wheat, 9.5% protein, soft kernels for production of cookies, cakes, pastry.

In addition, Sample Account Variety (previously Canada Feed) includes unlicensed, high-yielding, semi-dwarf red spring wheats that meet physical grade specifications of No. 3 Canada Western Red Spring Wheat. They are generally low in protein content and high in alpha amylase activity (as a result of sprout damage). Table 5-4 shows the present official standards for Canadian western red spring wheat. The main grading factors are bushel weight, variety, soundness, and foreign material. Protein content is not a grading factor. Grain is evaluated on the basis of visual assessment of factors related to quality. Grades are assigned according to specifications established under the Grain Act.

According to U.S. grain standards (Table 5-5), wheat is divided into seven classes on the basis of type and color: hard red spring, durum, hard red winter, soft red winter, white, and mixed (see Figure 5-1; note that at present there are no subclasses of hard red winter). Division of wheat into classes and subclasses effects a classification according to suitability of different wheats for specific uses. Hard red spring, hard red winter, and hard white wheats are valued for the production of flours used in making yeast leavened baked goods (bread, rolls). Amber durum wheat is

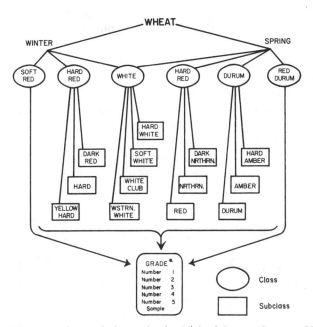

Figure 5-1. Wheat grades and classes in the United States (Source: USDA.)

prized for the manufacture of semolina and farina used to produce alimentary pastes (macaroni, vermicelli, etc). Red durum is used primarily as a poultry and livestock feed.

The wheat in each subclass is sorted into a number of grades on the basis of sound quality and condition. The tests that are generally used to determine the grade of grain and its conformity to the official U.S. Grain Standards include plumpness, soundness, cleanliness, dryness, purity of type, and the general condition of the grain.

Plumpness is measured by the weight per bushel (or kg/hl) test. It is supplemented by sizing of some grains. *Soundness* is indicated by the absence of musty, sour, or commercially objectionable odor and by the quantity of damaged kernels. *Cleanliness* is measured by the amount of foreign matter. *Dryness* is measured by the moisture content. *Purity* of type is provided for by classes of the same grain. *General condition* determining soundness is described above and is also designated by such terms as "smutty," "garlicky," "weevily," "stained," "bright," "tough," or "treated."

Other cereals are graded in a similar manner. Corn is divided into three classes: yellow, white, and mixed. These class designations apply to dent-type corn. If the corn consists of at least 95% of flint varieties, the word "flint" is used. If the corn consists of more than 5% but less than 95% flint varieties, the corn is designated as "flint and dent." Waxy corn is corn of any class that consists of at least 95% waxy corn. Moisture content has been for many years a grading factor for corn and sorghum but not for the other cereal grains. Moisture content, especially at the beginning of the harvest, is a very important determinant of corn quality and value. Broken corn and foreign material (BCFM) is determined by sieving through a 12/64 sieve.

Rough rice is divided into three classes: long, medium, and short. There are separate standards for rough, brown, and milled rice. All rough rice is graded on the basis of milled rice yields.

Barley is divided into three classes: six-rowed (subclasses: malting, blue malting, and barley), two-rowed (two-rowed malting and two-rowed), and barley. Plumpness is determined by sieving. The designation "malting" requires inclusion only of varieties that have been approved as suitable for malting purposes. Special grades and designations of oats are heavy and extra heavy (test weights of 38–40 and >40 lb/bu, respectively, thin, bleached, bright, ergoty, garlicky, smutty, tough (moisture >14% but <16%), and weevily.

Rye is classified into five regular grades and several special grades: plump, tough, smutty, garlicky, weevily, and ergoty.

The four classes of sorghum are brown, white, yellow, and mixed.

The five grades of triticale with special grade designations are: ergoty, light garlicky, garlicky, light smutty, smutty, and weevily.

According to the requirements of the European Economic Community, three factors determine standard quality of sound grain: moisture, Besatz

Table 5-6. Requirements for Cereal Grains According to the Standards of the EEC

Grain	Moisture (%, max)	Besatz[a] (%, max)	Test weight (kg/hl, min)
Wheat	16	5	75
Durum wheat	—	2	78
Rye	16	5	71
Barley	16	3	67
Corn	15	8	—
Oats	16	3	49
Sorghum	15	16	—
Millet	13	—	—

Courtesy: Commonwealth Secretariat, London.
[a]Damaged grains, grains of other cereals etc, according to specifications for each type of grain.

(admixture of any part of a sample that does not constitute faultless speciments of the basic grain) and test weight (expressed as kg/hl). The quantitative requirements of the EEC are listed in Table 5-6.

Australian wheat is marketed in seven general classes: Australian prime hard is limited to high-quality, hard-grained varieties of good milling quality and balance of dough properties. Australian prime hard is marketed on the basis of guaranteed minimum protein content, with levels set at 13%, 14%, and 15% protein. Australian hard consists of hard-grained varieties of proven bread-making quality. Australian hard is marketed at specified protein levels, usually within the range of 11.5–14%. Australian standard white is a multipurpose wheat of intermediate grain hardness with a protein level generally in the range of 9.5–11.5%. ASW wheat must satisfy a wide range of end-use requirements including conventional pan bread, steamed bread, Arabic and Indian style flat bread, and most types of noodles. Australian soft consists of low-protein wheat of specified soft varieties and is segregated in the southern part of Western Australia, Victoria, and New South Wales. Australian soft is suited to the production of sweet biscuits, cakes, and pastry goods. Australian general purpose and Australian feed relate to wheat that fails to conform to the standards of any of the above classes, usually in terms of test weight, weather damage, unmillable material, or contamination with foreign matter or seeds. Depending on the degree of weather damage and the test weight (74–68 kg/hl), some of the general purpose wheat may be suitable for milling purposes. Wheat of less than 68 kg/hl is marketed as Australian feed. Australian durum is limited to durum varieties and is suited to the production of semolina for pasta products.

Chapter 6

Storage

Cereal grains have been important to civilization because of their excellent keeping qualities. Postharvest losses, however, are a significant factor in the world food supply. Estimates of losses as high as 50% have been reported for some countries. Most losses result from infestation by insects, microorganisms, rodents, and birds. A smaller, but quite significant proportion of the total losses results from respiration and gradual deterioration of viability, nutritive quality, and end-use properties under commercial conditions. Nutrients are lost because of changes in carbohydrates, proteins, lipids, and vitamins. Functional properties including discoloration, caking, and abnormal odors also occur. Also, fungus-produced mycotoxins that elicit a toxic response when damaged grain is ingested may be produced.

Grain can be stored in flat warehouses or in reinforced concrete or steel vertical silos. The former are less expensive to build; the latter make possible better utilization of space, simple mechanical and/or automated loading and unloading, aeration and blending, and treatment against pests. Hermetically closed (under- or above-ground or even underwater) storage is useful for uninterrupted long-term preservation.

STORAGE INSECTS

Although there are several hundred species of insects associated with stored grains and their products, only a few of these cause in practice serious damage. They are confined to the groups of beetles (Coleoptera) and moths (Lepidoptera). The major beetle pests include the rice weevil *Sitophilus oryzae* (L.); maize weevil, *S. zeamais* Motschulsky; granary weevil *S. granarius* (L.); lesser grain borer, *Rhyzoperta dominica* (F.); red flour beetle, *Tribolium castaneum* (Herbst); confused flour beetle, *T. confusum* du Val; khapra beetle, *Trogoderma granarium* Everts; cadelle, *Tenebroides mauritanicus* (L.); saw-toothed grain beetle, *Oryzaephilus surinamensis* (L.); merchant grain beetle, *O. mercator* (Fauvel); flat grain beetle, *Cryptolestes pusillus* (Schonherr); rusty grain beetle, *C. ferrugineus* (Stephens); and the flour mill beetle *C. turcicus* (Grouvelle). The major moth pests are the *Angoumois* grain moth, *Sitotroga cerealella* (Olivier); Indian meal moth,

Plodia interpunctella (Hubner); almond moth, *Cadra cautella* (Walker); Mediterranean flour moth; *Anagasta kuehniella* (Zeller); and the tobacco moth, *Ephestia elutella* (Hubner). The grain mite *Acarus siro* (L.); the mold mite, *Tyrophagus putresentiae* (Shrank); and several other mites are often serious pests in cold climates.

The rice weevil, the maize weevil, the granary weevil, the lesser grain borer, and the Angoumois grain moth are internal feeders because they develop inside the grain kernel. All the other above-mentioned beetles feed primarily in the open or broken cereals and are externally developing insects. Examples of some important pests and kernel damage are shown in Figures 6-1 and 6-2.

STORAGE FUNGI

The most common genera of fungi in grains are *Fusarium*, *Alternaria*, *Cladosporium*, *Helminthosporium*, *Aspergillus*, and *Penicillium*. The last two are considered storage fungi; the others, field fungi. Figure 6-3 shows several of these fungi. Among the most common storage fungi are *Aspergillus flavus*, *A. ochraceus*, *A. glaucus*, *A. restrictus*, *A. candidus*, and several *Penicillium* species. These organisms grow at temperatures up to 50°C and at relative humidities above 65%. They cause most damage in grain stored above 25°C and 85% RH. The penicillia generally thrive best below 30°C, and the aspergillia tolerate higher temperatures. Both groups of fungi can exist in a wide range of oxygen tensions. *A. parasiticus* and *A. flavus* produce aflatoxins, a family of steroid-type carcinogens. *A. ochraceus* elaborates ochratoxin, which causes liver and kidney damage. Among the penicillia, citreoviridin is a toxin from *P. toxicarium* (mainly in rice) that may cause acute cardiac beriberi. Several penicillia may cause a swine kidney disease called fungus nephrosis. Several *Fusarium* species, generally considered as field fungi, may be present in stored wet grain and excrete toxins. *F. tricinctum* can cause alimentary toxic aleukia, and *F. roseum* may produce zearalenone, which causes an estrogenic syndrome of pigs.

STORAGE PROPERTIES OF GRAIN

The principal factors that control respiration and weight losses of stored grain are moisture, temperature, aeration, and previous history (condition). Moisture is by far the most important single factor. If the moisture content is sufficiently low and uniform throughout the bulk of the stored grain, it can be stored for many years with little deterioration. Moisture content at which a marked rise in the respiratory rates occurs is in the range in which heating and spoilage start in storage. Different cereal species exhibit some-what different critical moisture values—generally, at about 14%. A RH of

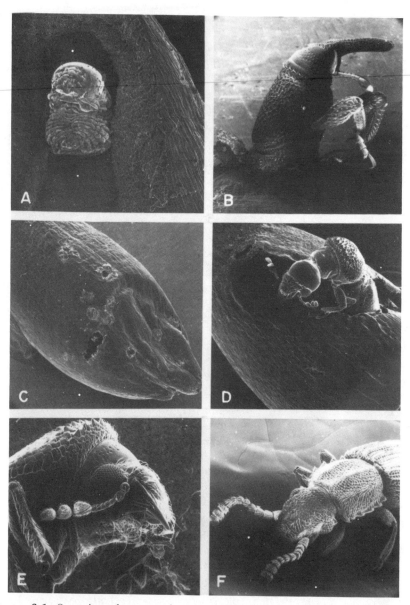

Figure 6-1. Scanning electron micrographs of beetles and wheat damage. (A) Section of wheat showing rice weevil larvae in endosperm tunnel (25×). (B) Rice weevil adult initially emerging from wheat kernel (12×). (C) Germ end of wheat kernel showing rice weevil egg plugs, feeding punctures, and beginning of emergence hole (6×). (D) Lesser grain borer adult emerging from wheat kernel, head extended (8×). (E) Head and prothorax of lesser grain borer, head retracted (25×). (F) Merchant grain beetle (15×). (Courtesy U.S. Grain Marketing Research Laboratory, Manhattan, Ks, G.D. White and J. L. Wilson.)

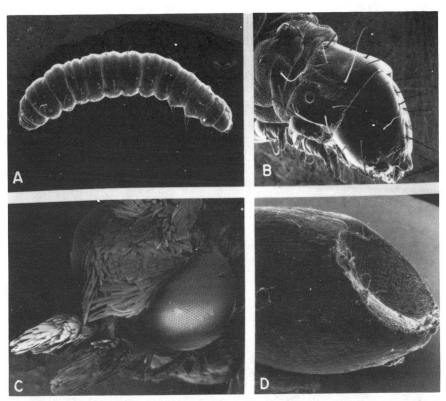

Figure 6-2. Scanning electron micrographs of Indian meal moth. (A) Larva (12×). (B) Larval head and first two thoracic segments (18×). (C) Adult head (22×). (D) Larval damage to wheat kernel (germ eaten away) (7×). (Courtesy U.S. Grain Marketing Research Laboratory, Manhattan, Ks, G.D. White and J. L. Wilson.)

75% is, generally, minimum for the germination of fungal spores at storage temperatures. It is now generally agreed that the so-called critical moisture level for any cereal species is the percentage moisture at which the seed is in equilibrium with an atmospheric RH of about 75%. That the marked increases in respiration for different grains occur at a rather constant RH of 75% in the interseed atmosphere, at which the equilibrium moisture content may vary widely among cereal grains, clearly indicates that the total moisture content of the grain is not the controlling factor. If the grain is cracked or broken, or if the storage is prolonged at high temperatures, the maximum moisture limit should correspond to a lower RH. For long-term storage (up to about 3 years), an RH as low as 65% is considered as a safe maximum. Respiration is accelerated by an increase in temperature until it is limited by such factors as the thermal inactivation of the enzymes involved, exhaustion of substrate, limitation in oxygen supply, or accumulation of inhibitory concentrations of carbon dioxide. In addition, the effect of temperature on respiratory rate depends on the moisture content of the seed

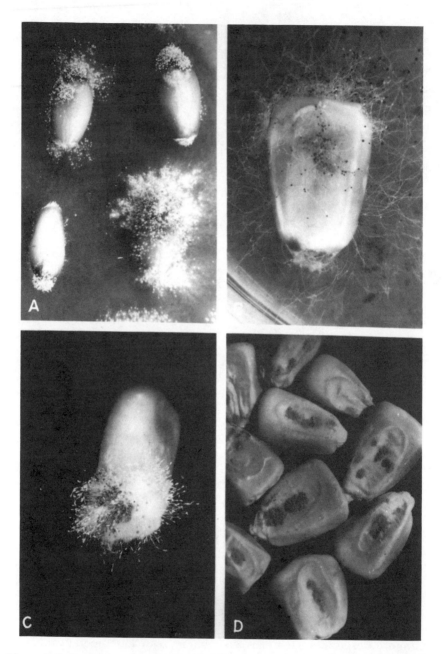

Figure 6-3. Representative members of the group of storage fungi. (A) *Aspergillus glaucus* and *A. restrictus* on wheat. (B) *Rhizopus* sp. on corn. (C) *A. clavatus* on corn. (D) *A. glaucus* on corn. (Courtesy: D.B. Sauer.)

and relative contributions of seeds, microorganisms, insects, and mites to total respiration. The interrelation of the different variables is so complex that empirical determination of optimum temperatures for grain respiration yield only approximate values and can be applied only to the conditions under which the values were determined. Since aerobic respiration of the grain and of the microorganisms associated with the grain involve consumption of oxygen and liberation of carbon dioxide, the process tends to be limited by the oxygen supply.

The storage properties of grain are influenced by environmental conditions during growth and maturation, by the degree of maturity at harvest, by methods of harvesting, and by the handling the seed has received until it is placed in storage. Softer grain types respire more rapidly than harder types at similar moisture levels and temperatures. Respiratory activity and the tendency of grain and its products to deteriorate in storage are considerably influenced by the "condition" or "soundness" of the product. This is one of the main reasons it is impossible to establish a safe maximum limit for the storage of any grain or grain product. Grain containing a high percentage of damaged kernels or showing other evidences of unsoundness is more likely to heat than sound grain of the same moisture content.

One of the poorly defined types of damage in wheat associated with storage deterioration is the condition known in the grain trade as "sick" wheat. It is manifested by kernels with dull appearance, in which the germs are dead and exhibit various degrees of darkening. Although storage under conditions that exclude mold damage may result in germ discoloration, the changes are not typical of sick wheat encountered in commercial storage. Under commercial grain storage, fungi are the primary cause of loss of germinability and germ damage.

Drying grain at elevated temperatures may reduce its nutritive value, bread-making potential, and germinative power and capacity. There is a relation between grain moisture content and the "critical temperature" at which it may be dried without damage. As the moisture content increases, the critical temperature decreases. The exact relationship depends on the grain, its history, drying conditions, and the criteria used to estimate damage (ie, germinability, nutritional value, functional properties in bread making or in starch production or in production of breakfast cereals, etc). To avoid damage to bread making, wheat temperatures should not exceed 35°C at 20%, 40°C at 18–20%, and 45°C at 18% moisture. This corresponds to air temperatures of 60–80°C. Damage to nutritive value of corn (losses of essential amino acids) is relatively small even at air temperatures of 120°C for about 30 minutes. Drying should be slow and gradual (reduction of about 4% per cycle) to avoid excessive cracks and fissures, especially in corn. Grain with moisture levels above 20% can be stored after application of 0.75–1.50% propionic acid or 0.90–1.75% of a 40:60 mixture of propionic and acetic acids. The grain can be used for feed purposes only. The treatment kills the embryo.

DORMANCY, VIABILITY, GERMINATION

Germination of seeds is a crucial step in the reproductive cycle; it is essential in malting, brewing, and several distilling industries and serves often as an index of grain soundness. Seeds of most species of cultivated plants germinate when they are morphologically mature and are placed under optimum conditions of moisture, temperature, and light. Seeds may fail to germinate as a result of quiescence or dormancy. Seed quiescence is an environmentally imposed temporary suspension of growth and reduced metabolic activity in viable seeds. It occurs under conditions that are unfavorable to germination. Seed dormancy, on the other hand, is an endogenously controlled and/or environmentally imposed temporary suspension of growth that is independent of immediate environmental conditions. Consequently, dormant seeds do not germinate under conditions that may be favorable to growth. Basically, dormancy is a physiological adaptation that delays germination until favorable environmental conditions are likely to occur. The causes of seed dormancy generally fall into five classes: (1) rudimentary embryos, (2) physiologically immature embryos (inactive enzyme system), (3) mechanically resistant seed coats, (4) impermeable seed coats, and (5) presence of germination inhibitors. Dormancy contributes to the survival of wild plants and reduces hazard of field sprouting and associated damage. On the other hand, prolonged dormancy is troublesome (and costly) if the grain is to be malted or sown shortly after harvest.

Museum botanists in Canada reported that lupine seed buried in a peat bog for about 10,000 years retained their viability and germinated. This is the longest record for safe seed storage. Germinating Indian lotus seeds from a Manchurian lake bed were found to be 1,000 years old. Barley and oat seeds stored in glass tubes in the foundation stone of a Nuremberg, Germany, theater were reported to have retained their viability after 123 years. Both internal and external factors affect the life-span of seeds. The internal factors include physical conditions and physiological state. The main external factors are relative humidity and temperature.

KEEPING QUALITY

The bread-making quality of freshly harvested wheat or freshly milled wheat flour tends to improve for a time depending on the nature of the product and storage conditions. Subsequently, a point is reached where further aging no longer improves baking potential and longer storage is accompanied by a gradual decline. In wet milling of corn, it is more difficult to obtain a good separation of starch from other constituents of the grain when corn has been stored under unfavorable conditions, even though no apparent damage to the corn has been observed, than when unquestionably sound corn is milled. Heat-damaged corn yields less starch than normal corn, the viscosity of the starch suspension is reduced, and the starch granules are mechanically damaged. Rice stored at elevated temperatures has impaired organoleptic properties. Brewers are reluctant to use as an adjunct milled rice that has been stored for long periods, as the rice requires high gelatinization temperatures and converts slowly, by the action of malt amylases, to fermentable sugars.

The keeping quality of breakfast cereals depends to a large extent on the content and keeping quality of the fat they contain. Thus, products made from cereal grains with low oil contents (wheat, barley, rye—oil content of about 2%) have an advantage over products made from oats (average oil content of about 7%). Whole corn has a relatively high oil content (about 4.5%), but most of the oil is contained in the germ and can be removed in the breaking of grits and flakes. The keeping quality of the fat depends on the degree of unsaturation, the presence of antioxidants or prooxidants, and the length of storage and storage conditions (temperature and moisture content). Several heat treatments, such as toasting or puffing, may destroy antitoxidants or induce formation of prooxidants. On the other hand, short temperature treatment at the surface may produce new antioxidants by the Maillard reaction. This may explain the improved antioxidant activity of steam-treated oat products.

Storage of breakfast cereals can cause bitter taste. In wheat and rye products, the bitter taste appears only after long storage and is not noticeable in a freshly baked product. In oat products, the danger of bitterness is noticeably higher. Bitter taste seldom develops in oat products that have been prepared by heat treatment, followed by kilning or dry heating. In addition to inactivating the enzymes and removing some bitter substances, heat treatment also facilitates separation of groats. Best results are obtained by treating raw oats with steam before kiln drying, by stabilization. In the stabilization process, the oats at 14–20% moisture are quickly raised to 205–212°F (96–100°C) by injection of live steam at atmospheric pressure and are maintained at that temperature for 2–3 minutes. The stabilization process also improves flavor and resistance to the onset of rancidity. Sterilized oats are dried to 4–8% moisture by kilning. Purposes of kilning are (1) to reduce the moisture content to a level satisfactory for the storage of milled products, (2) to facilitate the separation of groats by increasing the brittleness of the husk, and (3) to develop in the oats a characteristic nutty flavor. Excessive heat treatment should be avoided, as it reduces the thiamine content and damages the biological value of the proteins.

Involvement of enzymes in the formation of bitter substances has been postulated by several investigators. The participation of microorganisms seems excluded under normal storage conditions. Bitter taste is correlated with peroxidase activity. Peroxidase is apparently the most thermostable of the enzymes that could be involved in bitterness during storage and has been suggested as the best index of effective heat treatment.

Apparently, however, there are several bitter substances in oats. Bitter substances of the saponin type seem to form an adduct complex with other oat components that have no bitter taste. The formation of those bitter compounds could not be traced to enzymatic action. A second group of bitter substances is of the lipid type and is important in storage. Defatted oat products become bitter less readily than untreated oats. Formation of the bitter compounds is enhanced by free fatty acids, which are more susceptible to oxidation than triglycerides. Peroxides seem to act as precursors of the bitter substances. The extent to which the bitter compounds are formed is controlled by the concentration of peroxides. Peroxides could be produced by the enzyme lipoxygenase. That enzyme in oats, however, is low (about 10% of the lipoxygenase activity of rye, wheat, or barley). The low lipoxygenase activity seems to result from inhibition by natural antioxidants.

Several tests can be used to determine the commercial "condition" of grain and predict its future storage behavior. Although mold counts and grain viability tests are good indices of incipient deterioration, fat acidity and nonreducing sugar content are useful as measures of the actual damage that has actually taken place. High viability and germinability are probably the best and most meaningful indices of soundness especially in grain to be used in seeding and malting. In grain and milled grain products several biochemical tests can be used to estimate usefulness during processing into foods or feeds. None can be considered as the "tell-all" test, as each measures changes in a direction that is of greater consequence in some application than in others.

PEST CONTROL

Some of the most commonly used methods to control grain pests include sanitation and physical, chemical, and biological treatments. Physical treatments include cooling and heating (temperature is a key factor that influences growth and reproduction of insects and mites), irradiation and sonication (at present they cannot be easily and effectively applied), and use of inert dusts which repel or kill insects by abrasion, dehydration and cuticle delipidation, or antirespiration mechanism. Some problems in the use of dusts are lowered effectiveness at high RH, respiratory hazards to handlers, and abrasion to machinery.

Among chemical treatments, protectants and residuals are the most widely used and most effective supplements to sanitation. Potential hazards to mammals and increased resistance of insects are some of the concerns in application of chemical protectants. Controlled atmospheres can be lethal to storage pests; they include depletion of oxygen, use of certain fumigants, and use of sex attractants (pheromones). Insect growth regulators are compounds that interfere with a specific physiological system of a target pest (ie, morphogenesis and molting).

Biological treatments include the use of pathogens (various microorganisms exhibit pathogenicity to insects), parasites and predators (are of limited value), release of insects that have been sterilized by mutagenic chemicals or radiation, and conventional plant breeding or novel genetic engineering of cereal grains with varietal, genetically controlled resistance to pests. The best overall approach is integrated pest management that utilizes a judicious combination of several methods of pest control. It is in the long run most effective and eliminates continuous and excessive use of a single control method.

Chapter 7

Grain Quality

Cereal grains and their milling products are converted to many foods and used in many industries. Thus, wheat and its milled products are used in making bread, biscuits, pastry, cakes, and alimentary pastes. All-purpose wheat flour is used widely by the housewife; some wheat is converted into breakfast cereals and a large variety of dietary and specialty foods. By-products of the wheat flour milling industry are used in the manufacture of feeds, in the production of protein or enzyme concentrates, and by several organic and microbiological industries. The barley kernel is germinated for use in the malting and brewing industries. Wheat flour is fractionated to starch and gluten serving, respectively, as concentrated sources of carbohydrates and proteins. Wheat gluten can be used as a basis for manufacture of protein-rich foods, protein hydrolys-ates, or texturized protein substitutes. Corn (maize) starch is used as an ingredient in many foods, in the paper industry, in laundry starches, in oil drilling, and in the manufacture of various starch derivatives modified and tailored for specific industrial applications. Those are just a few examples of the various uses of cereals and cereal production. All these projects, processes, and uses require tests for determination of composi-tion, quality, and end-use properties. Information on methods used in testing cereals in the United States is available in the Official Methods of Analysis of the Association of Official Analytical Chemists, generally known as the AOAC (1980); in Cereal Laboratory Methods of the American Association of Cereal Chemists, known as the AACC (1980); in Official and Tentative Methods of the American Oil Chemists' Society, known as AOCS (1976); in Standard Analytical Methods of the Member Companies of the Corn Industries Research Foundation (Anon., 1967); and in Methods of Analysis of the American Society of Brewing Chemists (Anon., 1976). Information on analytical methods employed in England is available from the classical textbook on Modern Cereal Chemistry by Kent-Jones and Amos (1957), and on procedures used by many cereal chemists on the European Continent from Standard-Methoden fuer Getreide, Mehl, und Brot (Arbeitsgemeinschaft Getreideforschung 1978). The International Association for Cereal Chemistry (known as ICC) has issued (Anon., 1980) standard analytical procedures used in international grain trade.

This chapter outlines tests used to follow processing of cereals and discusses the significance of the tests. Quality control in processing of cereals involves testing the raw material, determining the best method of converting it into a processed product of uniform and established quality, and evaluating changes the product undergoes during storage.

Analytical chemistry and food analysis are becoming extremely sophisticated and are using elaborate instrumentation. Many of the new procedures provide valuable information that previously could not be obtained. Yet, no sound test should be discarded only because it has been used for many years. Nor does the value of a new method depend on the price of the instrument, complexity of the method, or degree of automation used. Many old and simple methods used by the experienced grain buyer are as useful today as they were years ago. The four "tests" used by experienced millers in evaluating a new shipment of wheat were (1) feeling the temperature of the grain in a sack to determine whether the wheat was heating, (2) visual inspection of a handful of grain to assess its purity and presence of foreign or objectionable material, (3) smelling the grain to establish off flavor or odor, and (4) chewing and masticating the grain to get some indication of its vitreousness, hardness, and approximate gluten contents.

Admittedly, the above "tests" could not establish the precise composition, potential use, and optimum processing. Yet they continue to have a place in many evaluation schemes. Those simple tests provide useful and important supplementary information. They are good screening tests and can, under certain conditions, save costly, lengthy, and laborious analyses. Consequently, though many modern methods and instruments are described, the time-proven, classical, and simple tests are included when applicable.

PHYSICAL TESTS

Grain Standards

As stated previously, some of the tests of grain quality important to producers and consumers are plumpness, soundness, cleanliness, dryness, purity of type, and general condition of the grain. The Official Grain Standards of the United States (USDA, 1978) function as commercial measures of quality and condition, and assist in marketing the grain in national and international trade channels. The Grain Grading Primer (USDA, 1957) explains the trading system and provides the background and details of methods used in sampling and grading cereal grains. Grain is inspected and graded by licensed inspectors who work under federal supervision. Standards are available for wheat, corn, barley, oats, rye, grain sorghum, flaxseed, soybeans, rice, triticale, and mixed grain.

The three factors that determine standard quality of sound grain in the regulation of the European Economic Community are moisture content, test weight, and Besatz (Schafer, 1963). Besatz covers all parts of a sample that do not constitute faultless specimens of the basic grain. Tolerances for the European standard are 5% total Besatz, 0.5% Schwarzbesatz, 1.0% sprouted kernels, and 2.0% broken kernels, spoiled kernels, smutty kernels, chaff, impurities, insect fragments, and weevils. Kornbesatz includes shrunken kernels, other grains, insect-attacked grains, and grains with discolored germs (see also page 61).

There is a basic difference in philosophy between testing wheat by mechanical devices (as in the U.S. grading system) and in determining its commercial value by using its Besatz content. In the first case, wheat is precleaned (Figure 7-1) to remove and determine impurities before being graded according to fixed standards. In the second case, the Besatz figure is used to determine a discount from the basic price for clean grain. The relatively long time required to perform the Besatz test (about 1 hour per analysis), its impreciseness, and its poor reproducibility in determining foreign matter from small samples by subjective judgment of technicians are reasons against its popularity or acceptance in the United States.

Figure 7-1. The Hart Carter Dockage Tester. (Courtesy, Hart Carter Co., Minneapolis MN.)

Test Weight

One of the most widely used and simplest criteria of grain quality is weight of the grain per unit of volume (Figure 7-2). In the United States and Canada, the test weight is expressed in terms of pounds per bushel; in most countries using the metric system, the weight is expressed in kilograms per hectoliter. The Winchester bushel used in the United States is 2,150.42 cu. in. The following factors can be used to convert from one system to another, provided containers of similar size are used (Zeleny, 1971):

From	To	Multiply by
lb/Winchester bu	lbl imperial bu	1.032
lb/Winchester bu	kg/hl	1.287
lb/imperial bu	kg/hl	1.247

Test weight is important to the farmer because grain is bought and sold on a minimum weight-per-bushel and grade basis. Test weight is also important to the miller because flour yield is usually related to weight-per-bushel wheat. Kernel shape, moisture content, wetting and subsequent drying, and even handling affect the test weight because they can modify the packing of the grain. Thus, the relationship between test weight and flour yield may lose significance under certain conditions (Miller and Johnson, 1957). Above 57 lb bu, test weight has relatively little influence on flour milling yield (Zeleny, 1971). At lower test weights, milling yields usually

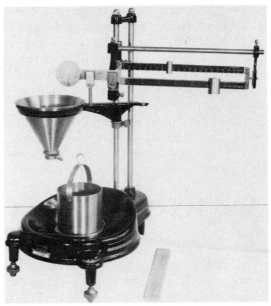

Figure 7-2. Standard apparatus for determining the test weight per bushel.

falls off rather rapidly with decreasing test weight of wheat. Average test weight per bushel of U.S. wheat is about 60 lb and may be as high as 65 lb. Badly shriveled kernels have test weights of 45 lb or less.

Kernel Hardness

Objective hardness measurements are mainly useful in differentiation between soft and hard wheats in plant-breeding programs (Miller and Johnson, 1954). Extremely hard milling characteristics are usually reflected in increased power requirements and reduced yields of flours of acceptable quality (ie, ash content or color). Alternatively, milling to a fixed extraction impairs mineral content and color. Excessive softness interferes with efficient bolting and increases sieving space.

Wheat hardness is a characteristic commonly used in wheat trade and classification. The term hardness means different things to different people. To some, it means physical hardness; to others, the way wheat mills; to still others, the amount of work required to grind grain. Those definitions may be related, but are not necessarily so, especially if the objective is to measure small but meaningful differences among cultivars or commercial samples from a single class of wheat. The major factors involved in wheat hardness are the physical hardness of the main components, starch and protein, the strength of their interaction within the cell, and the interaction of individual cells to produce the overall grain structure (Stenvert and Kingswood, 1977). Differences in hardness can be explained by differences in continuity of the protein matrix and the strength with which that matrix physically entraps starch granules. The primary determinant of wheat hardness is genetically controlled and relates to factors that influence compactness of endosperm cell components. Environmental factors and protein content also determine formation of an ordered structure (Stenvert and Kingswood, 1977). The relation of wheat hardness and protein contents has been studied by many investigators with various, often conflicting results. They are summarized in Table 7-1. In most cases, protein had no direct effect or the protein content varied with (or modified) varietal effects. Thus wheat hardness can be considered to denote a wheat class and variety characteristic that may be modified by environmental factors and protein contents (Symes, 1965).

Three types of factors affect hardness measurements in wheat: moisture and temperature, size and shape, and inherent hardness. We are interested in controlling the moisture and temperature effects by proper calibration. If this is impossible, the effects of moisture and temperature may be calculated from proper regression or calibration values. The effect of small differences in temperature is practically negligible; however, this is seldom the case for moisture effects. We prefer methods in which size and shape of kernel have no significant effects and look for methods that truly

Table 7-1. Wheat Protein and Hardness

Investigator(s)	Method(s)	Results
Newton et al (1927)	Cracking	No relation
Worzella (1942)	Particle size	No relation
Berg (1974)	Particle size	Varietal character, uninfluenced by protein
Fajerson (1950)	Particle size	Varietal character, influenced by protein
Symes (1961)	Particle size	Protein effect varies among varieties
Williams (1967)	Particle size index, starch damage	No relation
Symes (1969)	Particle size	No relation
Greenaway (1969)	Wheat hardness index	Relation with protein/m^2 of flour and protein
Seckinger and Wolf (1970)	Microscopy of endosperm particles	Protein particles for hard (unlike soft) wheat compact and hard to disrupt
Barlow et al (1973)	Penetrometer-starch granules and storage protein fragments	No varietal differences
Moss et al (1973)	Pearling resistance, particle size index	Negative relation for single cultivar
Trupp (1976)	Particle size index (protein by dye binding)	V. low relation, affected by variety and environment
Stenvert and Kingswood (1977)	Time to produce a fixed volume of ground wheat	Positive relation with protein content and formation of a continuous phase; cultivar dependent
Moss (1978)	Starch damage, particle size, resistance to abrasion, sp. vol. of whole meal	Optimum hardness and starch damage related to minimum protein
Obuchowski and Bushuk (1980a,b,)	Miscellaneous	No relation
Miller et al (1981b)	Work required to grind	No relation
Miller et al (1982a)	Time to grind, work to grind, particle size, NIR	No relation

Source: Pomeranz and Miller, 1982.

measure inherent hardness, including effects of growth conditions, maturation, and storage length and conditions.

Studies from the author's laboratories (Pomeranz and Miller, 1982) evaluated and compared four methods of hardness determination (work to grind, time to grind, particle size of ground wheat, and near infrared reflectance of ground wheat). The Brabender hardness tester was modified to measure the work required to grind 25–55 g of wheat and provide a digital readout of the data. Temperature, protein content, kernel size, and growth location had little effect on the results, but the work required to grind wheat increased with increasing moisture content. The assay is of limited value in distinguishing between hard and soft wheat (Miller et al, 1981a). In a subsequent study, a Brabender automatic hardness tester was used to measure the time required to grind 4 g of wheat (Miller et al, 1981b). Increasing moisture content increased grinding time of soft wheat much more than it increased the grinding time of hard wheat. The method is simple and rapid (less than 5 minutes), involves one basic semiautomated step, and requires no manual weighing. The tester can detect differences related to processing and end-use properties and in hardness among some wheat classes and among some varieties representing a single class of wheat. The instrument can also be used to evaluate hardness of plant breeders' samples so that selections with extreme hardness or softness can be discarded and to distinguish between hard and soft wheats in marketing channels. There was considerable variation in hardness within each marketing class (especially soft red winter). However, this is not likely to present problems in marketing channels because blends of many varieties are generally marketed.

The four methods mentioned previously were then used to measure the hardness of cultivars representing durum, hard red spring, hard red winter, white, white club, soft white winter, and soft red winter wheats (Miller et al, 1982a). Wheat protein content for samples within a class did not significantly affect hardness as measured by any of the four methods. Irrigation consistently increased protein content, reduced time to grind, and increased Near Infrared Reflectance (NIR) hardness values. The power to grind method had the poorest differentiation power. Durum wheats could be distinguished from the other classes by the other three methods. The time to grind and NIR values measured at 1680 nm distinguished between hard and soft wheats. None of the methods distinguished between hard red winter and hard red spring wheats, which indicates that there are no inherent major differences in hardness and milling between these two classes.

Hardness was then determined by the four methods in dark and vitreous (DHV) and yellow hard (YH) fractions from three hard red winter wheat cultivars (each from three locations) and five commercial hard red winter wheat samples ranging in DHV from 27% to 63% (Miller et al, 1982b). DHV kernels were harder than YH kernels. There was no consistent, significant difference in milling quality of commercial samples. There was a highly significant relationship between the percent DHV kernel content and the hardness of the commercial samples measured by time or power required to grind wheat and NIR of ground wheat.

To relate still further hardness characteristics with end-use properties, 25 Australian wheats that varied widely in hardness and showed relatively narrow range in protein contents were evaluated. Some of the results are shown in Table 7-2. The range in time to grind was twice that of PSI (3.11 vs 1.56); NIR at 1680 nm had an intermediate range (1.87). The flours differed little in density (range 1.37) and substantially in starch damage

Table 7-2. Characteristics of 25 Wheats

Type	No.	PSI (%)	Time to grind (sec)	NIR (1680 nm)
Very hard	10	28.5	36.3	321
Hard	7	33.5	43.9	290
Soft	5	39.8	76.2	197
Very soft	3	44.5	112.9	172

Source: Pomeranz and Miller, 1982.

Table 7-3. Characteristics of 24 Flours

Type	No.	Flour density (g/100 cc)	Flour protein (%)	Starch damage (%)
Very hard	10	38.4	13.7	8.5
Hard	7	36.4	14.3	6.7
Soft	5	29.8	13.2	5.5
Very soft	3	28.1	13.4	4.3

Source: Pomeranz and Miller, 1982.

Table 7-4. Correlation Coefficients

	Starch damage	PSI	Time to grind	NIR wheat
Flour density	0.903	0.924	0.893	0.972
Starch damage	—	0.939	0.840	0.910
PSI		—	0.923	0.967
Time to grind			—	0.933

Source: Pomeranz and Miller, 1982.

(1.98) (Table 7-3). Correlation coefficients among the five estimates (direct and indirect) of hardness are summarized in Table 7-4. All correlations are highly significant.

The method of choice for measuring hardness depends on the available equipment, the degree of discernibility required, and the wheat classes or types to be studied. Excellent differentiation can be obtained with the Brabender Automatic Micro Hardness Tester. However, if a near infrared instrument that will measure reflectance at 1680 nm is available, satisfactory differentiation between soft and hard wheats can be obtained. The reflectance measurement can be made easily in combination with other determinations such as moisture and protein content.

The Milling Test

The objectives of experimental milling are to prepare flour for evaluation of chemical, physical, or end-use properties, to screen wheats in plant breeding programs to eliminate selection of unsatisfactory milling proper-

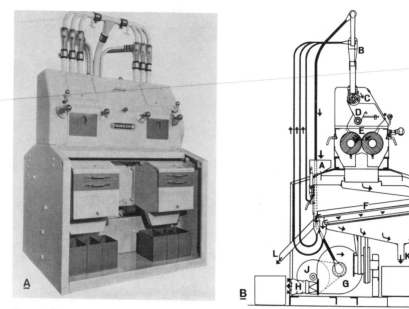

Figure 7-3. (A) Picture and (B) diagram of a semiautomatic experimental mill. A—grain inlet; B—cyclone; C—locks; D—feed; E—grinding; F—sifting; G—fan; H—filter bag; J—drive; K—outlets for flour; L—outlets for feed and bran. (Courtesy Buhler-Miag Co., Uzwil, Switzerland.)

ties, and to parallel results from commercial milling. Those three objectives present increasing levels of desirability and difficulty. Many types of experimental mills have been designed, manufactured, and used. Those used most extensively are the Allis Chalmers, Buhler (Figure 7-3), Brabender Quadrumat Junior and Senior (Figure 7-4), and Ross (Schellenberger and Ward, 1967). Both continuous (usually automatic) and batch-type mills are available. Other experimental flour mills sometimes used include the Hobart Micromill and the Miag Multomat. The Hobart Micromill is simple and requires small wheat samples; it is particularly useful in testing samples early in breeding programs. The Miag Multomat has a milling capacity of about 1,000 g/min and is close in performance to a pilot or commercial mill.

Results of experimental milling can be used to predict the milling operations of a specific commercial mill but fail to yield reliable results about the performance of wheats in the whole-milling industry. The Approved Methods of the AACC describe sample preparation, tempering (conditioning), and continuous and short-flow or long-flow batch milling. Tempering times of about 18 hours are recommended. No method for experimental milling is included in the Standard Methods of the International Association for Cereal Chemistry (ICC Standards). The Standard Methods of the West German Association of Cereal Research

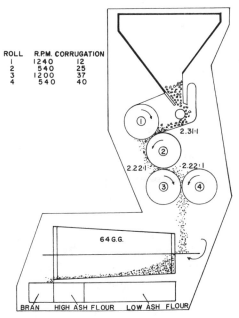

ROLL R.P.M. CORRUGATION
1   1240   12
2   540   25
3   1200   37
4   540   40

2.3:1

2.22:1   2.22:1

64 G.G.

BRAN   HIGH ASH FLOUR   LOW ASH FLOUR

Figure 7-4. Cross section view of a Quadrumat Junior Mill. (Courtesy Professor A.B. Ward.)

(Arbeitsgemeinschaft Getreideforschung, 1978) recommend the use of a Buhler automated mill (or equivalent) plus a bran separator. The wheat (2–4 kg) is conditioned for 24 hours to obtain a flour of about 74% milling yield and maximum ash of 0.47% (14% moisture basis).

The many attempts to standardize experimental milling procedures have met with limited success. Thus, the speed and ease of operating the Buhler mill make it possible to prepare a flour of about 65% extraction, which compares in bread-making properties with commercially milled flour. The Buhler mill also differentiates between wheats varying in milling properties. The rigid flow of the semiautomatic Buhler mill is, however, a disadvantage in evaluating mixed wheat grists or wheat varieties that cover a narrow, yet meaningful, and practical range of milling properties. Under such conditions, a mill that has a more flexible diagram provides more complete information in the hands of an experienced operator. Micromills capable of providing small samples of flour that can be evaluated by microquality tests are required in plant breeding programs (Hehn and Barmore, 1965). The Buhler mill requires about 1-kg samples and is therefore not useful for preliminary quality evaluation of early (F_3) generations.

The milling value of wheat is affected by wheat cleanliness, flour yield, wheat type (soft or hard), kernel uniformity and size, kernel weight, response to conditioning, thickness of the pericarp and aleurone layers, and behavior during milling (Shellenberger and Ward, 1967). Calibration of milling value range from the simple:

Milling value = flour extraction − Kent-Jones flour color
to the rather complex formula of

Milling evaluation figure =
$$\frac{\text{flour yield \% (1 − ash on dry matter basis)} + \text{reduction properties \% (100 − semolina \%)}}{100}$$

$$+\ \frac{10\ (1 − \text{power requirements in kw}) − \text{break bran \% (starch on dry- matter basis)}}{100}$$

According to Finney (1982) and Finney and Yamazaki (1967), milling properties of interest in the evaluation of bread wheats include wheat hardness; bolting properties; yields of break flour, middlings flour, and total flour; flour ash content; wheat-to-flour protein contents; and kernel plumpness. Neither flour ash nor flour yield can be properly assessed or evaluated without a knowledge of kernel plumpness. Flour yield decreases and flour ash increases with decrease in kernel plumpness or increase in shrivelled kernels.

Many attempts have been made to include various attributes in a single-value milling score, in some cases bringing in economic consider-ations such as prices of various milling products. Such attempts have not been uniformly successful; the biggest problem is whether to include some variables and their relative importance.

Pomeranz et al (1985) obtained flours by three variations of a micromil-ling method from 100-g samples of 12 hard red winter wheat varieties that varied widely in weight per bushel, kernel weight, protein and ash contents, and kernel hardness. In all three variations, the wheat was tempered to 15% moisture and was passed through two prebreak treatments (Figure 7-5) which reduced the flour ash by 0.005–0.10%. In variation A, the wheat was rested for 15 minutes after the first prebreak was made 1 1/4 hours after water addition. In variations B and C, the two prebreak treatments were made 2 and 24 hours after water addition, respectively, with practi-cally no intermediate rest. The three milling variations were also applied to three hard red winter wheats, each from four locations and of different hardness characteristics. Variation A produced slightly higher flour yields than those by variation B and slightly lower yields than those by variation C. The flour yields obtained by the three variations were 68–70% and flour ash contents were about 0.35%. Flours obtained by the three milling variations were comparable in protein content and bread-making charac-teristics. Yields of flour from the three milling variations were positively and highly significantly correlated.

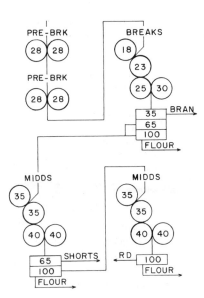

Figure 7-5. Micro experimental mill flow.

CHEMICAL TESTS

Duntley (1981) listed quality control tests performed in U.S. mills and times required to perform the tests in a mill laboratory. The tests determine moisture, ash, protein, fat, diastatic activity, presence and level of enrichment and oxidants, viscosity, pH, and particle size. The most commonly employed chemical tests in the cereal laboratory of a flour mill are moisture, ash, and protein—called the "tiresome triumvirate of routine tests" (Sullivan, 1949). A working diagram for the control of a mill is given in Table 7-5 (Anon., 1936).

Moisture

The percentage moisture that a food contains means the amount of water and other volatile compounds that will evaporate under specific drying conditions. Moisture contents are important for several reasons. Composition percentages are inversely related to moisture contents. Thus, results of chemical composition, milling, and baking tests should be reported on a common moisture basis. The 14% basis is most commonly used in the United States; in Europe results often are expressed on a dry-matter basis. Yield of flour and test weight are inversely related to moisture content. Efficient flour milling requires knowledge of the moisture content of the wheat before processing, and distribution of water in various parts of the kernel after conditioning and during milling. The amount of water that grain contains is economically significant and may result in certain advan-

Table 7-5. Working Diagram for the Control of the Mill

Stages of work	Methods of control
Raw wheat	Moisture content
	Test weight
	Specific weight
	Admixtures test
	Sedimentation
	Gluten content
	Condition of gluten (swelling test)
	Meal fermentation test
	Protein content
	Testing in the laboratory conditioner
	Grinding on the test mill
Wheat in silo	Moisture content
	Measuring temperatures in the wheat
Cleaning, washing, conditioning	Moisture content
	Preparation in the laboratory conditioner
	Grinding on the test mill
Wheat before first break	Moisture content
Grinding (break, etc)	Sieve analysis
Intermediate flours	Pekar test or other color tests
	Ash content
	If desired, gluten, protein, and diastatic power tests
Final flours	Moisture content
Flours of competitors	Ash content
	Sand content
	Protein
	Gluten content
	Condition of gluten
	Diastatic power
	Farinograph curves
	Baking test
	Test as to preparations added
	Sieve analysis
Old flours from store	Acidity and odor

Source: Anon., 1936.

tages or disadvantages at various stages of marketing. And finally, moisture content is of utmost importance in safe storage of grains. Grain that contains moisture in excess of the critical moisture level is subject to rapid deterioration from mold growth, heating, insect damage, and sprouting.

The basic official procedure to determine conformity of wheat with standard requirements is based on drying at 130°C for 1 hour in an air oven. The procedure is used in calibration of rapid moisture meters. Hunt and Neustadt (1966) described sources of error in determining moisture by

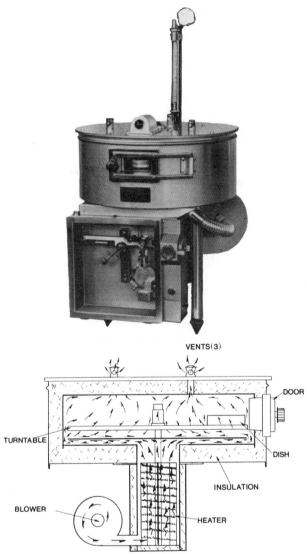

Figure 7-6. Picture and schematic of a semiautomatic moisture tester. (Courtesy Brabender Instruments.)

the oven method. The errors were traced to variations in grinding (effect of moisture level, screen size, sharpness of blades, rate of feed, and heating of mill), sample size, moisture dishes, and oven specifications and performance (controls, uniformity of heating, relative humidity of room air, and desiccant). The standard method calls for a forced-draft or a convection oven. A forced draft oven that is widely used is shown in Figure 7-6. The samples are placed in the oven at the loading door and moved on the hearth by a hand wheel. Ten-gram samples are dried by heat and forced draft and are reweighed on a built-in balance calibrated to give percentage of moisture. No cooling is necessary, and samples are weighed at the drying temperature. Generally, a temperature of 130°C and a time of 1 hour are used. However, in a standard type of force-draft oven, many more (about 50) samples can be dried at one time.

In the quest for a simple, rapid, and accurate method, many procedures for determining moisture in grain have been devised. They include rapid, high-temperature drying procedures, and electrical and chemical methods.

In practice there are problems in electrically measuring moisture. The main source of variation is the distribution and binding of water inside grain. Water can be present in a free or bound form. Free water is capable of acting as a solvent, contains dissolved salts, and conducts electricity, and its dielectric constant is near 80 when it does not contain excessive amounts of dissolved matter. Water bound to grain components has a dielectric constant similar to that of the molecule with which it is associated. Since in bound form it is not capable of dissolving mineral salts, it is a nonconductor of electricity. It cannot be measured by conductance-type meters. Determining moisture by electric meters is based on the assumption that the ratio of free to bound water is constant. While instruments measuring the dielectric value are subject to less error, large shifts in the ratio of bound to free water affect the readings. A dielectric moisture meter is used in grain inspection in the United States. Since 1959, an electric moisture meter (capacitance type) was made official in Canada to determine moisture in cereal grains. Among the chemical methods, the production of acetylene resulting from the reaction of water with calcium carbides and the Karl Fisher titration are sometimes used in mills (Anon., 1963).

Ash

Objectives of milling are to separate endosperm from bran and germ and, subsequently, to reduce endosperm particles to flour. The efficiency of separation can be judged by several empirical, indirect methods based on measuring any constituent that is concentrated to a larger extent in the bran or germ than in the endosperm. Since the mineral content of the bran is about 20 times that of the endosperm, the ash test fundamentally indicates the purity of the flour or thoroughness of the separation of bran and germ from the rest of the wheat kernel. The ash test has assumed greater importance in the milling trade than any other test for the control of the milling operation.

In the standard AACC procedure, a weighed sample is incinerated in a muffle furnace at 550°C for soft wheat flours and feeds and at 575–590°C for hard wheat flours. The amount of mineral residue or ash is expressed as the percentage of the original sample. The accelerated ash determination employs incineration at 600°C (for feedstuffs) or at 850°C in the presence of an alcoholic solution of magnesium acetate (for flours). In calculating ash by the second method, a blank must be subtracted. In Europe, ashing is carried out at 900°C ($\pm$ 10°C). That reduces the time of ashing but tends to melt the mineral constituents and to entrap carbon particles causing difficulties in cleaning and reusing crucibles.

Determination of Protein

Analysis of organic nitrogen is of great importance to cereal chemists as an indirect measure of protein content. The protein test, though not included as a grading factor in the U.S. standards for wheat, is accepted as a marketing factor. The relationship between crude-protein content and bread-making potential is so well established that today protein content is generally accepted as a sufficiently reliable indicator of the physicochemical properties of wheat flour to maintain protein's position as the commercial market criterion of quality supplementary to grade standards (Hehn and Barmore, 1965). Customarily, data on protein content are made available to buyers of wheat and flour.

The protein content of wheat and flour are extremely important because almost all flour properties (water absorption, mixing requirement, mixing tolerance, handling characteristics, oxidation requirement, loaf volume, and even bread crumb grain) are highly correlated with protein content. Highest protein content is required for wheat flour used in bread making, intermediate for family all-purpose flour, low for cookies, and lowest for cakes. Satisfactory malt is obtained, generally, from low-protein barleys. High protein content is generally desirable in rice, oats, rye, corn, and sorghum used as food or feed.

Determining Total Protein Content

A direct and accurate method of protein determination in cereals would require isolating and weighing all of the protein in a sample. Use of such a procedure for determining the protein content in a small number of samples would be difficult; for routine testing, it would be unsuitable.

In most procedures of protein determination, a specific element or functional group in the protein is determined and the protein content calculated by using an experimentally established factor. Methods based on analyses for constituents of proteins include those for determining nitrogen, certain amino acids, or the peptide linkage. Those methods assume that the particular constituent assayed for is present entirely in the protein fraction and that the ratio between the amount of that constituent and the protein content is constant. Those two assumptions are almost invariably in the realm of wishful thinking.

Kjeldahl Procedure

Three types of procedures for protein determination on the basis of nitrogen content are the Kjeldahl method, the Dumas method, and several physicochemical methods. The Kjeldahl procedure is the most widely accepted. Many hundreds of papers have described modifications of the original Kjeldahl procedure, which goes back to 1883. Basically, a sample is heated in sulfuric acid in the presence of a catalyst mixture and digested until the carbon and hydrogen are oxidized and the protein nitrogen is reduced and transformed to ammonium sulfate. Then sodium hydroxide is added and the digest heated to drive the liberated ammonia into a known volume of a standard acid solution. The unreacted standard acid is determined, and the results are transformed by applying a mathematical factor to percentage of protein in the original weighed sample. The factor generally used to convert nitrogen to protein in wheat and flour is 5.7; in most other foods and feeds a factor of 6.25 is used. The factors are based on average nitrogen contents of proteins in the tested foods.

The protein test is a reliable and rapid assay well adapted to routine determination. It is the most widely used and accepted single chemical test for wheat and flour bread-making quality. Quantity of protein in whole wheat and in flour are highly correlated. Generally, the flour protein is 0.8–1.8% less than the protein content of the wheat from which the flour was milled. The difference increases with refinement and purity of the flour.

In commercial practice, where large numbers of samples are run daily, many time-saving devices are used (Neill, 1962). Thus, digesting a 1-g sample, using a known amount of standardized 0.1253 N sulfuric acid, titrating with 0.1253 N sodium hydroxide and using an inverse reading biuret make it possible to report percentage crude protein directly from the biuret reading. Use of the automatic pipettes for dispensing the receiver acid solutions is certainly an advantage during the peak harvest season.

If the catalyst digestion mixture contains mercury, a mercury precipitant is incorporated with the caustic solution when the solution is prepared. The use of mercury (and selenium) has been discontinued in most laboratories because of health hazards. The indicator is added to the receiver acid solution when it is prepared. The antibumper or pumice stone is blended with the catalyst powder mixture, making the addition a "one-shot" procedure.

Heating levels are adjusted so that digesting 1 g with 25 ml concentrated sulfuric acid and a catalyst mixture (potassium sulfate, mercuric oxide, and copper sulfate) is completed in 35–40 minutes. The boric acid method to determine protein saves time, especially in a commercial laboratory where many routine tests are made daily. It is accurate and has the advantage of requiring only one standard solution of acid. Neither the amount (about 50 ml) nor the concentration (about 4%) of boric acid in the receiving bottle has to be precise.

If small samples are available (10–30 mg), a micro-Kjeldahl modification, employing the boric acid procedure and steam distillation of the liberated ammonia, is available.

Several rapid, semiautomated or automated methods for protein determination by the Kjeldahl procedure are available. They include the Kjeltec II and Kjel-Foss Automatic (Sepp, 1979). The digestion step in the Kjeldahl procedure can be excluded. Ronalds (1974) described a 15-minute alkaline distillation for determining the protein contents of wheat and barley. The sample is distilled with a solution containing sodium hydroxide, barium chloride, and an antifoaming reagent. The protein content is computed from the titration value of the distillate. It is also possible to exclude distillation and determine the ammonium ion in the digested liquid. This can be done colorimetrically, potentiometrically by acid-base titration, coulometrically, or with an ammonia electrode (Geissler, 1975; Sepp, 1979; Schmutz, 1978; Gruppe et al, 1978).

Dumas Procedure

In the Dumas method, nitrogen is freed by pyrolysis and the freed elemental nitrogen is determined volumetrically. In recent years, improvements in both the pyrolysis step and in the determination of elemental nitrogen have ensured precise and accurate analyses of nitrogen by the Dumas method. Based on those improvements, several rapid automated procedures have been described. The major problem in using the Dumas method was the heterogeneity of the plant material and the requirement for very finely ground and homogeneous materials for accurate analyses of small samples ranging from 5 to 50 mg. This problem has been solved in recent modifications that can test up to 1-g samples of cereal grains (Revesz and Aker, 1977; Vondenhof and Schule, 1979). In all modifications of the Dumas procedure, pure oxygen is used to speed up combustion. In a further modification, the nitrometer has been replaced by a thermal conductivity detector with helium acting as a carrier gas. In a rather new approach, both the reduction and volumetric stages are excluded and the nitrogen oxide gases are ozonized to form excited NO_2. The chemiluminescence produced is measured in the 650- to 900-nm range of a spectrophotometer (Sepp, 1979).

Some Physiochemical Methods

In recent years, several physiochemical methods of protein determination have been developed. They include (in addition to near infrared reflectance) neutron activation analysis, proton activation analysis, thermal decomposition analysis (Williams et al, 1978b), and electron spectroscopy (Sepp, 1979). Nitrogen content may be measured by means of particles or gamma rays emitted from nuclear reactions in N^{14}. If a cyclotron is available, one may irradiate the sample with a proton beam, count the emitted gamma rays, and use the results to compute the nitrogen content of the sample (Dohan et al, 1976). Systems for neutron activation analyses are available commercially (Andras et al, 1977, 1978; Doty et al, 1970). If a material is irradiated with X-rays, electrons are photoejected (Sepp, 1979). The kinetic energy of these electrons is characteristic of the atom level from which the electrons originated. The method employs samples of several milligrams only.

Dye-Binding Methods

Certain dyes (ie, Orange G) bind under acid conditions specifically to free amino groups, the imidazole group of histidine, or the guanidyl group of arginine. Each protein has a definite dye-binding capacity (DBC). To determine the protein content, the DBC is related qualitatively to the total nitrogen content, determined by the Kjeldahl method, and the regression equation between the two parameters is used to estimate protein content. Acid Orange 12 (AO-12) is structurally similar to Orange G but has only one sulfonic acid group for binding and the color change per protein binding site of AO-12 is about twice that of Orange G. The dye reacts with the protein and forms an insoluble complex. The concentration of the unbound dye can be plotted against Kjeldahl nitrogen content to give a straight regression line (Udy, 1956).

A commercial instrumental setup is available. The dye, acid orange in a phosphate-buffered system at pH 1.7, is added to flour or wheat that has been ground in an efficient mill disintegrator. The reaction mixture is mixed in a laboratory shaker or in a reaction chamber and is transferred to a squeeze-type polyethylene bottle fitted with a Fiberglas filter disc in the dropper cap. Light transmittance through the filtered dye solution is determined in a special colorimeter as the filtrate is transferred dropwise into a flow-through cuvette.

The DBC method can be used to determine protein contents of sound, unaltered agricultural products (ie, cereal grains, oilseeds, milk) or fractions of such products. The method is based on the assumption that there is a constant ratio between Kjeldahl nitrogen and available cationic groups (from the basic amino acids histidine, arginine, and lysine, and from the free amino end groups of the protein chains) for binding with the anionic, sulfonic acid dye. The method cannot be used to determine total protein contents if this ratio has been changed, as in processing of barley to malt and beer (resulting in a large increase in free alpha amino groups), in heat-treated or moldy foods (which may make epsilon amino groups of lysine unavailable owing to browning reactions), or in the assay of foods with unusual amino acid compositions. At the same time, changes in DBC resulting from those alterations can be used (after proper calibration) to detect damage in stored products or unusual composition (ie, high lysine content of cereal grains).

The Biuret Method

In the biuret method, proteins are peptized with potassium hydroxide and treated with a copper sulfate solution, and the intensity of color formed is proportional to protein concentration. The biuret procedure is simple and rapid, and the equipment and apparatus used are relatively inexpensive and readily available. The biuret test involves reaction with the peptide

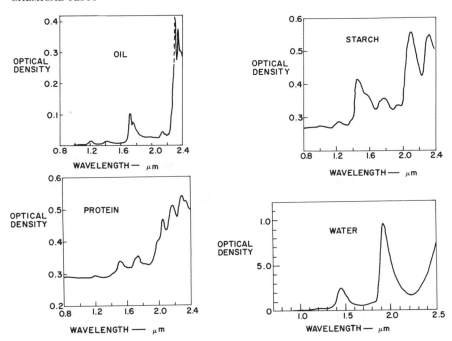

Figure 7-7. Infrared reflectance spectra of oil, starch, protein, and water. (Courtesy Neotec Instruments, Inc.)

linkage and measures protein-nitrogen only, whereas the Kjeldahl procedure measures total nitrogen and does not distinguish between protein and nonprotein nitrogen. The biuret method is widely used in biochemical laboratories to estimate small amounts of protein. It has been used to determine protein in cereals. The modified biuret reagent described by Pinckney (1961) contains potassium hydroxide, sodium potassium tartarate, and copper sulfate. Combining all the reagents in one solution simplifies and considerably shortens the determination.

The biuret procedure is not, however, an absolute method, and the color must be standardized against a known protein or against another method—ie, the Kjeldahl analysis of nitrogen. Several biuret procedures have been proposed for the determination of the protein content of cereals (Johnson and Craney, 1971; Noll et al, 1974). Although some of the rapid and automated procedures are quite attractive and precise, it is questionable how widely they are used in routine testing.

Infrared Reflectance

Extensive investigations in several laboratories have led to the development of instruments that utilize infrared reflectance to measure moisture, protein, and oil contents of foods and agricultural products. Typical spectra for protein, oil, starch, and water are given in Figure 7-7. Several methods of treating the data can be used to predict the composition from reflectance spectra.

Numerous investigations have shown that there is a good linear relationship between the chemical composition and the reflectance in the near infrared. However, the reflectance spectra are sensitive to extremes in particle size (reflectance increases as particle size decreases). A reflectance change of up to 50% can occur as particle size is reduced; the second derivative technique of data treatment lowers sensitivity to particle size effects.

As it is undesirable to extrapolate beyond the compositional limits of samples used for calibration and standardization, it is necessary to obtain samples that cover the entire expected range of composition. Finally, temperature changes of the instrument can cause both wavelength and reflectance sensitivity changes. Several companies—eg, Technicon Instruments Corp., Tarrytown, New York, and Neotec Instruments, Inc., Silver Spring, Maryland—market near infrared instruments to measure the composition of food and agricultural products. The following are from the descriptions of two instruments.

Neotec Grain Quality Analyzer

In the GQA, the light beam passes through a *tilting* filter system which changes angle in relation to a tungsten light source. In this manner, more than 300 measurement points are scanned 10 times every second using only three narrow band pass infrared filters. At all 300 points, the amount of IR light reflected by the grain sample is measured by a photodetector. Data are automatically fed into a built-in computer which solves third-order equations and displays protein, oil, and water content on the digital readout (Figure 7-8a).

The Dickey-John Instrument

The Infraanalyzer (Figure 7-8b) utilizes the difference in reflectance from a ground sample of six narrow-wave bands of energy in the near-infrared spectrum. The reflected energy is detected by a sensitive photocell, sampling the reflected energy level from each of the six different wave bands many times per second. This signal output is amplified and channeled through a synchronized device and then is processed through the computing electronic circuitry. In the computation process a series of mathematical equations are solved, resulting in presentation of direct readings on a digital readout device. For additional information see Norris (1974), Trevis (1974), Williams (1975), Pomeranz and Moore (1975), Watson (1977), Miller et al (1978), William et al (1978a,b), and Bolling (1979).

The use of infrared reflectance methods has been proposed for determination of wheat hardness (Williams, 1979), forage quality (Norris et al, 1976), amino acid composition of wheat proteins (Norris et al, 1977), and overall control of flour mill operations through process stream and

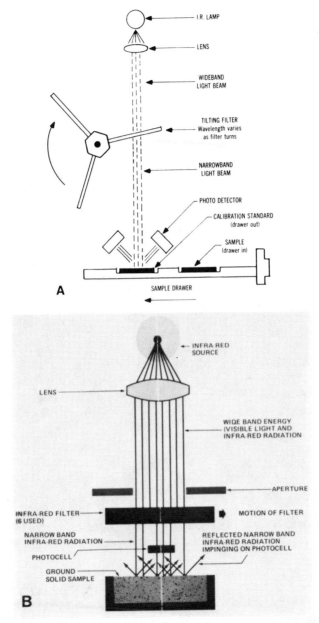

Figure 7-8. Diagrams of infrared commercial instruments. (A.) Grain Quality Analyzer (Courtesy Neotec Instruments, Inc.), (B.) Infra-Analyzer (Courtesy: Technicon Instruments, Inc.).

product assays (Anon., 1979). Various methods of protein determination were compared by Geissler (1975), Pomeranz and Moore (1975), Schmutz (1978), Williams et al (1978), and Sepp (1979).

Determination of Gluten

Many European cereal chemists determine gluten as an estimate of protein content. After drying, the gluten contains about 80% protein. Pelshenke and Bolling (1962) studied the relationship between Kjeldahl protein and gluten contents. The formula

$$\% \text{ wet gluten} = \frac{\% \text{ protein} - a}{b}$$

with a = 7.34 and b = 0.2271 was obtained for samples containing above 10.5% protein. The average deviation of a Kjeldahl-protein determination was ± 0.3% and for a gluten assay ± 1.5%. Dry gluten was about one-third of wet gluten, and the factor b was used to correlate dry gluten with protein 0.63.

Gluten determinations offer several advantages over the conventional Kjeldahl-protein test. The physical properties of the cohesive gluten ball can be tested by an experienced operator. Large differences in protein quality of various varieties or advanced stages of deterioration in storage, which cannot be detected by the Kjeldahl test, are brought out by the simple test of washing out a gluten ball. However, the gluten test is used little in the United States for the following reasons. (1) it is not precise; attempts to standardize the test by using salt solutions and a mechanical gluten washer have reduced the error somewhat. (2) Gluten can be washed easily from flour but not from wheat; consequently, it is of limited value in plant breeding programs. (3) The test is not suited for large-scale routine determinations. To overcome some of the limitations of the gluten test, a mechanical system for gluten washing has been developed (Figure 7-9). The wet gluten from the glutomatic system is dried quickly on a specially constructed Teflon hot plate. The dried cake is weighed and the results can be used to compute the protein content.

Reliability and Usefulness of the Methods of Protein Determination

In discussing the reliability and usefulness of methods of protein determination, it is important to consider the constraints imposed on the testing procedures with regard to the use of the method and to the objectives for which the tests are conducted. Thus, for instance, a consulting laboratory that conducts small numbers of tests on widely varying materials would probably find it difficult to justify the purchase of an expensive instrument which requires extensive calibration and standardization for each type of material. In all probability, such an instrument could not be used to test all

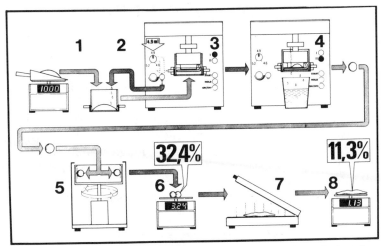

Figure 7-9. Glutomatic procedure. (1)Weighing; (2) dispensing; (3) dough making; (4) dough washing; (5) centrifuging; (6) result; (7) drying; (8) result. (Courtesy Falling Number AB, Stockholm.)

of the materials tested by the laboratory. Such an instrumental method is ideally suited for the routine testing of large numbers of samples of a similar nature and composition, such as in a terminal elevator, grain laboratory, or processing plant. The speed with which results are needed is an important consideration in selecting the method. In testing plant breeders' material large numbers of samples of similar nature and composition are evaluated, but the decision-making process does not require results as rapidly as it does in quality control or in grain marketing.

Finally, factors related to the reliability of the results (specificity, accuracy, precision, and sensitivity) must be considered. At the outset, it should be emphasized that none of the methods are truly accurate as none measure the true protein content of wheat. The biuret method is probably the most accurate, as it measures the peptide bond. On the other hand, it is not an absolute method, and the results vary with the material used for comparison and the standardization procedure. For all practical purposes, in developing new procedures for protein determination, we are interested in methods that show the best agreement with the Kjeldahl procedure rather than in methods that give an accurate estimate of protein. Insofar as the precision of methods for protein determination is concerned, we are concerned, in some instances, in high precision throughout the entire range of samples analyzed; in others, such a precision is essential for only part of the range. Thus, for example, in determining the composition of wheat for nutritive evaluation, in plant breeding work, or in plant quality control, we must have a precise measure at all protein levels. If the assay is to be used as a rapid screening test for binning wheat, we may tolerate small deviations at the extremes of high and low protein.

PHYSICAL DOUGH TESTING

Physical dough testing devices are used to evaluate bread-making potentialities (strength) and performance characteristics of flours under mechanized conditions. Such evaluation has assumed considerable importance as a result of high-speed mixers and continuous processing. Development of physical dough testing procedures was summarized by Brabender (1965). Comprehensive reviews on the basic considerations of dough properties were published by Bloksma (1971) and on dough testing equipment by Miller and Johnson (1954) and Kent-Jones and Amos (1957). Recommended procedures for three of the commonly used (in the U.S.) physical dough tests are given in Cereal Laboratory Methods (AACC, 1980).

Recording Dough Mixers

The two most widely used types of recording dough mixers are the farinograph and the mixograph. Both instruments record the power that is needed to mix dough at a constant speed. The record consists of an initial rising part which shows an increase in resistance with mixing time and is interpreted as dough development time. The point of maximum resistance (or minimum mobility) is generally identified with optimum dough development and is followed by a second part of more or less rapid decrease in consistency and resistance to mixing.

The Brabender farinograph was originally developed by Hankoczy, a Hungarian investigator. A picture and diagram of the instrument are given in Figure 7-10. The farinograph measures plasticity and mobility of dough subjected to a prolonged, relatively gentle, mixing action at constant temperature. Resistance offered by the dough to mixing blades is transmitted through a dynamometer to a pen that traces a curve on a kymograph chart.

The general practice has been to determine, by "titration curve" with water added from a biuret to the flour as it is mixed, the standard absorption [generally the amount of water needed for an optimum consistency of 500 Brabender Units (BU)-arbitrary]. Absorption increases as protein increases, and gluten quality improves. The farinograph has been most useful in determining water absorption. It should be noted, however, that some doughs tend to slacken and others to tighten during fermentation. Consequently, the consistency determined at the mixing stage may differ from that at the end of fermentation. Farinograms also provide information on optimum mixing time and dough stability. Figure 7-11 compares the farinograms of three types of flours and shows the main parameters calculated. Many investigators calculate the valorimeter value, a value expressing various characteristics of a farinogram as a single score. The larger the valorimeter value, the stronger the flour. The valorimeter

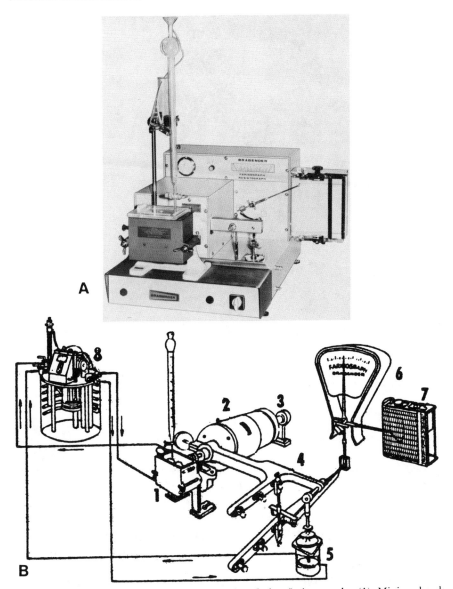

Figure 7-10. (A) picture and (B) schematic of the farinograph. (1) Mixing bowl; (2) free swinging dynamometer; (3) ball bearings; (4) lever system; (5) dash-pot; (6) scale system; (7) recording device; (8) thermostat. (Courtesy C.W. Brabender Instruments.)

value is determined by the time required to mix the flour to minimum mobility and by the descending slope.

The mixograph is a miniature, high-speed recording dough mixer. It has been found particularly useful in many laboratories in the United States to characterize and evaluate wheat varieties. The instrument was designed by Swanson and Working (1933). The mixer (Figure 7-12) has four vertical

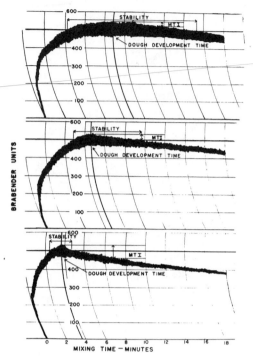

Figure 7-11. Comparison of farinograms of three flour types of decreasing strength; from top to bottom. (From Miller and Johnson, 1975.)

pins revolving about three stationary pins in the bottom of the bowl. As the gluten develops, a gradually increasing force is required to push the revolving pins through the dough. The increased force is measured by the tendency to rotate the bowl, which is in the center of a lever system. A record of the torque produced on the lever system is made on a chart moving at a constant rate. Mixograph mixing time is much more highly correlated with experimental bake mixing time than is farinograph mixing time. Unfortunately, the mixograph is less highly standardized than the farinograph. Finney and Yamazaki (1967) gave details of important specifications in standardizing the mixograph. Use of the mixograph in evaluating effects of proteolytic enzymes on bread flour properties has also been described (Pomeranz et al, 1966). Typical mixograms are given in Figure 7-13a. The effects of protein content and quality of mixograms are illustrated in Figure 7-13b.

Load Extension Meters

Several commercial instruments mix and shape dough under standardized conditions and stretch it until it ruptures, and a curve load versus

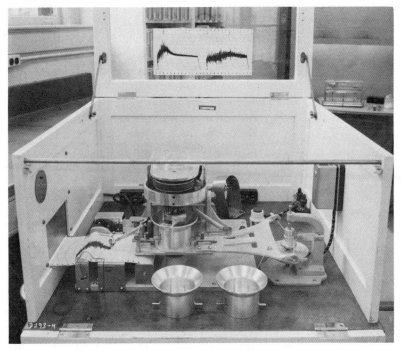

Figure 7-12. Mixograph for recording mixing properties of 35-g flour samples of wheat varieties. (From Finney and Yamazaki, 1967.)

elongation is recorded. From the curve, resistance to deformation, extensibility, and energy needed to rupture the dough are computed.

The extensigraph (Figure 7-14) was introduced in 1936 to supplement information supplied by the farinograph. The extensigraph is particularly useful in studying effects of oxidants on dough properties. The tested flour is mixed in the farinograph with a salt solution into a dough that reaches maximum consistency at 500 BU. It is taken from the mixer, and two 150-g pieces are shaped into balls by means of a rounder-homogenizer and then into cylinders by means of a roller. Each dough cylinder is clamped in a cradle and allowed to rest 45 minutes in a compartment maintained at 30°C. After being stretched, the dough is shaped, allowed to rest, and stretched again. Generally, three stretching curves are obtained (after 45, 90, and 135 minutes). It is usual to make the measurements:

1. Extensibility—length of the curve in millimeters.
2. Resistance to extension—height of the extensigram in BU measured 50 mm after the curve has started.
3. Strength value—area of the curve.
4. Proportionality figure—ratio between resistance and extensibility. The greater the proportionality figure, the "buckier" the dough.

The research extensometer designed by Halton and associates (Halton 1938, 1949) of the Research Association of British Flour Millers is similar in many respects to the extensimeter. The extensometer is part of a three-unit equipment system that also includes a water absorption meter and a mixer-shaper unit. The water absorption meter measures the extrusion time of doughs (generally yeasted) which have been made with known amounts of water and relaxed for standard periods. The doughs are packed into a cylindrical container and extruded at a fixed pressure through an opening in the bottom of the container. The rate of extrusion of a fixed amount of dough is determined by means of a micrometer dial gauge and stop watch. As the results are affected by relatively small differences in temperature, the whole determination is carried out in a temperature-controlled room. The small doughs tend to "skin" during the 3-hour relaxation period and should be kept in stoppered bottles. For routine determinations, doughs are prepared from a flour at three absorption levels. The log of extrusion time is plotted versus absorption, and the three points normally lie on a straight line. Optimum absorption was shown experimentally to correspond to an extrusion time of 50 seconds. Bennett and Coppock (1953) found that the absorption meter gave information that correlated with consistency evaluation by an experienced baker better than did farinograph information. The water absorption meter reflects better the effects of dough composition and flour

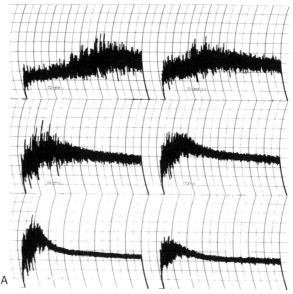

Figure 7-13. (A.) Typical mixograms of hard winter wheat flours (From K.F. Finney, USGMRL. ARS-USDA). (B.) Mixogram reference chart for wheat flours that vary in mixing time (increasing order from 1 to 8) and protein contents (Courtesy G.L. Rubenthaler, ARS-USDA, Pullman, WA.)

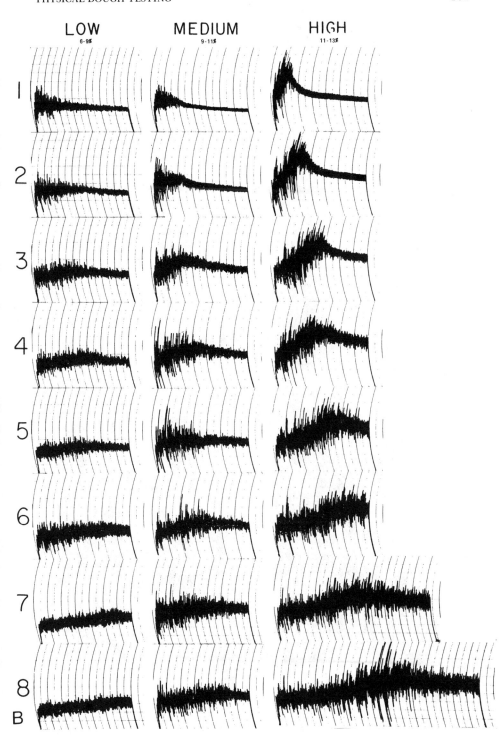

treatment on bread-making performance because it measures absorption of yeasted doughs at the critical molding stage, not at the mixing stage. For load-extension curves, doughs are mixed at optimum absorption and shaped in the mixer-shaper unit. The length of mixing is selected arbitrarily, depending on the expected strength of the flour. The shaped dough is impaled on a two-part peg of the research extensometer. This step requires considerable speed and skill. The upper half of the peg is fixed; the lower half moves at a fixed rate and stretches the impaled dough. The force that is exerted on the upper peg is transmitted and recorded on a graph paper. The stretching curves resemble extensometer curves.

The alveograph consists of three parts: mixer, bubble blower, and a recording manometer (Figure 7-15). The procedure involves using air pressure to blow a bubble from a disc of a flour-water-salt dough. The bubble is expanded to the breaking point, and the recording manometer, which is operated hydraulically, records a curve from which three basic

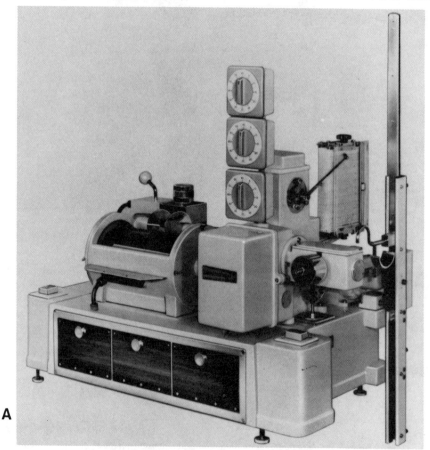

A

Figure 7-14. (A) Picture, (B) schematic of extensimeter and (C) diagrams of extensograms. (Courtesy C.W. Brabender Instruments.)

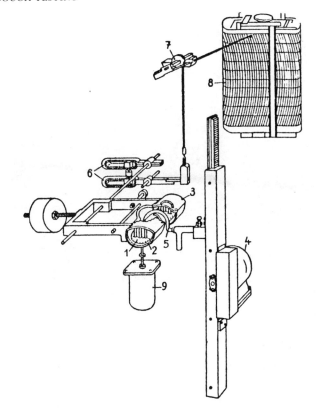

B

RESISTENCE TO EXTENSION
(RESISTENCE TO STRETCHING)

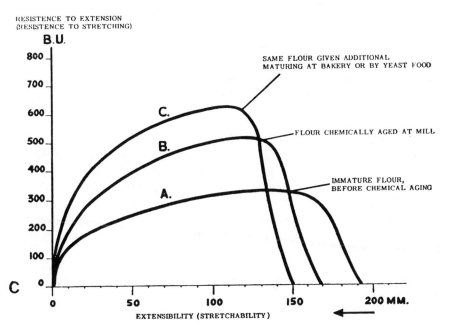

C

EXTENSIBILITY (STRETCHABILITY)

Figure 7-15. The alveograph. (Courtesy Tripette & Renaud France.)

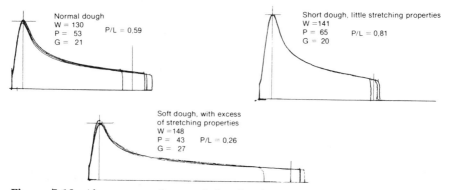

Normal dough
W = 130
P = 53   P/L = 0,59
G = 21

Short dough, little stretching properties
W = 141
P = 65   P/L = 0,81
G = 20

Soft dough, with excess
of stretching properties
W = 148
P = 43   P/L = 0,26
G = 27

Figure 7-16. Alveograms of normal dough, short dough, and soft dough. (G) Swelling index (calculated); (L) extensibility; (P) resistance to deformation; (W) work corresponding to deformation of dough sample. (Courtesy Tripette & Renaud, France.)

measurements (averages of testing 8 dough discs) are taken: distance (in millimeters) that the dough stretches before it ruptures, resistance to stretching at peak height (associated with dough stability), and area of curve. The area generally indicates bread-making potential and is substantially higher in strong than in weak wheat flours. Alveograms of normal, short, stiff, and soft extensible doughs are given in Figure 7-16. The alveograph is popular in several European countries and is the most commonly used physical dough testing device in France. Several modifications of the standard procedure make it possible to use the alveograph in studying effects of oxidizing improvers and in fundamental rheological research on structural relaxation of doughs.

A serious limitation of the alveograph is that it uses a fixed water absorption. That seems satisfactory with weak wheats but gives erratic results with strong wheats of high water absorption. Shogren et al (1963) modified the procedure to test small (15-mg) samples of hard winter wheat flours. Alveogram length was highly correlated with loaf volume and was particularly useful in predicting loaf volume when mixing and oxidation requirements were taken into account. The use of a modified alveograph in studies of hard red spring wheat flour was described by Chen and D'Appolonia (1985).

The Amylograph

The early oven stage is a neglected, yet important, phase of the bread-making process. During this phase, swelling, gelatinization, and liquefaction of starch take place. These changes affect considerably the grain and texture of the bread crumb, crust color, general appearance of the bread, and shelf life and freshness retention of the baked product.

The amylograph (Figure 7-17) is a torsion viscometer that provides a continuous automatic record of changes in viscosity of starch as the temperature is raised at a constant rate of approximately 1.5°C per minute (Anker and Geddes, 1944). The instrument consists of a cylindrical stainless-steel bowl that holds a suspension of 100 g flour in 460 ml of a phosphate citrate buffer (pH 5.30–5.35). The bowl is rotated at 75 rpm in an electrically heated air bath by a synchronous motor, which also operates the recording and temperature control devices. A stainless-steel arm that dips into the bowl is connected through a shaft to a pen that records changes in viscosity of the heated flour suspension in the bowl. Depending on the change in viscosity of the heated suspension, a torque is exerted on the steel arm and is recorded on the arbitrary scale (0 to 1,000 BU units). Viscosity of the slurry is generally recorded as the temperature rises from 30 to 95°C. Modifications of the amylograph that permit a record of viscosity changes at constant or at a uniformly rising or falling temperature (1.5°C per minute) are useful in studies of gelatinization characteristics of various starches. In Europe, the amylograph is used widely to predict baking performance of rye flours and to detect excessive amounts of flour from sprouted grain. Such flours tend to lower substantially the peak height of hot paste viscosity. In the United States, the amylograph is used primarily to control malt supplementation.

The starch-hydrolyzing enzymes of primary importance in cereal products are alpha- and beta-amylases. Alpha-amylase is present in very small concentrations in sound cereal grains or their milled products, but is synthesized abundantly when the grain germinates. Beta-amylase, on the other hand, is normally present in considerable amounts in normal wheat and flour and increases only moderately during germination. Supplementing sound flour with alpha-amylases from malt, fungi, or bacteria increases gas production during panary fermentation, improves crust color, and—

especially in low-sugar doughs—increases loaf volume. Alpha-amylase decreases viscosity of a gelatinized starch by liquefaction, and their supplementation can be measured by the amylograph.

In interpreting the results obtained with an amylograph, two points should be considered. The amylograph does not measure alpha-amylase activity alone, but the combined effects of alpha- and beta-amylases. While this may be of limited importance for alpha-amylase–rich supplements, it affects considerably the measurement of small amounts of alpha-amylase in the presence of high levels of beta-amylase (ie, in sound wheat flour). Secondly, alpha-amylases from various sources vary in thermolability. Malt and especially bacterial alpha-amylases are relatively thermostable and exert a much greater effect on amylogram characteristics than do

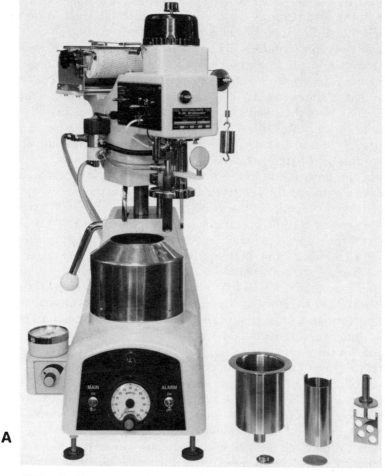

Figure 7-17. (A) Picture, (B) schematic of the amylograph and (C) diagrams of the amylograms. (Courtesy C.W. Brabender Instruments.)

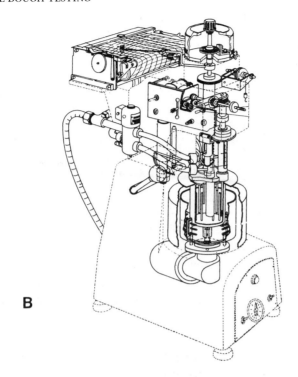

B

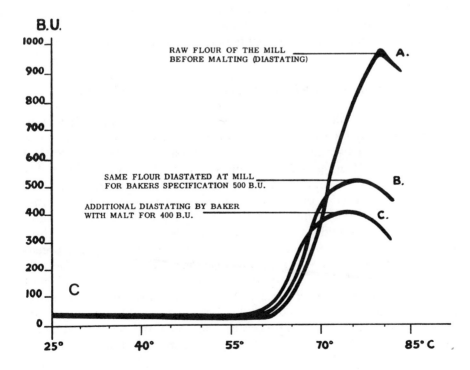

B.U.

RAW FLOUR OF THE MILL
BEFORE MALTING (DIASTATING) **A.**

SAME FLOUR DIASTATED AT MILL
FOR BAKERS SPECIFICATION 500 B.U. **B.**

ADDITIONAL DIASTATING BY BAKER
WITH MALT FOR 400 B.U. **C.**

C

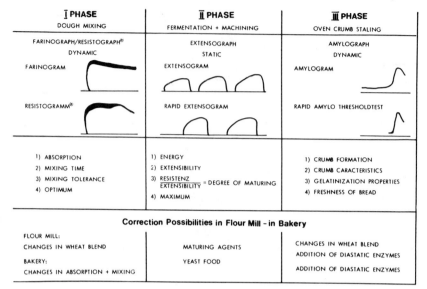

Figure 7-18. The three-phase concept of bread making. (Courtesy C.W. Brabender Instruments.)

heat-labile alpha-amylases from fungal sources. The thermolability of the fungal amylases, widely used in bread making, is so high that their activity cannot be determined by the conventional amylogram. To determine activity of the mold amylase, changes in viscosity resulting from liquefaction or pregelatinized starch held at 37°C are measured (Moro et al 1963).

Evaluation of Physical Dough-Testing Devices

The main value of physical dough-testing devices is to provide information on changes occurring during a short period of the bread-making process. This concept is illustrated in Figure 7-18. Physical dough-testing devices are useful in evaluating and predicting plant breeder samples, in quality control during flour milling and bread making, and in basic rheological studies. They provide precise information on specific properties that cannot be obtained by other means. Physical dough-testing methods have limitations that should be considered when interpreting results. Thus, rheological properties may differ in flours that vary little in their bread-making performance. Physical dough curves may be affected considerably by modification of flour constituents or by certain additives, neither of which may have much effect on bread making. Most rheological measurements are made on unyeasted doughs; fermentation is known to modify considerably both rheological properties and bread-making characteristics. Finally, although physical dough tests are useful analytical tools because they measure changes at certain stages of the bread-making process, they often fail to measure interactions among various factors in bread making.

PHYSICOCHEMICAL TESTS

Gas Production

Determining the rate of formation of sugars that can be fermented in the bread-making process is subject of numerous chemical tests and recording instruments (Seibel, 1964; Jongh, 1967). In a review of the various physical techniques that have been developed to measure gassing power of dough, Voisey et al (1964) described several volumetric and manometric methods, as well as automatic recording for those methods. They also designed an apparatus that measures and records gas production rates of a constant volume of dough. Rubenthaler et al (1980) described the Gasograph, an instrument designed to measure and continuously record the gas produced in 12 fermenting doughs (of about 10 g of flour). Both maltose and gassing power determinations are useful in controlling, to some degree, the proper level of gas production in panary fermentation. Both tests measure the effects of the combined activity of alpha-amylase and beta-amylase as well as susceptibility of mechanically damaged starch granules to amylolytic action. In the maltose determination, one measures the amount of reducing sugars, expressed as milligrams of maltose per 10 g of flour, formed after autolyzing a buffered (pH 4.7) flour dispersion for 1 hour at 30°C (AACC, 1980). In the gassing power determination, the amount of gas produced in a pressure meter (Sandstedt and Blish, 1934) over a prolonged period of autolysis is measured. Although both tests are useful in testing flours from the same mill and milled from a comparable wheat grist, they leave much to be desired in general evaluation of flour for bakery or household use.

For many years, cereal laboratories in the United States have used the amylograph to control malt supplementation for optimum baking results. The Hagberg Falling Number Test (Perten, 1964) was developed more recently and has had fairly wide acceptance in Europe. It is used, however, mainly to detect sprout damage. The Falling Number Test is also used in the United States. It is included in procurements by USDA of all-purpose flour and bread flour for distribution in the United States and overseas. The "falling number" is a simple, quick index of the amylase activity in grain. The falling number is the time in seconds required to stir and to allow a specified viscometer-stirrer to fall a fixed distance through a hot aqueous flour suspension being liquefied by the enzyme (see later in this chapter).

The Maturograph (Seibel, 1964) was designed to determine optimum proof conditions and fermentation tolerance. The instrument measures and records changes in dough elasticity in a fermented dough. The fermented dough is pressed at 2-minute intervals by a constant weight, and the relaxation is recorded until a plateau is attained.

Gas Retention

Several instruments are available to determine gas retention. The importance of gas retention at the baking stage cannot be overestimated. Determinations of gas retention at room temperature are, however, of limited value. Doughs are generally taxed little during the fermentation stage and show no significant differences in proof heights or gas retention capacities. Consequently most of the expensive, time-consuming, automatic, gas-production-retention measuring devices are of limited value.

An oven rise recorder (Ofentriebgeraet) has been described (Seibel and Crommentuyn, 1963; Marek and Bushuk, 1967). The instrument measures changes in a dough ball heated in an oil bath from 30 to 100°C within 22 minutes under conditions approximating those in a baked dough. The changes in volume of the ball are affected by gas retention capacity and seem to be useful parameters in predicting loaf volume potentialities. Effects of wheat conditioning, flour treatments, and oxidants on bread-making characteristics were correlated with effects of the variables on parameters of the oven rise recorder.

The Wheat Meal Fermentation Test

The wheat meal fermentation test originally proposed by Saunders and Humphries (1928) measures length of time (in minutes) elapsing before a dough ball made from wheat meal, water, and yeast disintegrates after it is placed in water. "Dough balls" made from weak wheats disintegrate rapidly, whereas those made from strong wheats remain intact a long time. The test has undergone many modifications (Cutler and Worzella, 1933; Pelshenke, 1933); it is used mainly to evaluate plant breeders' samples. The test is sensitive to even slight changes in procedure. Such changes are easily encountered in a procedure involving kneading dough in a mortar with an arbitrarily chosen amount of yeast suspension and shaping the dough ball in the palm of the hand. A number of investigators have therefore suggested that greater standardization of the method would permit it to be applied more widely in wheat-quality investigations. Pelshenke (1948) recommended using a standard grinding mill to obtain a meal of uniform fineness, a mechanical kneader, dough ball molder, and an accurately controlled temperature cabinet. A modified mixer and extrusion device developed by Seibel (1963) improved reproducibility of the test. The doughs were mixed for a fixed time at an arbitrarily selected absorption. Studies in the author's laboratory (Mamaril et al, 1962) showed that best results were obtained on samples ground on a Micro-Wiley mill to pass a 20-mesh sieve and mixed to the point of minimum mobility (at a consistency of 800–850 BU) on a farinograph, equipped with a microbowl. The doughs were made into balls by a simple mechanical device. As doughs with long meal fermentation times are affected by factors unrelated to strength (insufficient supply of fermentables), the doughs were supplemented with 5% sucrose.

Imbibitional Properties

The numerous methods that use imbibitional properties to evaluate bread-making potentialities of flour can be divided into those performed on gluten and those on the whole flour.

The Berliner-Koopman method (1929) measures the extent to which a known weight of moist gluten increases in volume when immersed under standardized conditions in a 0.1 N lactic acid solution. The increase in volume exhibited by 1 g of gluten after being immersed for 22 hours is termed the "specific swelling factor" of the gluten. The authors considered the method a test of gluten quality rather than a comprehensive test of flour quality. The test was modified to measure turbidity of the suspension (which is inversely related to strength) instead of recording actual swelling determinations. Separating gluten for the Berliner-Koopman test is time-consuming. It introduces an unknown factor; even when mechanical washers are used, the gluten is likely to vary and to be altered in composition. In addition, tests on gluten do not reflect the effects other flour constituents have on the contribution of gluten to breadmaking.

The method developed by Finney and Yamazaki (1946) measures water retention capacity of proteins in a flour suspended in a lactic acid solution. To evaluate soft wheat flours, an alkaline viscosity test using sodium bicarbonate instead of lactic acid was developed (Finney and Yamazaki, 1953). The alkaline water retention test (Yamazaki, 1953) has been useful in predicting performance of soft winter wheat flours in cookie manufacture.

The sedimentation test (Zeleny, 1947) was developed for estimating quality of small samples of wheat. Coarsely ground wheat is sifted to remove most of the bran, and a weighed portion of the crude white flour is suspended in water and treated with lactic acid in a graduated cylinder. The volume of the "sediment" after a 5-minute standing period is the sedimentation value. Values vary from about 5 for very weak wheat to about 70 for very strong wheat. Dividing the sedimentation value by the percentage of protein in the wheat gives a specific sedimentation value that can be used as an index of gluten quality. The sedimentation test has been suggested in early-generation wheat breeding work. The small sample size required and the speed and simplicity of the test have appealed to many plant breeders. The test is used widely in Europe. Several objections have been raised against the use of the sedimentation test. It can be performed on flour and not on ground wheat. Break flour that comprises only a fraction (about 12%) of the total flour is used. And, finally, the sedimentation test value decreases after harvest even though under optimum storage conditions breadmaking quality actually increases.

PERFORMANCE TESTS

The term "quality," as applied to cereals, means different things to growers, processors, and consumers. Yet the basic purpose in testing the grain, the intermediate raw materials, and the processed foods is to help produce an attractive, nutritious, and uniform end product at a competitive price under conditions imposed during the various processing stages. Many physical, chemical, and physicochemical tests can predict with varying degrees the suitability of raw materials and processes to produce an acceptable end product. Usefulness of the test is generally measured by the extent to which they correlate with a performance test—the final criterion of quality.

Bread Making

Although Cereal Laboratory Methods (AACC, 1980) describes two standard bread-baking tests, it is doubtful that any flour chemist employs either without modification. And so, after many years of experience and research, there is no generally accepted laboratory bread-baking formula and procedure (Durham, 1965).

In plant breeding programs, a straight dough procedure involving baking loaves from 100 g (or less) of flour is used. A layout of a dough-testing and baking laboratory in the Grain Research Laboratory in Canada was given by Kilborn and Aitken (1961) and in the Hard Winter Wheat Quality Laboratory (ARS-USDA), Manhattan, KS (Courtesy K.F. Finey) (Figure 7-19). The baking test is regarded primarily as a method of evaluating protein quality. Characteristics such as protein content, color, mixing time, absorption, etc are determined separately, though the baking supplies information on some of them. As the baking test is regarded primarily as a measure of protein quality, an optimized system with regard to dough composition and baking procedures has been designed. None of the baking ingredients (yeast, sugar, shortening) are permitted in a limiting capacity, and optimum mixing time, water absorption, and oxidant level are used. Such a baking test serves as a scientific analytical tool to determine bread-making potentialities rather than performance of a wheat flour under a set of fixed and arbitrarily selected conditions. The test gives a good measure of quality characteristics inherent in wheat or flour, as judged principally by loaf volumes (Figure 7-20).

In commercial baking operations, a wide variety of formulas and procedures is used. The commercial bakery test is generally intended to simulate actual production conditions. Such a test serves several purposes,

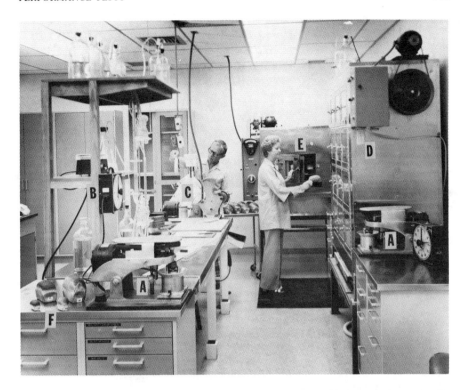

Figure 7-19. View of the experimental bakery in the Hard Winter Wheat Quality Laboratory at the U.S. Grain Marketing Research Laboratory, Manhattan, Kansas. Dough mixers are at (A). After wheat flour, soy flour and shortening are added to mixing bowl, solutions of sugar-salt, malt syrup, yeast, ascorbic acid, and potassium bromate are added from the dispensing stand (B). At C are sheeting (punching) rolls and a dough about to be molded and panned, prior to placing in the fermentation and proofing cabinet (D). A loaf has been removed from the baking oven at E, after which loaf volume is determined by seed displacement with a volumeter (not visible, on a table at the end of the fermentation cabinet). In the lower left-hand corner above F are 100-g and 10-g (flour basis) loaves baked from good- (right) and poor- (left) quality flours. Wheat breeders throughout the Great Plains of the United States send samples of standard varieties and experimental strains to the laboratory for evaluation of bread-making quality, which must be good if the new ones are to be distributed to the farmer for commercial production.

of which three appear to have the most merit: (1) assuring production of a flour of uniform quality; (2) determining the best manner to handle a flour to produce a given product; and (3) characterizing flours to predict commercial usefulness (Bradley, 1950). The common test in bakeries uses a sponge dough procedure for baking 1-lb loaves. In a survey of significance attached to different baking test factors, these were given as most important: loaf volume and internal loaf characteristics, handling properties, and mixing characteristics (Nelson, 1950).

Some bread is produced by continuous bread making in which dough

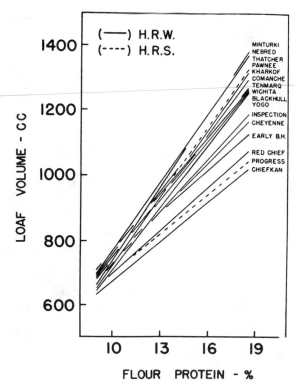

Figure 7-20. Loaf volume-protein content regression lines for hard red winter wheats and hard red spring wheats. (From Finney and Barmore, 1948.)

ingredients are mixed continuously at high speed and under pressure and where extruded dough is deposited directly into the baking pan. The importance of and interest in the process are demonstrated by five complete monthly issues of Cereal Science Today devoted to continuous bread making during 1963–67. The need for tools to evaluate flours used in the process resulted in the development of special equipment to be used in laboratory and pilot-scale continuous-process bread making. Such equipment is relatively expensive. A similar setup, suited to testing small samples, was described by Bushuk et al (1965).

Thus, in the United States wheat testing methodology differs for evaluation of plant breeding material and for samples in marketing channels. The cereal chemist working with the plant breeder or miller is concerned with the milling properties of the wheat. This is of indirect concern to the baker. Preparation of wheat to determine end-use properties may be required primarily to produce flour for each evaluation (various tests including baking), to distinguish among varieties that differ widely in milling properties, or to predict commercial milling yield. Under some circumstances the three objectives may not be compatible, and

Table 7-6. Milling, Compositional, Physical, and Bread-making Properties of Hard Red Spring Wheats and Flours

Wheat data	Flour data	Bread-making data
Test weight	Extraction	Absorption
Moisture	Ash	Dough character
Vitreous kernels	Protein	Loaf volume
1,000-kernel weight	Wet gluten	Grain and texture
Protein	Falling number	Crumb color
Ash	Amylograph peak	Crust color
Falling number	Farinogram	Symmetry
Alpha-amylase	Extensogram	

Regional hard red spring wheat quality: Montana, North Dakota, South Dakota, Minnesota.

Table 7-7. Milling and Bread-making (Functional) Properties of Bread Wheats and Flours

Milling	Bread-making
Bolting properties	Flour protein
Break flour yield	Mixing requirement
Middlings flour yield	Mixing tolerance
Flour yield (total)	Dough handling properties
Relative hardness	Water absorption
Flour ash	Oxidation requirement
Wheat-to-flour protein	Loaf volume potential
	Crumb grain and color

Source: K.F. Finney, Hard Red Winter Wheat Quality Laboratory, U.S. Grain Marketing Research Laboratory (USGMRL), Manhattan, Kansas, 1981.

ranking of bread-making potential may be affected by the method of material preparation.

Bread-making potential should be evaluated in reference to both protein content and protein quality. Kjeldahl protein determination may not provide the needed information to determine varietal effects or to evaluate wheat damaged by adverse climatic conditions during growth or damage during growth or storage.

Evaluation for Plant Breeders

Milling, compositional, physical, and bread-making properties of hard red spring (HRS) wheats and flours are evaluated in the United States according to the outline in Table 7-6. The functional (milling and bread-making) properties of wheats and flours that are evaluated for plant breeders in the hard red winter wheat area of the United States are listed in Table 7-7. The tests that are run on those samples are listed in Table 7-8.

Evaluation for Millers and Bakers

At more advanced stages of plant breeding and in mill and bakery laboratories, assays of protein and mixing properties are used in conjunc-

Table 7-8. Chemical, Milling, and Baking Properties of HRW Wheats

Wheat	Flour	Dough	Bread
Test weight	Ash		Loaf volume
Ash	Protein	Mix time	as Received
Protein	Absorption	Mixogram	Constant protein
Flour yield			Crumb grain
(also as index of hardness)			

Source: Hard Red Winter Wheat Quality Laboratory, U.S. Grain Marketing Research Lab, Manhattan, Kansas, 1983.

Table 7-9. Baking Formulations—Ingredients

Ingredient[a]	Regular	Short-time	No sugar
Flour	100	100	100
Sugar	6	6	0
Salt	1.5	1.5	1.5
Shortening	3.0	3.0	3.0
Yeast (adjusted)	2	5.3 (or 2 dry)	5.3 (or 2 dry)
Malt (120°L)	0.25	0.25	0.75
Nonfat dry milk or soyflour	4	0	4

[a]Parts per 100 parts flour.
Source: Hard Winter Wheat Quality Laboratory (USGMRL), Manhattan, Kansas, 1983.

Table 7-10. Baking Formulations—Procedures

Ingredient or procedure	Regular	Short-time	No sugar
Mixing time	Optimum	Optimum	Optimum
Absorption	Optimum	Optimum	Optimum
Oxidant	KBrO$_3$—optimum	10 ppm KBrO$_3$ + 100 ppm ascorbic acid[a]	10 ppm KBrO$_3$ + 100 ppm ascorbic acid[a]
Fermentation (min)	180	90	90
Proof time (min)	55	About 32	About 32
Bake time (min)	24	24	24
Bake temp (°C)	215	215	215

[a]Or 100 ppm ascorbic acid, or 50 ppm ascorbic acid.
Source: Hard Winter Wheat Quality Laboratory (USGMRL), Manhattan, Kansas, 1983.

tion with baking tests. The baking tests used in the Hard Winter Wheat Quality Laboratories (USGMRL) of the USDA are compared in Table 7-9 and 7-10. They are straight dough procedures that use 100 g flour and are optimized with regard to water absorption, mixing time, and oxidation level. They differ in yeast and sugar levels and resulting fermentation schedules. A rich formula in which wheat flour components are the only known limiting components is used.

Table 7-11. Selected Trade Methods—Test Baking Size and Ingredients

Size or ingredient	Milling (straight)	Baking (sponge 65:35)	Supplier (sponge 70:30)	Baking institute (sponge 70:30)
Size	1 lb	1½ lb	1 lb	Variable
Flour	100	100	100	100
Yeast[a]	2.5	2.0	2.5	2.5
Salt[a]	2.0	2.0	2.0	2.0
Sugar[a]	5.0	6.0	7.0	6.0 High fructose corn syrup (HFCS)
Nonfat dry milk solids (NFDM[a])	3.0	2.0	—	2.0 (replacer)
Shortening[a]	2.0	2.0	2.0	3.0
Oxidant[a]	—	Yeast food (fixed)	Yeast food (fixed)	Yeast food (fixed)

[a]Parts per 100 parts flour.

Baking tests for evaluation of wheats from state programs in the hard red winter and hard red spring areas are modeled after the American Association of Cereal Chemists (AACC) methods. Evaluation of wheats milled for at least 2 years, each a composite of several locations, on a semicommercial mill is required for approval by the Hard Winter Wheat Quality Council. Milling, composition, physical properties, and instrumental bread-making properties are conducted in a single laboratory. Actual baking results, by a method of choice of 35–40 collaborators and as compared with the check control, are then used to compute relative values. A similar procedure is used in the hard red spring area.

Finally, a comparison is presented for methods in a mill laboratory, a bakery laboratory, a laboratory for a manufacturer of flour improvers, and an institute affiliated with the baking industry. Consulting laboratories report results of wheat surveys on the basis of AACC straight dough pup loaves and of testing flours for mills and bakeries on the basis of AACC sponge dough loaves; occasionally, flour is tested by the continuous process. Although the methods compared in Tables 7-11 and 7-12 are specifically for four laboratories only, they are believed to be typical in practice in bake laboratories in U.S. flour mills and bakeries.

The mill-affiliated laboratory uses the straight dough procedures; the others insist on a sponge dough procedure. Comparison among the baking company, the supplier of improvers, and the baking institute indicate that insofar as testing in, for, and on behalf of the baking industry, there is a uniformity of approaches and philosophy. The straight dough procedure is satisfactory for preliminary evaluation (screening) for use by plant breeders and mills. For determination of mixing and fermentation tolerances, the sponge dough procedure, which produces the best texture, must be used. Fermentation tolerance can be established seasonally; mixing tolerance must be established for each batch of flour.

Physical dough testing can be used to assist in determining mixing time

Table 7-12. Selected Trade Methods—Test Baking Procedures

Procedure	Milling (straight)	Baking (sponge 65:35)	Supplier (sponge 70:30)	Baking institute (sponge 70:30)
Absorption	Optimum	Optimum	Variable	Variable
Mixing time	Optimum	Optimum	Variable	Variable
Fermentation	3 h, 86°F	4½ h or variable, 85°F	4½ h, 83°F	3½ h, 88°F
Recovery time	—	30–45 min	45 min, 83°F	20 min, 88°F
Proof	To height (2″) 86°F	To height (1½″) 105°F	To height (1½″) 105°F	To height (⅝″) 100°F
Bake	25 min 425–435°F	20 min 450°F	20 min 435°F	20 min 430°F
Loaf volume	As baked	Inconsequential	Cool	Cool
Bread score	Cool	Next day	After 48 h	Cool

but cannot predict it accurately. To predict mixing time, a series of tests using equipment specifically modified to duplicate the results in the plant must be used. Bread produced by no-time methods is unsatisfactory with regard to texture and flavor. Buns can be produced by some no-time methods, provided a preferment is used. Insofar as bakeries are concerned, the method of choice is sponge and dough. Flour that produces satisfactory bread by this method will produce satisfactory bread by any other method: straight, continuous, no-time, etc. We may conclude that the single, universal bread-baking test is the sponge and dough test. This is not always the case, however. While anyone who conducts the baking test does so (or should do so) to obtain information, the information sought varies widely. The baker uses the test to establish the best conditions to produce a uniform and acceptable bread under the constraints of plant equipment, formulation, time, and cost. He expects a flour (or blend of flours) from the right wheats, milled in a manner that will satisfy his needs. Potential bread-making properties (under optimized or variable conditions) are of no practical consequence or interest. A rather narrow range in protein content and quality is expected to result from a blend of flours produced under specific conditions and cultural practices, following an extensive evaluation program of plant breeders' material. Not surprisingly, if those requirements are met, loaf volume is of no real concern. It must be of concern, however, to the laboratory that evaluates material for the plant breeder. That laboratory evaluates a very wide range of material both in terms of protein content and protein quality. The bake test conducted by that laboratory is an analytical test of the wheat or the flour in terms of bread-making potential.

What then is the answer to the need for a single, universal bread-baking test? The answer is a reserved yes, as long as the test is compatible with the purpose of testing, including test baking to provide useful information and not merely to generate data. To meet those compatibility requirements, several baking tests may be required: for the plant breeder, the researcher, the miller, the supplier of additives, and the baker. Once the needs are defined and proper formulations and schedules are made known and laid down, two criteria must be met: (1) that users of the test can obtain results that are in agreement within experimental error, and (2) that all users know the meaning of the results, so that proper interpretations can be made.

Bread Staling

The problem of staling developed with the wholesale baking industry that distributes bread produced in a single large bakery of retail outlets in a wide area. Returns of stale bread are over 5% of production (Bechtel, 1955). Staling of bread crust is considered due mainly to its absorbing moisture from the air and the crumb. Crumb staling, which is more important in causing loss of palatibility, is a rather poorly defined, complex phenomenon that is difficult to measure. The following are characteristics of the staling process in bread crumbs: changes in taste and aroma; increased hardness, opacity, crumbliness, and starch crystallinity; and decreased absorptive capacity, susceptibility to attack by amylase, and soluble starch content (Bice and Geddes, 1949; Herz, 1965). Nearly every one of those changes has been, and can be, used as a basis for a test of the degree of staling (see Chapter 13).

Using crumb compressibility to evaluate staleness is justified by many consumers frequently using softness as a criterion of freshness. The methods used to measure loaf softness are based on three principles (Thomas and Thuger, 1965); (1) The crumb is subjected to a constant load for a fixed time, and the deformation is measured to give a softness index. (2) The force required to give a fixed deformation is measured, and crumb firmness is determined. (3) The crumb is subjected to shearing or squeezing forces.

The latest edition of Cereal Laboratory Methods (AACC 1980) gives two tests to measure staleness of bread. One is based on organoleptic evaluation by a trained panel, and one uses a compressimeter (Platt and Powers, 1940) that measures crumb firmness.

An instrument that quantitatively measures the effects of squeezing a loaf of bread was described by Hlynka and Van Eschen (1965). Compressibility of bread can also be measured by universal testing machines such as the shear-press, Instron, or Texturometer. The various physical tests are widely used as convenient ways to obtain measurements of changes in crumb properties as bread ages. It should be realized, however, that the various physical tests record different rates of change, so they are not entirely consistent. It has also been shown that such methods frequently do not agree with human judgment of freshness (Bechtel, 1955).

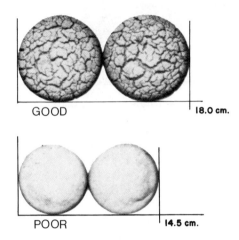

GOOD   18.0 cm.

POOR   14.5 cm.

Figure 7-21. Good and poor cookies baked by the Micro-Cookie Test. (From Finney and Yamazaki, 1967.)

Soft Wheat Flour Products

To assess milling quality of soft wheats, the following tests are performed: test weight, wheat ash, wheat protein, pearling index, particle size index, break flour yield, and total flour yield (Yamazaki, 1962). The pearling and particle size indices give an idea of kernel hardness, break flour yield, endosperm friability, and total flour yield of the wheat.

In spite of the variety of pastry products baked from soft wheat flours, evaluation of breeding lines in USDA laboratories is directed toward two end products—cookies and cakes (Finney and Yamazaki, 1967). To measure usefulness in cookie making, straight-grade flours from soft wheats are tested for moisture, ash, protein, no-time acid viscosity, mixogram area, and alkaline water retention capacity. The alkaline water retention capacity test (Yamazaki, 1953) is probably the best available chemical test to evaluate performance of soft wheat flours. Of all the tests used, however, the most meaningful is the cookie-baking test. Cookies are baked from 40 g of flour. Ingredients are mixed in a pin-type micromixer, and the dough is divided in half, rolled on a sheet provided with riders, cut with a standard cutter, and baked. The criterion used in evaluating the quality of the sugar-snap cookie is the diameter of the product (Figure 7-21). Cookie spread potential appears to be a wheat varietal characteristic. Within a variety, spread is somewhat affected by protein content, and some workers adjust cookie spread to a uniform protein content in making intervarietal comparisons.

The large number of cake formulas, special ingredients, and operational variations complicate the process of standardizing tests to evaluate cake

Figure 7-22. Good and poor cakes baked by the research formula. (From Finney and Yamazaki, 1967.)

flours whether in a commercial laboratory or as part of a breeding program. At the USDA Soft Wheat Quality Laboratory, finely milled 50% extraction flours are impact-milled to reduce average particle size and are treated with chlorine gas to a pH range of 4.6–4.8. Chemical tests are similar to those used to evaluate cookie flours. In addition, original pH and change in pH after treatment with chlorine are determined. Cakes are baked by a formula that was developed to subject flours to maximum strain through high sugar content and lack of milk solids and egg whites (Kissell, 1959) (Figure 7-22). Data for optimum liquid level, cake volume, and internal score are then recorded. The no-egg, no-milk formula gives responses of wider range from cake-making flours than does a formula of optional ingredients. None of the available chemical or physicochemical tests seem to be highly correlated with cake quality. Similarly, it is questionable whether a single cake-baking test can be used to predict performance of a flour for various types of cakes (Yamazaki, 1962).

Alimentary Pastes

For manufacture of alimentary pastes (macaroni, spaghetti, and noodles), semolina from durum is the preferred raw material (Irvine, 1971). Semolina, a coarsely ground, purified, middlings stock is mixed with water, and sometimes with salt, in a stiff dough, pressed or extruded into the desired shape, and dried. Highest yields of best semolina are produced from clean, large, vitreous, uniform kernels. The wheat should be high in protein, its gluten medium strong, the concentration of yellow pigments high, and lipoxygenase activity low. Particle size distribution and granulation of semolina are highly important in the manufacture of macaroni. Gluten from a good-quality durum semolina is clear, light yellow, extensible, but also considerably elastic; when stretched, it forms thin, stable membranes. High-quality semolina should be a clear, bright yellow, imparted by xanthophyll and practically free of a brownish cast. Pigment loss from processing ranges from 15% to 60%; low enzyme activity is desirable. Measurements of rheological properties (rate of water absorption, dough development, and dough stability) determined at an absorption of 27–35% on a farinograph provide useful information. Soznicov and Skuratova

(1963) report semolina mixed in a farinograph with 36.5% water satisfactory if its reading is over 550 BU and of medium quality with consistency at the point of minimum mobility 450–550 BU.

Good semolina should process readily, yielding a strong, clear, bright-yellow product, which when cooked retains its shape, has a firm bite, and does not become soft or mushy (Irvine, 1971). According to Holliger (1963), the quality of macaroni products is determined by the following factors: (1) color and appearance; (2) mechanical properties of the uncooked product, of special interest in packaging and transportation; and (3) behavior during cooking (cooking loss, texture of cooked product). Macaroni products are judged visually before being cooked, and by appearance and form, color, flavor, tenderness, and stickiness of the cooked product (Hoskins and Hoskins, 1959). Schneeweiss (1959) proposed to evaluate alimentary pastes by this point system:

	Points
Exterior appearance (length, shape, break, crumbliness)	4
Surface (color, virtreousness, cracks, smoothness)	4
Mechanical strength (breakability, elasticity, appearance of break)	4
Cooking properties (weight loss in cooking)	10
Flavor	3
Taste	5
Total points	30

Excellent macaroni has 27–30 points; good, 24–26; satisfactory, 20–23; and unsatisfactory, 19 points or lower. Experimental-scale macaroni-processing equipment is available commercially for testing 1000-g samples; instruments for smaller samples (about 50 g, semolina) have been designed cooperatively by laboratory workers and plant breeders (Fifield, 1934; Fifield et al, 1937; Harris and Sibbitt, 1942). An apparatus that permits mechanical properties of uncooked and cooked spaghetti to be measured was described by Holliger (1963).

CRITERIA AND SPECIFICATIONS

A cereal analyst in a well-equipped laboratory faces two problems: which method to select from many to solve a specific problem, and how to interpret the results of his analyses. Chapter appendices A, B, and C give details of the test used by the Grain Research Laboratory in Canada to evaluate bread wheats, durum wheats, and barley.

Surveying wheat in Germany, Pelshenke et al (1964) determined a

Table 7-13. Quality Characteristics of Bread, Pastry, and Macaroni Flour Measurable by Micromethods

	Bread	Pastry	Macaroni
Protein content (%)	11–13	7–9	11.5–13
Flour yield	High	High	High
Ash	Low	Low	Medium
Flour color	White	White	Light yellow
Sedimentation	High	Low	—
Flour absorption	High	Low	Low
Mixing properties	Strong	Very weak	Medium to weak
Loaf volume	Large	Small	Small
Viscosity	High	Low	—
Cookie diameter	Small	Large	—

Source: Hehn and Barmore, 1965.

number of external characteristics (kernel size and extent of damage) and functional bread-making parameters (protein, gluten, sedimentation value, swelling power, maltose value, and falling number). Paquet (1964) and Broekhuizen (1966) reviewed methods used to evaluate baking quality of wheat in several European countries. In France, the alveograph W value (energy in 10^3 erg to inflate dough sheet until it bursts) is the common test and, generally, the only criterion; a value of 200 or above indicates strong wheat; 70–95 to 200, medium; and below 70–95, low strength. In Germany, for many years a single score was computed from determinations of gluten contents, wheat meal fermentation time, and swelling power of gluten in dilute lactic acid. Today the Rapid-Mix Baking Test for hard rolls is used widely.

$$\text{Wertzahl} = \frac{\text{loaf volume index} \times \text{crumb grain}}{100} \pm \text{crumb texture}$$

Crumb grain is determined by comparison with standard pictures of fine (8 points) to coarse (1 point) bread crumbs. Strong wheats have Wertzahl values of 115–130 or above, inferior wheats 80 or below. In Austria, the Wertzahl figure is derived from the sum of 2 times the wet gluten content plus 3 times the gluten swelling value. Flours from strong wheats should have at least 28% wet gluten and a Wertzahl of 118 or above. Criteria used in evaluating bread, pastry, and macaroni flours by plant breeders in the United States are given in Table 7-13. Pratt (1971) prepared comprehensive tables of criteria generally used to measure quality and approximate values of flours manufactured in the United States for various end uses. Larsen (1959) presented specifications in terms of six parameters for bread, cookie, pastry, cake, cracker, and biscuit flours. The moisture should be 14.5% maximum, and Kjeldahl protein should be at least 11.5% in bread flour, below 9.5% in cookie and cracker flours, below 8.5% in pastry and cake flours, and 8.0–10.5% in biscuit flours. The ash should not exceed 0.50% in bread flour and should be somewhat lower in flours used for other purposes. Desirable maltose figures (milligrams maltose per 10 g flour) are up to 450 in bread flour and range from 150 to 250 for the other flours. Farinograph parameters were given for bread flours, and viscosity measurements by the MacMichael viscometer for soft flours. Particle size of flour used in bread making should be substantially larger than that of flours used to manufacture other baked products.

According to Schiller (1984), typical flour for white pan bread should be a dry, free-flowing powder, free of hard lumps, and should have a clear, bright, white color and a natural odor with no unusual off odors. The moisture content should be 14% (± 0.25%), the protein (14% mb) 11.4% (± 0.20%), the ash 0.46% (± 0.01%), the color 56–60 Agtron units. The enrichment should meet FDA specifications, the average particle size should be 16–22 FU (Fisher units, average diameter as measured by a Fisher subsieve sizer); and starch damage should be 7% (± 1.5%). Determined by the farinograph, the absorption should be 60.5% (± 2.5%), arrival time 2 minutes (± 0.5), peak time 7.5 minutes (± 1.5), stability time 12 (± 1.5), and mixing tolerance index (MTI) 30 BU (± 10). Hot paste viscosity, determined by the amylograph, should be 450–550 BU; the flour should be practically free of extraneous matter, rope spores, and other objectionable bacteria and/or molds.

Minor (1984) described specifications of straight-grade untreated cookie flours produced in various parts of the United States. The maximum moisture is 14%, the ash 0.43–0.45% (except for soft white flours, which may be somewhat higher in ash; about 0.46–0.50%), the protein around 8–8.5%, and MacMichael viscosity about 40–50 (in some flours 30–40). Farinograph specifications vary widely, depending on the protein content and quality. The absorption range is 50–60%, peak time 1.5–4.0 minutes, stability 2–8 minutes, and MTI 50–150 BU. The AACC bake test spread should be 8–10 BU.

There are large differences in the raw materials used in the production of baked goods in the United States and Germany. This affects the type and methods of processing those raw materials into consumer-acceptable products. Consequently, testing methodology and evaluation of the raw materials and final products differ. Pomeranz et al (1984) milled the six most widely grown commercial West German wheats on an experimental mill (extraction 73–78%) and characterized them by several chemical analytical methods (moisture, protein, ash, sedimentation value, and falling number), physical dough testing methods (farinograph, mixograph, and extensograph), and baking tests (bread making, rolls by the Rapid-Mix Test, and cookie baking). Three soft (cv Caribo, Okapi, and Diplomat) and three relatively hard (cv Monopol, Disponent, and Maris Hutsman) wheat cultivars were evaluated. Mixograms provided better evaluation of physical properties than farinograms. Extensograms were useful in meaningful characterization of the flours and their potential use. The correlation between loaf volume in a modified AACC baking test and the Rapid-Mix Test was highly significant. For cookie making, flours from low-protein soft wheats were best.

According to Gedye et al (1981), appearance, purity, moisture content, visible sprouting, and falling number as an index of excessive alpha-amylase are important selection criteria for all wheats. In grain for use as animal feed, the feeding value should be considered. In whole grain for use in breakfast cereals, additional criteria are protein content and variety identification. In grain used to produce flour for baking, alpha-amylase, protein content and quality, wheat hardness and variety, milling characteristics, starch damage, flour color, and baking characteristics should be considered. Interestingly, 1,000-kernel weight was considered of little importance in grain processed for animal feed, breakfast cereals, or seed, or milled to produce flour for bread making, other baked products, or industrial uses.

Degree of Milling of Rice

Brown (dehusked) rice is milled to separate the bran from the endosperm. Commercial rice bran contains the pericarp (including the tegmen), the embryo, and, in well-milled rice, most of the aleurone layer. The Food and Agriculture Organization Revised Model System for Grading Rice in International Trade (FAO, 1972) defines four types of milled rice: (1) undermilled, in which part of the germ and most or all of the pericarp have been removed; (2) reasonably well milled, in which the germ and pericarp and most of the aleurone have been removed; (3) well milled, in which the germ, the pericarp, and practically all the aleurone have been removed; and (4) extra well milled rice, which contains the starchy endosperm only. Those definitions pertain to the average of a sample, so small variations among individual kernels can be tolerated. There is a large price difference between whole-kernel milled rice and broken pieces. Because the amount of broken kernels increases rapidly with the degree of milling, it is important to control the milling process and develop quick and reliable tests for degree of milling.

Three types of methods are available for estimating degree of rice milling: visual, optical, and chemical. The visual methods, most widely accepted and used, have the usual errors of subjective tests. The methods have been discussed by Hogan and Deobald (1965), Barber (1972–74, 1975–76), Pomeranz et al (1975), and Stermer et al (1977). Most of the optical methods measure reflectance or transmittance properties in the visible range of the spectrum. The results are influenced by color of the rice as affected by variety, growth conditions, texture, and history of the grain. Those limitations can be reduced somewhat by visual (or optical) observation of kernels immersed in dye solutions that preferentially stain the bran layers. Determining optical properties in the far red (660 nm) or near infrared (850 nm) eliminates some of the interferences encountered in measurements in the visible range, but the results are affected by moisture, heat damage, chalkiness, and age of the grain (Stermer, 1968).

Chemical methods are based on measurement of a component that is more concentrated in bran tissues than in the starchy endosperm. Components proposed include crude fiber, ash and ash components, proteins, thiamine, and oil. Determination of either surface oil or total oil has been recommended. Determination of surface oil is an empirical method that measures only the exposed portion of oil in the germ and aleurone rather than the total amount of residual oil. The total oil content is a satisfactory index of rice milling because of the high concentration of oil in the germ and aleurone, but the classical determination of total oil is even more complex and time-consuming than that for surface oil.

The two methods used most involve extracting surface lipids (Hogan and Deobald, 1961) and measuring the colored bran index (FAO, 1972). In the first method, fat is extracted from the whole rice by a refluxing flammable solvent, the solvent is evaporated from the extract, and the lipid residue is weighed. Because the bran (mainly aleurone) and germ portions contain more lipid than the endosperm, extractable fat decreases progressively with milling. In the second method, the bran is stained with the May-Gruenwald reagent (1% methylene blue and 1% eosin in methyl alcohol), and the stained area is measured by planimetry. The arbitrary colored bran index ranges from 100 for brown rice to 5 for commercial samples and 0 for well-milled rice. This method measures the degree of milling of rice and also permits evaluation of the homogeneity of milling. The method is tedious, however, and analysis of one sample requires 1 1/2 hours. Neither method is entirely suitable for grading rice.

Wide-line nuclear magnetic resonance (NMR) spectroscopy has been used to analyze oil in seeds and has opened a new opportunity for geneticists and plant breeders. The method is nondestructive and rapid (about 2 minutes excluding drying time), and it can be used to determine the oil content of a single seed or of a bulk sample. The measured NMR value is related to the total hydrogen in the nonoil fraction. The oil content is calculated from calibration tables or curves. The method has been used successfully to assay oil in corn, soybeans, and other oil-containing seeds.

Brown rice samples (9 cultivars from 3 locations) were milled on an experimental mill (Pomeranz et al, 1975). Oil in 26 brown rices and 173 milled subsamples was determined by NMR. Oil contents of long- and medium-grain cultivars were not consistently different, but in each group there were consistent varietal differences. Weight loss during milling was significantly correlated at the 1% level ($r = -0.617$) with oil content of the milled samples. Correlation coefficients for samples from individual varieties were generally higher (up to $r = -0.874$) than for all samples combined.

Computerized spectroscopy was used to study the infrared (IR) spectra of milled rice. The reflectance at three principal oil absorption bands—0.928, 1.215, and 1.725 μm—was highly correlated with residual bran content (degree of milling). IR reflectance was compared to surface and total (NMR) lipid contents and to weight loss due to milling (Stermer et al, 1977).

A rapid, objective method was developed to rate the degree of milling of rice (Miller et al, 1979C). It involves a 5-minute extraction of 10 g of milled rice with 40 ml of isopropyl alcohol and water (1:1, v/v) and measurement of the electrical conductivity of the extract. An apparatus was built to submerge five rice samples automatically and simultaneously in a solvent contained in 100-ml centrifuge tubes and to drain the solvent from the samples. The baskets, made of 16-mesh stainless-steel screen and containing 10 g of rice, were lowered into and raised from 40 ml of solvent as a group every 6 seconds for 5 minutes. Extraction temperature was 25°C. The objective and subjective ratings for the degree of rice milling were highly correlated, and the correlations of both types of ratings with ash, surface lipid, and Agtron color were higher for experimentally than for commercially milled samples. The objective test was sensitive to the percentage of broken kernels in the sample.

Rice Hardness

Rice hardness is important from the standpoint of changes after harvest, drying to a safe moisture level in a manner that reduces hazards of cracks and fissures and prevents losses in milling, resistance to insects, milling, and processing.

Rice cultivars differ in hardness distribution in the endosperm. Nagato and Kono (1963) classified varieties on the basis of the ratio of hardness at halfway between the center and the periphery and at the central point. In *Indica* cultivars the hardness ratio is below 1.0, the central core is hardest, and hardness decreases toward the periphery. *Japonica* cultivars tend to have a ratio above 1.0; hardness is highest in the middle region and decreases toward the central core and peripheral region. In samples with a ratio of 1.20, the cells of the central core are uneven, flattened, and disorganized.

Saio and Noguchi (1983) studied the structure of polished rice and its milled and air-classified fractions under a scanning electron microscope. The cells of rice endosperm, showing a long, rectangular column shape, were distributed radially from the center to the outer layer. The vitreous region near the outer layer is harder than the central region, and the flour derived from the outer layer can be collected as a coarse fraction rich in protein.

Varietal differences in resistance to grain breakage are known but rare. Two varieties in Sri Lanka have shown exceptional resistance to breakage, but no differences in chemical composition could be established (Breckenridge, 1979). Kongseree and Juliano (1972) determined physicochemical properties of rice grain and starch from lines differing in amylose content and gelatinization temperature. Grain hardness was estimated from 20 individual kernels of brown rice with a Kiya-type hardness tester with a tapered plunger and from the percentage by weight of flour coarser than 80-mesh after grinding duplicate 10 grains for 20 seconds in a special mill. Grain hardness was not correlated with either amylose content or final gelatinization temperature.

Rice undergoes changes during the first 3–4 months after harvest, particularly if kept above 15°C, when stored in rough, brown, or milled form (Barber, 1972; Moritaka and Yasumatsu, 1972; Villareal et al, 1976; Indudhara Swamy et al, 1978). Brown rice becomes progressively harder, as reflected in tensile strength (Kunze and Choudhury, 1972), resulting in an increase in total and head-rice yields because of lower grain breakage on milling.

Fissuring of grains during drying is caused by stress from uneven contraction or expansion of the surface caused by sudden changes in relative humidity (Nagato et al, 1964). A high birefrigerence end-point temperature (BFPT) variety was more resistant to fissuring than a low-BEPT rice (Kunze and Hall, 1965). Varietal differences in fissuring during drying and wetting of rough rice have been reported (Srinivas, 1975; Srinivas et al, 1977).

Nguyen and Kunze (1984) used an Instron universal testing machine with a three-point bending test cell to determine the breaking force in rough rice. Investigations were conducted on breaking strength and on percentage of fissuring grains that developed in rice stored in different environments after being subjected to drying temperatures of 25, 40, and 60°C. Breaking strengths were generally related to the percentage of grains that fissured in the treated samples.

Varietal resistance to stored-rice insects is related to grain hardness. Rout et al (1976) found that susceptibility of brown rice to rice weevil *Sitophilus oryzae* L. in eight varieties with high-amylose content was negatively related to crushing hardness values.

The rice industry is interested in rices with bright, clear, translucent milled kernels that are free of chalky spots. They are typical of nonwaxy (common) varieties of rice of varying degrees of apparent hardness (Webb and Stermer, 1972). Chalky kernels occur when the rice is harvested too early and under certain environmental and cultural conditions. Glutinous, sweet, or waxy rice is characterized by a completely opaque endosperm. Chalky rice detracts from the general appearance and is usually weak, breaks up during milling, and reduces the yield of milled rice (Gariboldi, 1973).

Opaqueness in rice is caused by micropores on the surface of the granule and by cavities inside the polyhydral granules (Tashiro and Ebata, 1975; Utsunomiya et al, 1975b). Nonwaxy rices with about 10% amylose have a tombstone-white appearance. The grain and starch granules of waxy rice have lower densities than those of nonwaxy rice.

Chalky portions of nonwaxy grains are caused by loose arrangement of the cell contents, with air spaces (Del Rosario et al, 1968, Tashiro and Ebata, 1975; Utsunomiya et al, 1975a,b). Starch granules in the chalky portions tend to be spherical and loosely packed, in contrast to the tightly packed polyhedral granules in the translucent portions. The chalky areas contribute to grain breakage during milling, because they are softer than the translucent portions (Nagato, 1962).

According to Blakeney (1979), chalky rice grains are characterized by loose, irregularly packed, rounded-compound starch granules, and air spaces. There is an abrupt change from chalky to translucent tissue. Often one endosperm cell is chalky and the adjacent one is translucent. The sudden change from chalky to normal translucent cells may be caused by environmental influences during endosperm cell division.

Srinivas et al (1984) found that complete chalkiness, expressed uniformly in the endosperm, confers on the grain outstanding physicochemical and technological characteristics which include high resistance to cracking and breakage. The possibility cannot be excluded that the resistance was imparted, in part at least, by an increase in the thickness of the pericarp and aleurone layers.

The physical properties of the rice grain are determined by the dimensions, shape, and weight of the grain, and the hardness of the endosperm (Chang and Somrith, 1979). High milling recovery is generally associated with hardness and an absence of chalky spots in the endosperm. The degree of chalkiness varies among cultivars and among environments within a cultivar. While in some studies chalky spots appeared as monogenic recessive traits, in others they appeared to be a dominant trait. A multigenic system interacting with environmental factors seems plausible. Chalky kernels are more frequently associated with bold grain shape than with slender shape of comparable length. Additive effects appear to be a major component of F_2 variation. An extremely opaque and floury type of "crumbly" endosperm was shown to be controlled mainly by one gene and modifiers (Chang and Somrith, 1979).

There is at present no reliable empirical method to predict milling behavior, other than actual test milling (Barber and De Barber, 1979). In addition, there is no reliable or common conversion factor for extrapolating laboratory tests to industrial large-scale conditions.

According to Suzuki et al (1979), rice grain quality governs milling properties and the commercial value. Brown rice "grain quality" is affected by a great number of factors that include percentage of white-belly and white-core grains, the thickness of the seed coat, the depth of ridges, the percentage of whole-grain brown rice, color and gloss of grains, and size and shape of the grains. Potential milling properties of rice are evaluated

Table 7-14. Average Grain Size and Shape Measurements among Typical U.S. Commercial Rice Types

Grain type[a]	No. samples	Length (mm)	Ratio length/width	Thickness (mm)	1,000-grain wt (g)
Long	9				
Rough		9.3	3.9:1	1.8	22
Brown		7.3	3.6:1	1.7	18
Milled		6.8	3.5:1	1.6	17
Medium	6				
Rough		8.0	2.6:1	2.0	25
Brown		6.0	2.3:1	1.9	20
Milled		5.7	2.2:1	1.8	19
Short	3				
Rough		7.5	2.2:1	2.2	28
Brown		5.5	1.9:1	2.1	23
Milled		5.3	1.8:1	2.0	22

[a]Rough, unhulled grain; brown, grain with hull removed; milled, whole grain milled kernels with hull, bran, and germ removed.
Calculated from Webb (1975).

on the basis of grain size (extra long, long, medium, and short), shape (length-to-width ratio: slender, medium, bold, round), and appearance (percentage of area with chalkiness; none <10%; 10–20%; >20%) (Kush et al, 1979).

It is difficult to compare the milling quality of rices, which differ in size and shape. Optimum milling pressure increases with increase in grain thickness, because the aleurone layer is thicker in coarser samples. In addition, the dorsal portion of the aleurone layer is thicker than the ventral portion and is thinnest in the lateral portions (Hoshikawa, 1967).

Bhashyam and Srinivas (1984) have shown that milling characteristics of rice are affected by morphological characteristics of the grain and that milling breakage can be reduced if shallow-grooved varieties are used.

Rice Quality

Characteristics that influence rice quality include the genetic makeup, environmental effects, and cultural practices (Webb and Stermer, 1972). Virtually all U.S. commercial rice varieties have nonpigmented pericarp and are straw-hulled, translucent, nonscented, nonwaxy, and mild-flavored. Traditionally, rice varieties in the United States are classed as long-, medium-, and short-grain types. Varieties of each grain type are generally associated with specific physical, processing, cooking, and eating characteristics.

Average grain size and shape measurements among typical U.S. commercial long-, medium-, and short-grain types are given in Table 7-14. Average test weights are 42–45, 44–47, and 45–48 lb/bu for long-,

Table 7-15. Average Milling Yields for Selected Rice Varieties in the United States

| Grain type | No. samples | Average milling yields[a] | |
		Whole-kernel rice (%)	Total milled rice (%)
Long	8	58	69
Medium	5	66	71
Short	3	66	74

[a]Total milled rice is the whole-kernel rice and all sizes of broken kernels. Mill yields are based on clean, mature, rough rice samples.
Calculated from Webb (1975).

medium-, and short-grain rice types, respectively. Average milling yields for selected rice varieties in the United States are given in Table 7-15. Average chemical and physical characteristics of selected rice varieties in the United States are given in Table 7-16.

High-quality U.S. long-grain rices cook dry and fluffy, and the cooked grains remain separate. Medium- and short-grain rices after cooking are rather moist and chewy, and the grains clump together.

Breakage Susceptibility of Shelled Corn

Many buyers and processors want whole-kernel corn that is not susceptible to breakage during handling. The increase in broken corn between place of origin and destination is of particular concern. Broken corn is defined by the Official United States Standards for Grain (1978) as broken kernels and fine material that will pass through a screen with 4.76-mm- (12/64 in.) diameter round holes. Since grain may be handled several times from harvest to export, the amount of breakage and foreign material may exceed 7%, which classifies corn as sample grade.

Methods to measure breakage susceptibility can be classified into two groups: (1) subjective methods that measure the extent to which whole kernels show stress cracks, and (2) methods that measure the amount of cracked grain formed when whole grains are impacted or ground. The latter methods are preferred because of their simplicity, objectivity, and reproducibility.

Keller et al (1972) described a pneumatic system to accelerate grain toward an impact surface. Sharda and Herum (1977) used a centrifugal impeller to throw corn kernels against a steel surface and reported that it indicated grain damage susceptibility. Stephens and Foster (1976) compared results from normal handling in an elevator with results from a commercially available, experimental breakage tester (Stein tester). They found that the experimental device showed the relative breakage susceptibilities of different lots of corn but that the testing procedure would have to be standardized for use in official grain grading.

It is clear that for testing large numbers of samples, either in plant

Table 7-16. Average Chemical and Physical Characteristics of Selected Rice Varieties in the United States

Grain type	No. samples	Amylose content (%)	Alkali reaction (avg.)	Water uptake at 77°C (ml/100 g)	Gelatinization temperature (BEPT) (°C)	Class	Amylographic viscosity Peak	After 10 min at 95°C (BU)	Cool to 50°C (BU)	Protein ($N \times 5.95$) (%)	Parboiled rice: Parboil-canning stability (solids) loss (% db)
Long	9	24.6	3.9	128	73	Intermediate	800	425	809	6.7	19
Medium	6	16.7	6.5	324	66	Low	950	401	728	6.4	33
Short	4	18.5	6.8	335	67	Low	845	385	685	6.0	32

Calculated from Webb (1975).

breeding programs or in marketing channels, a simple and empirical method such as the Stein breakage tester can be used. Until recently, there were two options of how to calibrate the Stein breakage tester: (1) comparison with performance in a grain elevator or (2) evaluation by an experienced operator. The first one is very expensive and difficult to control; the second one is subjective. Miller et al (1979a) built a grain acceleration device that impacts corn against corn at velocities both above and below that attained by corn falling vertically 30.5 m (100 ft). Subsequently, Miller et al (1979b) described a standard procedure for measuring the breakage susceptibility of commercial and plant breeder's corn with a Model CK2 Stein breakage tester. A diagram of the Stein breakage tester is given in Figure 7-23. The Stein breakage tester has a removable chamber in the shape of a cup. A steel impeller fits into this cup with a small clearance at the bottom and on the sides. The impeller rotates at about 1,800 rpm. This impacts and slings the grain against the inside of the cup.

Miller et al (1979b) found that the results of the grain acceleration device and a model CK3 Stein breakage tester were highly correlated. The effects of several variables on the results were studied. Breakage increased as the temperature and/or moisture decreased. Small kernels were more resistant to breakage than large kernels. Breakage susceptibility of a mixture of corn samples could be calculated from the proportion of breakage susceptibility of the components of the mixture.

In determining the breakage susceptibilities of commercial samples, it was found that neither total dockage nor broken kernels and foreign material reflected the percent breakage measured by the Stein breakage tester. Five samples that ranged in breakage from 2% to 21% all graded No. 2 yellow corn; similarly, there was no consistent relationship between grade and percentage of breakage of six other samples. Those findings indicated that the present assignment of grade values does not reflect susceptibility of corn to breakage.

It was concluded that the model CK2 Stein breakage tester can measure the breakage susceptibility of corn (and probably of other grains, such as soybeans) and can serve as grain-grading device. A standard procedure was proposed and it was emphasized that the instruments must be manufactured and maintained according to rigorous specifications to assure uniformity of results by different machines and to avoid spurious results.

Corn Hardness

Corn hardness is of great significance to producers, processors, and workers in the grain trade. Hardness is related to kernel density; bulk density; storability; attack by storage insects; breakage susceptibility caused by drying, storage, handling, or processing (milling characteristics, power requirements, dry and wet milling yields); production of special foods; and classification.

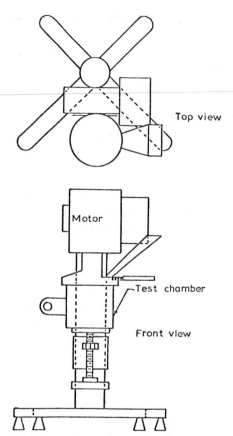

Top view

Motor

Test chamber

Front view

Figure 7-23. Diagram of a Stein grain breakage tester. (From McGinty, 1970.)

Pomeranz et al (1984) determined breakage susceptibility and kernel hardness as measured by density, near-infrared reflectance, and average particle size of ground material for four groups of corn samples. The groups were isogenic pairs with regard to hardness (dent and flint), commercial dent hybrids, dent corn heat-dried under various conditions, and a group that varied in starch composition (waxy, regular, high-amylose) and in protein, oil, and ash contents. Density, near-infrared reflectance, and average particle size values were highly, clearly, and positively correlated, provided homogeneous groups were analyzed and evaluated. In samples that were highly susceptible to breakage, correlation coefficients of hardness determination increased when calculations were made on the basis of constant breakage susceptibility. The three methods of hardness determination were equally sensitive and useful in routine analyses.

EVALUATION OF MALTING AND BREWING QUALITY

The assessment of malting quality has been of interest almost ever since barleys were first malted and malt was used for beer production. Evaluation of barley varieties for malting quality was described by Dickson and Burkhart (1956). Burkhart (1953) compared pilot brewing of commercial and laboratory malts. Bendelow et al (1971) reviewed assessment of malting and brewing quality for Canadian barley breeding programs. Table 7-17 compares mean values, over five stations, of five varieties that included two accepted cultivars and three experimental selections; data for barley, malt, wort, enzymatic, and beer properties are included. Kernel plumpness of selection C is unsatisfactory. Under malt extract are given four figures obtained under various experimental conditions. The extracts in selections

Table 7-17. Mean Values, Over Five Stations, of Varieties Grown in 1967 for Barley, Malt, and Beer Properties

	Unit	Parkland	Conquest	A	B	C	Necessary difference
Barley							
Plump	%	87.5	92.4	86.3	82.7	78.3	8.0
1,000-kernel wt	g	33.1	33.0	32.3	32.8	32.3	—
Nitrogen	%	2.18	2.28	2.21	2.20	2.19	—
Malt							
Extract, fine	%	76.1	76.5	77.1	75.5	75.1	0.9
coarse	%	75.0	75.3	75.6	74.2	73.2	—
70°	%	73.6	73.8	74.4	72.6	71.6	1.3
cold	%	23.2	24.8	24.5	21.7	22.7	—
Wort							
Nitrogen, fine	%	1.03	1.08	0.96	1.06	1.02	0.08
70°	%	0.94	0.98	0.88	0.85	0.84	0.07
cold	%	3.17	3.05	2.87	3.17	3.03	0.15
Perm sol nitrogen, fine	%	84.6	79.7	85.1	79.9	91.5	7.6
70°	%	79.4	81.7	78.1	87.6	99.7	4.8
cold	(%)	70.5	63.2	63.3	62.8	67.5	—
Enzymatic activity							
Saccharogenic	°L	151	157	166	166	164	—
Alpha-amylase	°L	22.2	24.6	20.0	15.8	18.8	4.8
Cytolytic	units	23.8	27.6	25.3	20.5	36.1	7.5
Beer							
Brewhouse yield	%	76.1	75.8	76.9	74.9	73.5	1.4
Fermentability	%	63.2	64.5	64.6	64.1	62.5	—
Alcohol	%	3.65	3.73	3.70	3.76	3.63	—
Protein	%	0.83	0.84	0.77	0.79	0.76	0.06
Chill haze		13.4	10.7	11.2	10.6	20.8	3.7

Source: Bendelow et al, 1971.

B and C, and especially C, are below the expected. The difference between fine- and coarse-grind extract in selection C was higher than in the others, indicating unsatisfactory modification during malting. Percent permanently soluble nitrogen in selection C is high. Selection B was low in alpha-amylase and cytolytic activity. The brewing properties of selections B and C were substandard. Selection C was relatively low in brewhouse yield and high in chill haze. In summary, selection A is good and selections B and C are unsatisfactory.

SPROUT DAMAGE

Germination, malting, and sprouting are related phenomena from a plant physiological standpoint. Germination of grain in the field to produce a new plant and malting in the malthouse to produce a commercial product with specified properties for specific food uses are desirable processes. Sprouting, on the other hand, almost invariably implies an undesirable process of germination of rain-damaged grain in the field or of water-damaged grain in storage under adverse conditions. Sprout damage, especially in wheat, may be associated with impaired functional properties.

A large number of changes takes place during germination-malting-sprouting. They include physical changes (including acrospire and rootlet formation), gross chemical changes (including weight loss), increase in solubility of major grain components (mainly starch and proteins), and enzymic activities. Immature, developing grain contains substantial amounts of hydrolytic enzymes; their amounts reach a minimum at maturity. Many hydrolytic enzymes are synthesized de novo after germination-malting-sprouting. Levels of any of those enzymes can be used to assess degree of sprouting. Generally, however, determination of alpha-amylase is the simplest, because it has a specific action, can be used on defined and readily available substrates, and is easy to measure.

Initially, sprout damage was evaluated by visual inspection of the grain for the presence of an acrospire. The subjective and laborious method is still widely used. The need for a more precise and sensitive evaluation led to the development of objective analytical methods such as the amylograph and falling number devices. The amylograph has found use as a research tool and in mills or bakeries to evaluate malt supplementation of bread flours. The falling number was developed by Hagberg (1960, 1961) as an index of alpha-amylase in grain; it is used in several countries to detect sprout damage. The falling number method is more rapid and better suited to determine sprout damage (alpha-amylase) than the amylograph paste viscosity method. Both methods cook a slurry of flour or wheat and water under arbitrarily selected conditions and measure viscosity of the hot paste. Falling number is the time (in seconds) required to stir and allow the viscometer stirrer to fall a fixed distance through a hot aqueous flour or

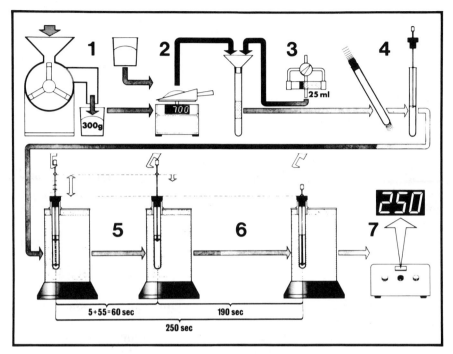

Figure 7-24. Falling number procedure. (1) Sample preparation; (2) weighing; (3) dispensing; (4) shaking; (5) stirring; (6) measuring; (7) result. (Courtesy Falling Number AB, Stockholm.)

wheat dispersion; the time is inversely related to the concentration of the liquefying alpha-amylase that "thins" the hot suspension. Perten (1964) has shown a relationship between falling number, amylograph peak viscosity, and dextrinogenic alpha-amylase. The falling number procedure is shown disgramatically in Figure 7-24. Falling number values around 250 are typical of flour that produces bread of satisfactory loaf volume and crumb gain; values much below 250 produce dark and sticky bread; values much above 250 produce dry, dense bread.

Though the falling number test is used extensively in European grain markets, it has not been used extensively in the United States. Viscometric methods require time-consuming, sophisticated equipment and measure the effect of, rather than actual, alpha-amylase activity. The results are influenced by several factors such as inherent variations in the viscosity of native starches. Recently, chromogenic substrates consisting of dye-labeled starch fractions were developed. These substrates, some of which are commercially available, are convenient and sensitive, require little equipment, and are simple to use. Most work with these substrates has concerned clinical medicine, but agricultural applications are receiving increasing attention. Mathewson and Pomeranz (1977) developed a rapid, simple assay for alpha-amylase in sprout-damaged wheat using a commercially available, dyed amylose substrate. The test is suitable for minimally equipped laboratories, such as those at large grain elevators. The range of the test was selected to evaluate bread wheats. For evaluation of wheat used in producing Japanese noodles or certain types of cakes, a more sensitive test may be required. Such a test was developed by using a different substrate and optimizing pH and temperature (Mathewson and Pomeranz, 1979).

Subsequently, the colorimetric alpha-amylase assay was simplified and shortened by combining the extraction and reaction steps (Mathewson et al, 1979). The procedure includes weighing the sample, extracting and reacting in one step, and a rapid filtration. The resulting colored solution is read directly in a colorimeter. Within-laboratory collaborative studies indicated that the procedure is sufficiently simple that untrained personnel can routinely determine amounts of alpha-amylase below levels that would be detrimental in bread, cake, or cookie making.

The above test cannot, however, be used to evaluate sprout damage in the field or in other locations where no laboratory equipment is available. A comparable test that could be performed entirely without laboratory facilities would enable personnel at small-elevator locations to evaluate sprout damage and afford the grower the opportunity to monitor the alpha-amylase content of wheat and make informed decisions on when to harvest, thus reducing losses from excessive sprout damage. Mathewson et al (1978) describe a compact self-contained unit that can be used to determine alpha-amylase under field conditions.

Colorimetric determination of alpha-amylase in solid products requires five basic steps: grinding, extraction of the enzyme, filtration, enzymatic hydrolysis of an amylose-dye complex, and comparison of color. A field unit that performs these operations was developed. The outside dimensions of the unit (a box) are 39.0 cm (width), 39.5 cm (height), and 28.5 cm (depth). The total weight of the unit, constructed mostly of hardwood and steel, is 18 kg. The main components include a specially designed mill operated by a 12-V vehicle battery, a spoonlike device for sample weighing, a special filtration assembly, a Thermos bottle equipped for temperature control and mixing, and a test tube colorimetric comparator.

This self-contained, compact unit could have wide application in determining alpha-amylase in a variety of commodities and foods. In addition, the unit contains several basic components (ie, mill, filtration assembly, temperature control, and mixing device) that could be incorporated in other self-contained units for evaluating composition, quality, and properties of agricultural products and foods.

BREEDING FOR NUTRITIONAL IMPROVEMENT

Cereal and legume breeders in many parts of the world are currently developing new varieties with improved agronomic characteristics. Since cereals and legumes are a major source of protein, the breeders also seek to improve the quantity and/or quality of the protein in the grains. This requires screening many samples. To meet those requirements, special schemes and methods were developed. They were described by Villegas and Mertz (1971) for chemical screening methods for maize protein quality, by Mertz et al (1975) for simple chemical and biological methods to evaluate cereals for protein quality, by the Protein-Calorie Advisory Group of the United Nations (1976) for protein methods for cereal breeders as related to human nutritional requirements, and by Hulse et al (1976) for nutritional standards and methods of evaluation for food legume breeders. A schematic representation for screening cereal grains for protein content and nutritive quality is given in Figure 7-25.

Complete amino acid analyses in F_5 and later generations can be used to predict the amino acid score. The amino acid score can be calculated when total nitrogen and concentration of the first limiting amino acid is known. The amino acid score provides a chemical prediction of the nutritive value of a protein; it does not take into account availability of the amino acid. To determine availability, biological methods are required. The score compares the level of an amino acid in the test protein and in a reference protein (FAO, 1972; FAO/WHO, 1973). The pattern in Table 7-18 is derived from estimates of the amino acid requirement of young children.

The amino acid score (previously known as chemical score) is calculated from

$$\frac{\text{content of limiting amino acid in test protein}}{\text{content of same amino acid in reference protein}}$$

If the complete amino acid profile is known or if the limiting amino acid is not known, the chemical score is calculated from

$$\frac{\text{content of amino acid in test protein}}{\text{content of same amino acid in reference protein}}$$

The lowest score for any essential amino acid of the test protein is a rough approximation of its utilization. The general outline of screening methods for food legume breeders is basically similar to that for cereal breeders. In view of the relatively low content of sulfur-containing amino acids, there is increased emphasis on sulfur, methionine, and cystine determination. Included are methods for raffinose and stachyose, as these sugars might be responsible for a large part of the flatulence effect of legumes. The use of near-infrared reflectance methods in screening is recommended. Analytical methods recommended for screening legumes include methods to determine cooking quality (seed size, hydration coefficient, cooking time, seed hardness, and thickness of cooking broth) and processing quality (damaged starch, amylose, viscosity, water-absorption capacity, viscosity or thickness of bean puree, and emulsifying capacity).

In recent years, methods were developed for reliable variety identifica-

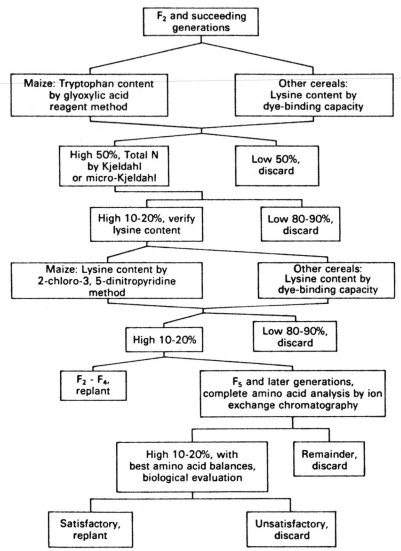

Figure 7-25. Schematic representation for screening cereal grains for protein content and nutritive quality. (From Protein-Calorie Advisory Group, 1976.)

tion. Most are based on determination of prolamins by polyacrylamide gel electrophoresis or high-pressure-liquid chromatography. Methods for routine determinations of several dozens of samples within a working day are available. For more reliable identification combinations of the two methods can be performed on the non-germ part of the kernel and retain a viable germ portion for seeding. Some methods employ separation of total proteins and especially gliadins, by either of the methods.

Table 7-18. Amino Acid Scores Calculated for Selected Cereals[a]

Amino acid	Scoring pattern amino acid levels	Whole wheat		Wheat flour (60–70% extraction)		Maize		White rice		Sorghum	
		Amino acid content (mg/g N)	Amino acid score	Amino acid content (mg/g N)	Amino acid score	Amino acid content (mg/g N)	Amino acid score	Amino acid content (mg/g N)	Amino acid score	Amino acid content (mg/g N)	Amino acid score
Lysine	340	179	53	113	33	167	49	226	66	126	37
Threonine	250	183	73	153	61	225	90	207	83	189	76
Methionine and cystine	220	253	115	229	104	217	99	229	104	181	82
Leucine	440	417	95	400	91	783	178	514	117	832	189
Isoleucine	250	204	82	217	87	230	92	262	105	245	98
Valine	310	276	89	240	77	303	98	361	116	313	101
Phenylalanine and tyrosine	380	469	123	423	111	544	143	503	132	473	124
Tryptophan	60	68	113	58	97	38	63	84	140	63	105

[a]Scoring pattern amino acid levels are those suggested by the FAO/WHO joint committee (FAO/WHO, 1983). Amino acid levels of cereal are from FAO data (FAO, 1972).

LABORATORY AUTOMATION

According to Munson (1974), laboratory automation through the years has concentrated mainly on automation of actual analyses. Today, increasingly efforts are made to automate the whole flow from sampling through reporting and evaluation of data.

Munson (1974) listed the following possible steps in such comprehensive automation: (1) sample inspection and entry into the laboratory; (2) sample preparation; (3) assignment of priorities; (4) scheduling of analytical work; (5) aliquot weighing; (6) addition of reagents; (7) digestion, extraction, ashing, etc; (8) dilution or concentration; (9) value determination; (10) calculation of analytical results; (11) report writing and distribution; and (12) data storage and retrieval.

Trevis (1974a) defined automation as it applies to any computerization, instrumentation, or advanced equipment that speeds or eliminates the manual processes currently employed. Trevis (1974b) listed seven automated instruments available to the cereal-processing industry.

The Kjel-Foss Macro Automatic combines the basic Kjeldahl protein method with an automatic control that produces the results on the first sample within 12 minutes of the start of the test. Subsequent test results are obtained at 3-minute intervals. The key to accelerated analysis is speeding up digestion by using hydrogen peroxide in the initial stages and steam distillation. Titration is automated and the results are displayed on a digital panel. There is no flask cleaning.

The Kaman Sciences Neutron Activation System involves the activation of nitrogen by fast neutrons, followed by the detection of the radiation from the activated samples as they decay. The Udy Protein Analyzer measures protein by the dye-binding method. The Neotec Grain Analyzer and the Dickey-John Grain Analysis computer employ near-infrared reflectance to measure directly many gross components including protein, moisture, starch, fiber, and fat and, indirectly, mineral components, kernel hardness, biological value of forage components, amino acid composition, etc.

The Technicon Auto Analyzer is an automated modular system capable of performing up to three determinations simultaneously at rates of up to 60 samples per hour. It has been adapted for automatic determination of protein, riboflavin, niacin, thiamine, iron, phosphorus, nicotinamide, and amylose in various cereal products. It is widely used to measure enzymic activities in many biological systems.

The Agtron model CA-1 monitors material that passes along a conveyor. It records the ratio of reflectance against a desired color. It provides programmable calibration for each of the four modes (red, blue, green, and yellow) and permits investigation of samples in four areas of the spectrum.

McGinty et al (1974) described an automatic system for weighing individual kernels. The system is controlled by a minicomputer that counts the kernels and calculates weight distribution. The following parameters can be determined and recorded: minimum, maximum, and average kernel weight; number of kernels within a certain range; and weight of individual kernels. The results can be used to evaluate variability within a lot and potential end-use properties—ie, flour milling yield.

Williams et al (1978a) described the automated digital analyzer system for the automated analyses of up to 2,500 samples per day of cereal grains for protein and moisture. The system operates through near-infrared reflectance spectroscopy. Unlike other equipment of this type, the instrument is not hard wired to any particular wave length. Instead, wavelength areas that are most significant for the material being analyzed can be selected. An on-line digital computer performs the necessary calibration calculations. Samples are presented by means of microswitch-activated conveyor chains at 9-second intervals. Standard deviations of 0.27 with respect to Kjeldahl protein and 0.24 with respect to oven moisture have been achieved for hard red spring wheat under on-line, high-volume conditions. The system is well suited to large-scale testing of plant breeders' samples or in marketing channels for protein segregation programs.

Schur et al (1978) compared several systems for automated beer analyses. Equipment for determining specific gravity, alcohol, total extract, and degree of fermentation was described.

CONTINUOUS QUALITY CONTROL

The trend of thought and practice in the milling industry favors more automation, so use of continuous and automatic controls is likely to increase. Mechanized, large-scale baking, especially by a continuous process, along with the need to meet specified pressing schedules and low tolerances, makes automatic, analytical methods much more urgent than before.

If automation extends to process control, continuous sampling and analyses are essential. In addition, high-speed computing elements are necessary if rapidly and continuously produced information is to be translated into appropriate corrective action. Pomeranz and Ward (1967) outlined present and possible future uses of automated analyses in grain processing. Several of the procedures are described in detail here. The Bushelmaster consists of a specially designed container with continuous flowing wheat actuating a precise automatic weigher, so the scale movements are recorded on a chart. The instrument provides continuous information on the test weight of the grain as well as on any changes in feed rate of the wheat stream.

Efficient milling of wheat depends on the moisture content and its distribution in the grain. Mambis and Lapickaja (1961) described equipment for automatic control of wheat flow, automatic moisture determina-

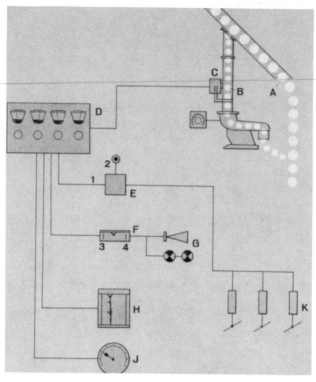

Figure 7-26. Schematic of continuous color intensity recorder of flour. A. Main flour stream; B. By-pass flour stream for measuring; C. Photo-electric receiver; D. Amplifier with indicators; E. Regulator; F. Limit indicator; G. Alarm system; H. Recorder; J. Indicator; K. Remote control of flour valves; 1. Recorded index 2. Pre-set index 3. Indicates darker flour 4. Indicates brighter flour. (Courtesy Buhler-Miag Co., Uzwil, Switzerland.)

tions, and regulation of water added during wheat conditioning. Water additions are controlled by results of moisture determinations in the dampened wheat. The control and regulation of rolls while wheat is milled into flour requires up to 60% of a flour mill's employee time.

Mambis and Sibirjakov (1963) have designed instruments to automatically sample ground grain, test by sieving, and adjust roll distance according to sieving results. The system increased productivity and improved quality and uniformity of milled products.

Large advances have been made in control of flour color as an index of extraction. The Simonitor was developed by modifying the Kent-Jones and Martin Color Grader; it uses a flour paste to eliminate the effect of particle size and shape. For automatic and continuous control determinations dry flour is preferred (Kent-Jones and Brown, 1966). The instrument employs a continuous sampling unit that compresses flour and presents it to an optical unit that detects differences in brightness between the flour and a standard. The difference is electronically amplified and recorded on a chart. Significant deviations from a desirable standard can be monitored by a device that operates warning lights or alarms. A diagram of a continuous color intensity recorder is shown in Figure 7-26. In addition to optical or accoustical warning signals, the system is equipped to automatically divert flow of flour that deviates from the preset color value.

The advantages of automated quality control are numerous. It reduces cost, is faster, and is more reproducible. The advantages are greatest when many highly repetitive, time-consuming, and tedious but not too complex tests are necessary. Automated control provides the only way to uniform quality, rather than correcting poor quality after an inferior product is obtained. However, it should be constantly borne in mind that an automated quality control system itself needs routine inspection to ensure consistently uniform results.

Instruments for on-line, real-time analysis of product monitoring and dough testing were described by Anon. (1985). They include several instruments for measuring, monitoring, and controlling color, moisture, and water activity, and dough temperature. Continuous near infrared reflectance spectroscopy in flour testing was described by Kummer (1985).

The flour milling industry was one of the first to process foods in an automated and continuous process. Development of instrumental and automatic control methods permits the cereal industry to adapt available tools to continuously measure and control quality of the raw materials and their conversion to uniform and high-quality products.

BIBLIOGRAPHY

AACC. "Cereal Laboratory Methods," 7th ed. Am. Assoc. Cereal Chemists; St. Paul, Mn; **1980**.

Anderson, R.A.; Pfeiffer, F.F.; Peplinki, A.J. *Cereal Sci. Today* **1966** *11*, 204–209.

Andras, L.; Bakos, L.; Csoke, A. *Hung. Acad. Sci. Cent. Res. Inst. Phys.* KFKI-1977-94, **1977** 10 pp.

Andras, L.; Balint, A.; Csoke, A., Nagy, A.Z. *Hung. Acad. Sci. Cent. Res. Inst. Phys.* KFKI-1978-38, **1978**, 6 pp.

Anker, C.A.; Geddes, W.F. *Cereal Chem.* **1944**, *21*, 335–360.

Anon. "The Miller's Handbook." Miag: Braunschweig, West Germany; **1936**.

Anon. "Methods of Analysis of the American Society of Brewing Chemists," 6th ed. Am. Soc. Brewing Chemists; Madison, WI; **1958**.

Anon. *Cereal Prod. Control* **1963**, *1*, 389–344.

Anon. "Standard Analytical Methods of the Member Companies of the Corn Industries Research Foundations, Inc." Corn Ind. Res. Found.; Washington, D.C.; **1967**.

Anon. *Food Eng.* **1979**, *59* (2) 122.

AOAC. "Official Methods of Analysis," 13th ed. Assoc. Offic. Anal. Chemists; Washington, D.C.; **1980**.

AOCS. "Official and Tentative Methods of the American Oil Chemists' Society," 2nd ed. Am. Oil Chemists' Soc.; Chicago; **1967**.

Arbeitsgemeinschaft Getreideforschung. "Standard Methods for Grain, Flour, and Bread." Schafer Publ. Co.; Detmold, Germany; **1958**.

Balow, K.K.; Buttrose, M.S.; Simmonds, D.H.; Vesk, M. *Cereal Chem.* **1973**, *50*, 443–454.

Barber, S. In "Rice: Chemistry and Technology," Houston, D.F., Ed. Am. Assoc. Cereal Chemists; St. Paul, MN; **1972**, pp. 215–263.

Barber, S. ICC Study Group Report No. 21, **1972–74**.

Barber, S. ICC Study Group Report No. 40, **1975–76**.

Barber, S.; De Barber, C.B. In "Proc. Workshop on Chemical Aspects of Rice Grain Quality." Int. Rice Research Inst., Los Banos, Laguna, **1979**, p. 209.

Bechtel, W. G. *Trans. AACC* **1955**, *13*, 108–121.

Bendelow, V.M.; Meredith, W.O.S.; Sisler, W.W. In "Barley Genetics II." Nilan, R.A., Ed. Washington State Univ. Press; Pullman, WA; **1971**, pp. 575–581.

Bennett, R.; and Coppock, J.B.M. *Trans. AACC* **1953**, *11*, 172–182.

Berg, S.O. *Cereal Chem.* **1947**, *24*, 274.

Berliner, E.; Koopman, J. *Z. Ges. Muhlenwesen* **1929**, *6*, 57–63, 75–82, 91–93.

Bhashyam, M.K.; Srinivas, T. *J. Food Sci.* **1984**, *49*, 393.

Bice, C.W.; Geddes, W.F. *Cereal Chem.* **1949**, *26*, 440–465.

Blakeney, A.B. In "Proc. Workshop on Chemical Aspects of Rice Grain Quality." Rice Research Inst., Los Banos, Laguna, **1979** p. 115.

Bloksma, A.H. In "Wheat Chemistry and Technology," Monograph 3. Y. Pomeranz, Ed. Am. Assoc. Cereal Chemists; St. Paul, MN; **1971**.

Bolling, H. *Getreide Mehl Brot* **1979**, *33,* 264–268.

Breckenridge, C. In "Proc. Workshop on Chemical Aspects of Rice Grain Quality." Int. Rice Research Inst., Los Banos, Laguna, **1979**, p. 175.

Bruinsma, B.L.; Rubenthaler, G.L. In "Proc. 8th Technicon Int. Congr." Technicon Instruments Co., Ltd.; London; **1978**.

Burkhart, B.A.; Dickson, A.D.; Hunt, W.N. *Proc. Am. Soc. Brewing Chem.* **1953**, *1953*, 34–35.

Bushuk, W.; Kilborn, R.H.; Irvine, G.N. *Cereal Sci. Today* **1965**, *10*, 402–405.

Chang, T.T.; Somrith. In "Proc. Workshop on Chemical Aspects of Rice Grain Quality." Int. Rice Research Inst., Los Banos, Laguna, **1979**, p. 49.

Cutler, G.H.; Brinson, G.A. *Cereal Chem.* **1935**, *12*, 120–129.

Cutler, G.H.; Worzella, W.W. *Cereal Chem.* **1933**, *10*, 250–262.

Del Rosario, A.R.; Briones, V.P.; Vidal, A.J.; Juliano, B.O. *Cereal Chem.* **1968**, *45*, 225.

Dickson, A.D.; Burkhart, B.A. *Proc. Am. Soc. Brewing Chem.* **1956**, *1956*, 143–155.

Dohan, D.A.; Standing, K.G.; Bushuk, W. *Cereal Chem.* **1976**, *53*, 91–100.

Doty, W.H.; Munson, A.W.; Wood, D.E.; Schneider, E.L. *J. Assoc. Offic. Anal. Chem.* **1970**, *53*, 801–803.

Duntley, J.M. *Cereal Sci. Today* **1961**, *6*, 282–287.

Durham, R.K. *Cereal Sci. Today* **1956**, *10*, 253–256, 359.

Fajerson, F. *Agric. Hortique Genet. Landskrona* **1950**, *8*, 109.

FAO. "Degree of Milling of Rice: The Standard Method Adopted by the Food Agency, Government of Japan. Sub-Group on Rice Grading and Standardization, Intergovernmental Group on Rice," FAO Document CCP:RI/GS C.R.S.1, FAO: Rome; **1972a**.

FAO. "Amino Acid Content of Foods and Biological Data on Proteins," FAO Nutr. Stud. 24. FAO; Rome; **1972b**.

FAO/WHO. "Energy and Protein Requirements. Reports of a Joint FAO/WHO Ad Hoc Expert Committee. WHO Tech. Rep. Ser. 522. WHO: Geneva. Also as FAO Nutr. Rep. Series 52. FAO: Rome; **1973**.

Fifield, C.C. *Cereal Chem.* **1934**, *11*, 330–334.

Fifield, C.C.; Smith, G.S.; Hayes, J.F. *Cereal Chem.* **1937**, *14*, 661–673.

Finney, K.F.; Barmore, M.A. *Cereal Chem.* **1948**, *25*, 291–312.

Finney, K.F.; Yamazaki, W.T. *Cereal Chem.* **1946**, *23*, 416–427.

Finney, K.F.; Yamazaki, W.T. *Cereal Chem.* **1953**, *30*, 153–159.

Finney, K.F.; Yamazaki, W.T. In "Wheat and Wheat Improvement," Monograph 13. Quisenberry, K.S.; Reitz, L.P., Ed. Am. Soc. Agron.; Madison, WI; **1967**.

Gariboldi, F. "Rice Testing Methods and Equipment," Agric. Services Bull. 18. FAO; Rome; **1973**.

Gedye, D.J.; Doling, D.A.; Kingswood, K.W. "A Farmer's Guide to Wheat Quality." NAC Cereal Unit, National Agricultural Centre; Stoneleigh, Warwickshire, England; **1981**.

Geissler, C. *Die Nahrung* **1975**, *19*, 363–377.

Greenaway, W.T. *Cereal Sci. Today* **1969**, *14*, 4.

Gruppe, G.A.; McDonnel, W.G.; Robe, K. *Food Process.* **1978**, *39* (11, Part I), 78–80.

Hagberg, S. *Cereal Chem.* **1960**, *37*, 218–222.

Hagberg, S. *Cereal Chem.* **1961**, *38*, 202–203.

Halton, P. *Cereal Chem.* **1938**, *15*, 282–294.

Halton, P. *Cereal Chem.* **1949**, *26*, 24–25.

Hardy, A.C. "Handbook of Colorimetry." Technology Press, MIT: Cambridge, MA; **1936**.

Harris, R.H.; Sibbitt, L.D. *Cereal Chem.* **1942**, *19*, 388–402.

Hehn, E.R.; Barmore, M.A. *Adv. Agron.* **1965**, *17*, 85–114.

Herz, K.O. *Food Technol.* **1965**, *19*, 90–103.

Hlynka, K.; Van Eschen, E.L. *Cereal Sci. Today* **1965**, *10*, 84–87.

Hogan, J.T.; Deobald, H.J. *Cereal Chem.* **1961**, *38*, 291–293.

Hogan, J.T.; Deobald, H.J. *Rice J.* **1965**, *68*(10), 10.

Holliger, A. *Cereal Chem.* **1963**, *40*, 235–240.

Hoshikawa, K. Proc. Crop Sci. Soc. Jpn. **1967**, *30*, 221.

Hoskins, C.M.; Hoskins, W.G. In "The Chemistry and Technology of Cereals as Food and Feed," Matz, S.A., Ed. Avi Publishing Co.; Westport, CT; **1959**.

Hulse, J.H.; Rachie, K.O.; Billingsley, L.W. "Nutritional Standards and Methods of Evaluation for Food Legume Breeders." International Development Research Centre; Ottawa, Canada, **1977**, 100 pp.

Hunt, W.H.; Neustadt, M.H. *J. Assoc. Offic. Anal. Chem.* **1966**, *49*, 757–763.

Indudhara Swamy, Y.M.; Sowbhagya, C.M.; Bhattacharya, K.R. *J. Sci. Food Agric.* **1978**, *29*, 627.

Irvine, G.N. In "Wheat Chemistry and Technology," Monograph 3, 2nd ed. Pomeranz, Y., Ed. Am. Assoc. Cereal Chemists; St. Paul, MN; **1971**.

Johnson, R.M.; Craney, C.E. *Cereal Chem.* **1971**, *48*, 276–282.

Jongh, G. *Getreide Mehl* **1967**, *17*, 1–4.

Juliano, B.O. In "Rice Chemistry and Technology." Houston, D.F., Ed. Monograph Series Vol. IV. Am. Assoc. Cereal Chem.; St. Paul, MN; **1972**, p. 16.

Juliano, B.O. In "Proc. Workshop on Chemical Aspects of Rice Grain Quality." Int. Rice Research Inst.; Los Banos, Laguna, **1979**, p. 69.

Katz, R.; Collins, N.D.; Cardwell, A.B. *Cereal Chem.* **1961**, *38*, 364–368.

Keller, D.L.; Converse, H.H.; Hodges, T.O.; Chung, D.S. *Trans. Am. Soc. Agr. Eng.* **1972**, *15*(2), 330–332.

Kent-Jones, D.W.; Amos, A.J. "Modern Cereal Chemistry," 5th ed. Northern Publishing Co.; Liverpool, England; **1957**.

Kent-Jones, D.W.; Brown, T.J. *Cereal Sci. Today* **1966**, *11*, 88–90, 119–120.

Khush, G.S.; Paule, C.M.; De la Cruz, N.M. In "Proc. Workshop on Chemical Aspects of Rice Grain Quality." Rice Research Inst., Los Banos, Laguna, **1979**, p. 22.

Kilborn, R.H.; Aitken, T.R. *Cereal Sci. Today* **1961**, *6*, 253–259.

Kissell, L.T. *Cereal Chem.* **1959**, *36*, 168–175.

Kongseree, N.; Juliano, B.O. *J. Agric. Food Chem.* **1972**, *20*, 714.

Kummer, E. "Kontinuierliche NIR-Messung am Mehl." Ber. 35 Tagung Muellerei Technologie 8–12. Granum Verlag, Detmold, **1985**.

Kunze, O.R.; Choudhury, M.S.U. *Cereal Chem.* **1972**, *49*, 684.

Kunze, O.R.; Hall, C.W. *Trans. Am. Soc. Agric. Eng.* **1965**, *8*, 396.

Lai, F.S.; Afework, S.; Pomeranz, Y. *Cereal Foods World* **1983**, *28*, 572.

Larsen, R.A. In "The Chemistry and Technology of Cereals as Food and Feed." Matz, S.A., Ed. Avi Publishing Co.; Westport, CT; **1959**.

Mamaril, F.P.; Pomeranz, Y.; Shellenberger, J.A. Unpublished data, **1962**.

Mambis, I.W.; Lapickaja, L.F. *Documenta Cerealia* **1961**, *2*, 18.

Mambis, I.E.; Sibirjakov, V.A. *Arb. Unions-Forsch. Inst. Getreide Produkte* **1963**, *47*, 91–101.

Marek, C.J.; Bushuk, W. *Cereal Chem.* **1967**, *44*, 300–307.

Mathewson, P.R.; Pomeranz, Y. *J. Assoc. Off. Anal. Chem.* **1977**, *60*, 16–20.

Mathewson, P.R.; Pomeranz, Y. *J. Assoc. Off. Anal. Chem.* **1979**, *62*, 198–200.

Mathewson, P.R.; Rousser, R.; Pomeranz, Y. *Cereal Foods World* **1978**, *23*, 717–719, 722–724.

Mathewson, P.R.; Pomeranz, Y.; Miller, B.S.; Farenholz, C.H. *Cereal Foods World* **1979**, *24*, 469.

McGinty, R.J. "Development of a Standard Grain Breakage Test (A Progress Report)." U.S. Agr. Res. Serv. ARS 51-34; Washington, D.C.; **1970**, 13 pp.

McGinty, R.J.; Watson, C.A.; Sanders, D.E.; Kurtenbach, A.J. *Cereal Sci. Today* **1974**, *19*, 196–199.

Mertz, E.T.; Jambunathan, R.; Misra, P.S. "Simple Chemical and Biological Methods Used at Purdue University to Evaluate Cereals for Protein Quality." Int. Programs Agric., Agric. Exp. Sta., Purdue Univ.; W. Lafayette, IN; **1975**, Bull. No. 70, 25 pp.

Miller, B.S.; Johnson, J.A. *Kansas Agric. Exp. Sta. Tech. Bull.* **1954**, *76*.

Miller, B.S.; Johnson, J.A. "Testing Wheat for Quality," USDA Production Res. Rep. 9. USDA: Washington, D.C.; **1957**.

Miller, B.S.; Pomeranz, Y.; Thompson, W.O.; Nolan, T.W.; Hughes, J. W.; Davis, G.; Jackson, N.G.; Fulk, D.W. *Cereal Foods World* **1978**, *23*, 198–201.

Miller, B.S.; Hughes, J.W.; Rousser, R.; Pomeranz, Y. *Cereal Chem.* **1979a**, *56*, 213–216.

Miller, B.S.; Hughes, J.W.; Rousser, R.; Pomeranz, Y. "Standard Method for Measuring Breakage Susceptibility of Shelled Corn." Am. Soc. Agric. Eng. Paper No. 79-3087: Winnipeg, Canada; **1979b**.

Miller, B.S.; Lee, M.S.; Pomeranz, Y.; Rousser, R. *Cereal Chem.* **1979c**, *56*, 172–180.

Miller, B.S.; Hughes, J.W.; Afework, S.; Pomeranz, Y. *J. Food Sci.* **1981a**, *46*, 1851.

Miller, B.S.; Afework, S.; Hughes, J.W.; Pomeranz, Y. *J. Food Sci.* **1981b**, *46*, 1863.

Miller, B.S.; Afework, S.; Pomeranz, Y.; Bruinsma, B.L.; Booth, G.D. *Cereal Foods World* **1982a**, *27*, 61–64.

Miller, B.S.; Afework, S.; Pomeranz, Y.; Bolte, L. *Getreide Mehl Brot* **1982b**, *36*, 114.

Miller, B.S.; Pomeranz, Y.; Afework, S. *Cereal Chem.* **1984**, *61*, 201.

Milner, M.; Shellenberger, J.A. *Cereal Chem.* **1953**, *30*, 202–212.

Minor, G.K. *Cereal Foods World* **1984**, *29*, 659–660.

Moritaka, S.; Yasumatsu, K. *J. Jpn. Soc. Food Nutr.* **1972**, *25*, 59.

Moro, M.A.; Pomeranz, Y.; Shellenberger, J.A. *Phyton* **1963**, *20*, 59–64.

Moss, H.J. *Aust. J. Agric. Res.* **1978**, *29*, 1117.

Moss, H.J.; Edwards, C.S.; Goodchild, N.A. *Aust. J. Exp. Agric. Anim. Husb.* **1973**, *13*, 233.

Munson, A.M. *Cereal Sci. Today* **1974**, *19*, 178, 179, 206.

Nagato, K. *Proc. Crop Sci. Soc. Jpn.* **1962**, *31*, 102.

Nagato, K.; Kono, Y. *Nippon Sakumotsu Gakkei Kiji.* **1963**, *32*, 181 (cited in Juliano, 1972).

Nagato, K.; Ebata, M.; Ishikawa, M. *Proc. Crop Sci. Soc. Jpn.* **1964**, *33*, 82.

Neill, C.D. *Cereal Sci. Today* **1962**, *7*, 6–12.

Newton, R.; Cook, W.H.; Malloch, J.G. *Sci. Agric.* **1927**, *8*, 205.

Nguyen, C.N.; Kunze, O.R. *Cereal Chem.* **1984**, *61*, 63–68.

Noll, J.S.; Simmonds, D.H.; Bushuk, W. *Cereal Chem.* **1974**, *51*, 610–616.

Norris, K.H. *Agric. Eng.* **1974**, *45*(7), 370.

Norris, K.H.; Barnes, R.F.; Moore, J.E.; Shenk, J.S. *J. Anim. Sci.* **1976**, *43*, 889–897.

Norris, K. H.; Williams, P.C.; Mattern, P. *Cereal Foods World* **1977**, *22*, 461.

Obuchowski, W.; Bushuk, W. *Cereal Chem.* **1980a**, *57*, 421.

Obuchowski, W.; Bushuk, W. *Cereal Chem.* **1980b**, *57*, 426.

"Official United States Standards for Grain." U.S. Department of Agriculture, Federal Grain Inspection Service, Inspection Division; Washington, D.C.; **1978**.

Paquet, J. *Getreide Mehl* **1964**, *14*, 73–78, 89–96.

Pelshenke, P. *Cereal Chem.* **1933**, *10*, 90–96.

Pelshenke, P.F. *Getreide Mehl Brot* **1948**, *15–16*, 119–127.

Pelshenke, P.F.; Bolling, H. *Getreide Mehl* **1962**, *12*, 29–33.

Pelshenke, P.F.; Bolling, H.; Springer, F.; Zwingelberg, H. *Getreide Mehl* **1964**, *14*, 121–128.

Perten, H. *Cereal Chem.* **1964**, *41*, 127–140.

Pinckney, A.J. *Cereal Chem.* **1961**, *38*, 501–506.

Platt, W.; Powers, R. *Cereal Chem.* **1940**, *17*, 601–621.

Pomeranz, Y.; Miller, B.S. In "Proc. 7th World Cereal and Bread Congress," Prague, **1982**.

Pomeranz, Y.; Moore, R.B. *Baker's Dig.* **1975**, *49*(1), 44–48, 58.

Pomeranz, Y.; Ward, A.B. *Cereal Sci. Today* **1967**, *12*, 159–163.

Pomeranz, Y.; Rubenthaler, G.L.; Finney, K.F. *Food Technol.* **1966**, *20*, 95–98.

Pomeranz, Y.; Stermer, R.A.; Dikeman, E. *Cereal Chem.* **1975**, *52*, 849–853.

Pomeranz, Y.; Bruinsma, B.L.; Zwingelberg, H. *Bakers' Digest* **1984**, *58*(6), 12, 14–16.

Pratt, D.B., Jr. In "Wheat Chemistry and Technology," Monograph 3, 2nd ed. Pomeranz, Y., Ed. Am. Assoc. Cereal Chemists; St . Paul, MN; **1971**.

Protein-Calorie Advisory Group (PAG) of the United Nations System. *Adv. Cereal Sci. Technol.* **1976**, *1*, 378–408.

Revesz, R.N.; Aker, N. *J. Assoc. Offic. Anal. Chem.* **1977**, *60*, 1238–1242.

Ronalds, J.A. *J. Sci. Food Agric.* **1974**, *25*, 179–1285.

Rout, G.; Senapati, B.; Ahmed, T. *Bull. Grain Technol.* **1976**, *14*, 34.f
Rubenthaler, G.L.; Finney, P.L.; Demaray, D.E.; Finney, K.F. *Cereal Chem.* **1980**, *57*, 212–216.
Saio, K.; Noguchi, A. *Nippon Shokuhin Kogyo Gakkaishi*, **1983**, *30*, 331.
Sandstedt, R.M.; Blish, M.J. *Cereal Chem.* **1934**, *11*, 368–383.
Saunders, H.A.; Humphries, S. *J. Natl. Inst. Agric. Bot.* **1928**, *2*, 34.
Schafer, W. *Cereal Sci. Today* **1963**, *8*, 160–164, 183.
Schiller, G.W. *Cereal Foods World* **1984**, *29*, 647–651.
Schmutz, W. *Monatschr. Brauerei* **1978**, *31*(1), 10–16.
Schneeweiss, R. *Getreide Mehl* **1959**, *9*, 101–102.
Schneeweiss, R. In "Proc. Technical and Executive Committee," Int. Assoc. Cereals IGV-Mitt. **1965**, *1*, 150–154, 181–184.
Schur, P.; Anderegg, P.; Pfenninger, H. *Brauerei Rdsch.* **1978**, *89*(2), 229–234.
Seckinger, H.L.; Wolf, M.J. *Cereal Chem.* **1970**, *47*, 236.
Seibel, W. *Getreide Mehl* **1963**, *13*, 43–45.
Seibel, W. *Muhle* **1964**, *101*, 659–663.
Seibel, W.; Crommentuyn, A. *Brot Gebaeck* **1963**, *17*, 139–144.
Seitz, L.M.; Yamazaki, W.T.; Clements, R.L.; Mohr, H.E.; Andrews, L. *Cereal Chem* **1985**, *62*, 467–469.
Sepp, R. *Starch* **1979**, *31*(2), 57–63.
Sharda, R.; Herum, F.L. "A Mechanical Damage Susceptibility Tester for Shelled Corn." Am. Soc. Agric. Eng. Paper No. 77-3504, Chicago; **1977**.
Shellenberger, J.A.; Ward, A.B. In "Wheat and Wheat Improvement," Monograph 13. Quisenberry, K.S.; Reitz, L.P., Eds. Am. Soc. Agron.; Madison, WI; **1967**.
Shogren, M.D.; Finney, K.F.; Bolte, L.C.; Hoseney, R.C. *Agron J.* **1963**, *55*, 19–21.
Smeets, H.S.; Cleve, H. *Milling Production* **1956**, *21*(4,5), 12–14, 16.
Soznicov, A.A.; Skuratova, O.L. *Selekcija Semenov.* **1963**, *28*, 63–64.
Srinivas, T. *J. Sci. Food Agric.* **1975**, *26*, 1479.
Srinivas, T.; Bhashyam, M.K.; Mahadevappa, M.; Desikachar, H.S.R. *Indian J. Agric. Aci.* **1977**, *47*, 27.
Srinivas, T.; Singh, V.; Bhashyam, M.K. *Rice J.* **1984**, *87*(1), 8.
Stenvert, N.L.; Kingswood, K. *J. Sci. Food Agric.* **1977**, *28*, 11.
Stephens, L.E.; Foster, G.H. "Breakage Tester Predicts Handling Damage in Corn." U.S. Agric. Res. Serv. ARS-NC-49, Washington, D.C., **1976**, 6 pp.
Stermer, R.A. *Cereal Chem.* **1968**, *45*, 358–364.
Stermer, R.A.; Watson, C.A.; Dikeman, E. *Trans. Am. Soc. Agric. Eng.* **1977**, *20*, 547–551.
Sullivan, B. *Trans. AACC* **1949**, *7*, 63–70.
Suzuki, H.; Ikehashi, H.; Kushibuchi, K. In "Proc. Workshop on Chemical Aspects of Rice Grain Quality." Int. Rice Research Inst., Los Banos, Laguna, **1979**, p. 149.
Swanson, C.O.; Working, E.B. *Cereal Chem.* **1933**, *10*, 1–29.
Symes, K.J. *Aust. J. Exp. Agric. Anim. Husb.* **1961**, *1*, 17.
Symes, K.J. *Aust. J. Agric. Sci.* **1965**, *16*, 113.
Symes, K.J. *Aust. J. Agric. Res.* **1969**, *20*, 971.
Tanda, Y. *Nippon Sakumotsu Gaikkai Kiji.* **1966**, *35*, 38 (cited in Juliano, 1972).
Tashiro, T.; Ebata, M. *Proc. Crop Sci. Soc. Jpn.* **1975**, *44*, 205.
Taylor, J.W.; Bayles, B.B.; Fifield, C.C. *J. Am. Soc. Agron.* **1939**, *31*, 775–784.
Thomas, B.; Thunger, L. *Brot Gebaeck* **1965**, *19*, 65–74.
Trevis, J.E. *Cereal Sci. Today* **1974a**, *19*, 180, 181.
Trevis, J.E. *Cereal Sci. Today* **1974b**, *19*, 182–189.
Trupp, C.R. *Crop Sci.* **1976**, *16*, 618.
Udy, D.C. *Cereal Chem.* **1956**, *33*, 190–197.
USDA. "Grain Grading Primer," USDA Misc. Publ. 740. USDA: Washington, D.C.; **1957**.
USDA. "Official Grain Standards of the United States," USDA Publ. SRA-C8US-177. USDA: Washington, D.C.; **1964**.
Utsunomiya, H.; Yamagata, M.; Doi, Y. *Bull. Fac. Agric. Yamaguti Univ.* **1975a**, *26*, 1.
Utsunomiya, H.; Yamagata, M.; Doi, Y. *Bull. Fac. Agric. Yamaguti Univ.* **1975b**, *26*, 19.
Villareal, R.M.; Resurreccion, A.P.; Suzuki, L.B.; Juliano, B.O. *Starke* **1976**, *28*, 88.
Villegas, E.; Mertz, E.T. "Chemical Screening Methods for Maize Protein Quality at CIMMYT," Res. Bull. No. 20. Int. Maize and Wheat Improvement Center; Mexico City; **1970**, 14 pp.
Voisey, P.W.; Miller, H.; Kloek, M. *Cereal Sci. Today* **1964**, *9*, 393–396, 412.

Vondenhof, T.; Schulte, K. *Dtsch. Lebensm. Rundschau* **1979**, *75*(6), 187–189.

Watson, C.A. *Anal. Chem.* **1977**, *49*(9), 835A–840A.

Webb, B.D. In "Six Decades of Rice Research in Texas," Research Monograph No. 4, Texas Agric. Exp. Station, **1975**, pp. 97–106.

Webb, B.D.; Stermer, R.A. In "Rice Chemistry and Technology." Houston, D.H., Ed. Am. Assoc. Cereal Chem.; St. Paul, MN; **1972**, pp. 102–139.

Williams, P.C. *Cereal Chem.* **1967**, *44*, 383–391.

Williams, P.C. *Cereal Chem.* **1975**, *52*, 561–576.

Williams, P.C. *Cereal Chem.* **1979**, *56*, 169–172.

Williams, P.C.; Stevenson, S.C.; Irvine, G.N. *Cereal Chem.* **1978a**, *55*, 263–279.

Williams, P.C.; Norris, K.H.; Johnson, R.L.; Standing, K.; Fricioni, R.; MacAffrey, D.; Mercier, R. *Cereal Foods World* **1978b**, *23*, 544–547.

Worzella, W.W. *J. Agric. Res.* **1942**, *65*, 501.

Yamazaki, W.T. *Cereal Chem.* **1953**, *30*, 242–246.

Yamazaki, W.T. *Cereal Sci. Today* **1962**, *7*, 98–104, 125.

Zeleny, L. *Cereal Chem.* **1947**, *24*, 465–475.

Zeleny, L. In "Wheat Chemistry and Technology," Monograph 3, 2nd ed. Pomeranz, Y., Ed. Am. Assoc. Cereal Chemists; St. Paul, MN; **1971**.

Chapter 8

Wheat—Processing, Milling

Over two-thirds of the annual U.S. harvest of wheat is processed for food. The limited use for industrial purposes is due mainly to its high price in relation to other cereal grains. The main use of wheat for food is the manufacture of flour for making bread, biscuits, pastry products, and semolina and farina for alimentary pastes. A small portion is converted into breakfast cereals. Large quantities of flour are not sold in the form in which they come from the mill but are utilized as blended and prepared flours for restaurants, cafeterias, and schools and as all-purpose flours for the private household. Industrial uses of wheat include the manufacture of malt, potable spirits, starch, gluten, pastes, and core binders. Because of the relatively high price, wheat malt is used little in the brewing and distilling industries. It is used mainly by the flour milling industry to increase the alpha-amylase activity of high-grade flours. In the USA, small quantities of wheat flour (mainly low-grade clears) are used to manufacture wheat starch as a by-product of viable (functionally in bread making) gluten. The gluten is used to supplement flour proteins in specialty baked goods (hamburger buns, hot-dog buns, hearth-type breads, specialty breads, etc) and as a raw material for the manufacture of monosodium glutamate, which is used to accentuate the flavors of foods. Some low-grade flours are used in the manufacture of pastes for bookbinding and paper hanging, in the manufacture of plywood adhesives, and in iron foundries as a core binder in the preparation of molds for castings. In Australia, the starch is a by-product of wheat gluten manufacture. Low-grade flours are also used in Australia as an adjunct in brewing (as a source of fermentable sugars). The high yields of wheat in western Europe (compared to those of corn) make attractive production of starch and gluten, provided both products can be marketed economically.

WHEAT AND FLOUR QUALITY

In wheat and flour technology, the term quality denotes the suitability of the material for some particular end use. It has no reference to nutritional attributes. Thus, the high-protein hard wheat flour is of good bread-making quality but is inferior to soft wheat flours for chemically leavened products such as biscuits, cakes, and pastry (see also Chapters 7 and 9).

The miller desires a wheat that mills easily and gives a high flour yield. Wheat kernels should be plump and uniformly large for ready separation of foreign materials without undue loss of millable wheat. The wheat should produce a high yield of flour with maximum and clean separation from the bran and germ without excessive consumption of power. Since the endosperm is denser than the bran and the germ, high-density wheats produce more flour. In production of bread flours, the reduction in protein content from wheat to flour should be minimum (not above 1%). The test weight is affected by kernel shape, moisture content, wetting and subsequent drying, and even handling, because these characteristics and operations affect the grain packing. Above 58 lb per bushel, the test weight has relatively little influence on flour milling yield. At lower test weights, the milling yield usually falls off rapidly with decreasing test weight. Average test weight per bushel of U.S. wheat is about 60 lb. Badly shriveled kernels have test weights of 45 lbs or less. Weathering lowers the test weight by swelling kernels, but the proportion of the endosperm remains the same. Some environmental factors influence the ease of milling. Bran of weathered and frosted wheats tends to pulverize, and it is difficult to secure clean separation of flour from bran.

ROLLER MILLING

Milling grain as food for man has been traced back more than 8,000 years. Flour milling has advanced from a primitive and laborious household task to a vast and sophisticated, to a large extent automated industry. In the production of white flour, the objective is to separate the starchy endosperm of the grain from the bran and germ. The separated endosperm is pulverized. A partial separation of the starchy endosperm is possible because its physical properties differ from those of the fibrous pericarp and oily germ. Bran is tough because of its high fiber content, but the starchy endosperm is friable. The germ, because of its high oil content, flakes when passed between smooth rolls. In addition, the particles from various parts of the wheat kernel differ in density. This makes possible their separation by using air currents. The differences in friability of the bran and the starchy endosperm are enhanced by wheat conditioning, which involves adding water before wheat is actually milled. The addition of water toughens the bran and mellows the endosperm. The actual milling process comprises a gradual reduction in particle size, first between corrugated break rolls and later between smooth reduction rolls. The separation is empirical and not quantitative. The milling process results in the production of many streams of flour and offals that can be combined

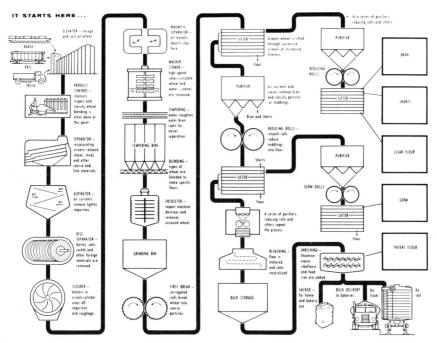

Figure 8-1. Milling process of flour. (Source: Wheat Flour Institute, Chicago.)

in different ways to produce different grades of flour. Still, the offals contain some of the starchy endosperm particles, and some of the flour streams have little bran and germ particles.

Wheat flour production involves wheat selection and blending, cleaning, conditioning, breaking, bolting or sieving, purification, reduction, and treatment (bleaching, enrichment, supplementation). An outline of the wheat milling process is shown in Figure 8-1. (See also CIGI, 1975; Inglett, 1974; Kent-Jones and Amos, 1967; Lockwood, 1960; Pomeranz, 1971; Scott, 1951; Smith, 1944, and Swanson, 1938.)

Selection and Blending

The miller must produce a flour of definite characteristics and meet certain specifications for a particular market. The most critical requirement is maintaining a uniform product from a product (wheat) that may show a wide range of characteristics and composition. Consequently, selection of wheats and binning according to quality for proper blending are essential phases of modern milling. An adequate supply of wheat, binned according to quality characteristics, makes it possible to build a uniform mix to meet some of the most stringent specifications. The availability of rapid, nondestructive, near-infrared reflectance instruments has made this task substantially easier.

Cleaning

Wheat received in the mill contains many impurities. Special machines are available to remove those impurities. Preliminary cleaning involves the use of sieves, air blasts, and disc separators. This is followed by dry scouring in which the wheat is forced against a perforated iron casting by beaters fixed to a rapidly revolving drum. This treatment removes foreign materials in the crease of the kernel and in the brush hairs. Some mills are equipped with washers in which the wheat is scrubbed under a flowing stream of water. The washed wheat is then passed through a "whizzer" (centrifuge), which removes free water. In practice, little wheat is washed today, because the process is relatively ineffective, may actually increase microbial populations, and creates problems of disposing large amounts of polluted water with a high biological oxygen demand (BOD).

Conditioning

In this process water is added and allowed to stand for up to 24 hours to secure maximum toughening of the bran with optimum mellowing of the starchy endosperm. The quantity of water and the conditioning time are varied with different wheats to bring them to the optimum conditioning for milling. The quantity of added water increases with decreasing moisture content of the wheat, with increasing vitreousness, and with increasing plumpness. Generally, hard wheats are tempered to 15–16% moisture and soft wheats to 14–15% moisture. In the customary conditioning, the wheat is scoured again, after it has been held in the tempering bins for several hours. A second small addition of 0.5% water is made about 20–60 minutes before the wheat goes to the rolls.

Breaking

The first part of the grinding process is carried out on corrugated rolls (break rolls), usually 24–30 inches long and 9 inches in diameter. Each stand has two pairs of rolls, which turn in opposite directions at a differential speed of about 2.5:1. In the first break rolls there are usually 10–12 corrugations per inch. This number increases to 26–28 corrugations on the fifth break roll. The corrugations run the length of the roll with a spiral cut, which is augmented with an increase in the number of corrugations. As the rolls turn rapidly toward each other, the edges of the corrugations of the fast roll cut across those of the slow roll, producing a shearing and crushing action on the wheat, which falls in a rapid stream between them. The first break rolls are spaced so that the wheat is crushed lightly and only a small quantity of white flour is produced. After sieving, the coarsest material is conveyed to the second break rolls. The second break rolls are set a little closer together than the first break rolls so that the material is crushed finer and more endosperm particles are released. This process of grinding and sifting is repeated up to six times. The material going to each succeeding break contains less and less endosperm. After the last break, the largest fragments consist of flakes of the wheat pericarp. They are passed through a wheat bran duster, which removes a small quantity of low-grade flour.

Sieving

After each grinding step, the crushed material, called stock or chop, is conveyed to a sifter, which is a large box fitted with a series of sloping sieves. The break sifters have a relatively coarse wire sieve at the top and progressively finer silk sieves below, and end with a fine flour silk at the bottom. The sifter is given a gyratory motion so that the finer stock particles pass through the sieves from the head (top) to the tail (bottom). Particles that are too coarse to pass through a particular sieve tail over it and are removed from the sifter box. The process results in separation of three classes of material: (1) coarse fragments, which are fed to the next break until only bran remains; (2) flour, or fine particles, which pass through the finest (flour) sieve; and (3) intermediate granular particles, which are called middlings.

Purification

The middlings consist of fragments of endosperm, small pieces of bran, and the released embryos. Several sizes are separated from each of the break stocks; individual streams of similar size and degree of refinement result from the sieving of several break stocks and are combined. Subsequently, the bran-rich material is removed from the middlings. This is accomplished in purifiers. Purifiers also produce a further classification of middlings according to size and thereby complete the work of the sifters. In the purifier, the shallow stream of middlings travels over a large sieve, while shaken rapidly backward and forward. The sieve consists of a tightly stretched bolting silk or grits gauze, which becomes progressively coarser from the head to the tail end of the purifier. An upward air current through the sieve draws off light material to dust collectors and holds bran particles on the surface of the moving middlings so that they drift over to the tail of the sieve.

Reduction

The purified and classified middlings are gradually pulverized to flour between smooth reduction rolls, which revolve at a differential of about

1.5:1. The space between the rolls is adjusted to the granulation of the middlings. The endosperm fragments passing through the rolls are reduced to finer middlings and flour. The remaining fibrous fragments of bran are flaked or flattened. After each reduction step, the resulting stock is sifted. Most of the bran fragments are removed on the top sieve while the flour passes through the finest bottom sieve. The remaining middlings are separated according to size, are moved to their respective purifiers, and are then passed to other reduction rolls. These steps are repeated until most of the endosperm has been converted to flour and most of the bran has been removed as offal by the reduction sifters. What remains is a mixture of fine middlings and bran with a little germ; this is called feed middlings. Impact mills have been used in reduction grinding, especially with soft wheats. Close grinding using clean middlings on reduction rolls, followed by a pin mill or detacher, increases the yield of flour from a reduction step. This process has been used more for soft than for hard wheats.

The embryos are largely released by the break system and appear as lemon-yellow particles in some of the coarser middling streams. These streams are called sizings. The embryos are flattened in reduction of the sizings and are separated as flakes during sieving. Germ may be separated also without reduction of the sizings by gravity and regular air currents. Previously, all the germ was mixed with the shorts as feed. Some special uses of germ in foods and as a source of pharmaceuticals have been developed.

Flour Grades

Each grinding and sieving operation produces flour. In addition to the various break and middlings flours, a small quantity of flour is obtained from dust collectors and bran and shorts dusters. With each successive reduction, the flour contains more pulverized bran and germ. The flour from the last reduction, called "red dog," is dark in color and high in components originating from the bran and germ, such as ash, fiber, pentosans, lipids, sugars, and vitamins. Such flour bakes into dark-colored, coarse-grained bread but is mostly sold as feed flour.

In a large mill there may be 30 or more streams that vary widely in composition. If all the streams are combined, the product is called straight flour. A straight flour of 100%, however, does not mean a whole wheat flour. It means, generally, a 75% flour, because wheat milling yields about 75% white flour and about 25% feed products. Frequently, the more highly refined (white) streams are taken off and sold separately as patent flours; the remaining streams, which contain some bran and germ, are called clear flours. A diagram of flours and milled feed products is given in Figure 8-2. Some clear flours are dark in color and low in bread-making quality. Some of the better, lighter, clear flours are used in blends with rye and/or whole wheat flours in the production of specialty breads. The darker grades of

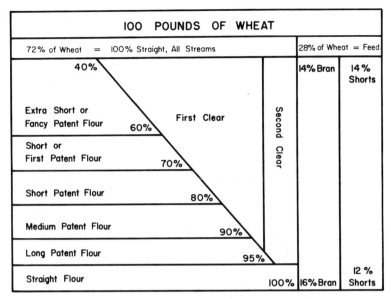

Figure 8-2. Grades of flour. (Source: C. O. Swanson, Kansas State Univ., Manhattan, KS.)

clear flours are used in the manufacture of gluten, starch, monosodium glutamate, and pet foods.

Yields of Mill Products

The plump wheat grain consists of about 83% endosperm, 14.5% bran, and 2.5% germ. These three structures are not separated completely, however, in the milling process. The yield of total flour ranges from 72% to 75%, and the flour contains little bran and germ. In ordinary milling processes only about 0.25% of the germ is recovered. Bran ranges from 12% to 16% of the wheat milled. The remaining by-products are shorts. The low-grade flour and feed middlings may be sold separately as feed by-products. The objective of efficient milling is to maximize the monetary value of the total mill products, generally by increasing the yield of flours.

Flour Fractionation

Wheat flour produced by conventional roller milling contains particles of different sizes (from 1 to 150 μm), such as large endosperm chunks, small particles of free protein, free starch granules, and small chunks of protein attached to starch granules (see Figure 8-3). The flour can be ground, pin-milled to avoid excessive starch damage, to fine particles in which the protein is freed from the starch. The pin-milled flour is then passed through an air classifier. A fine fraction, made up of particles about 40 μm

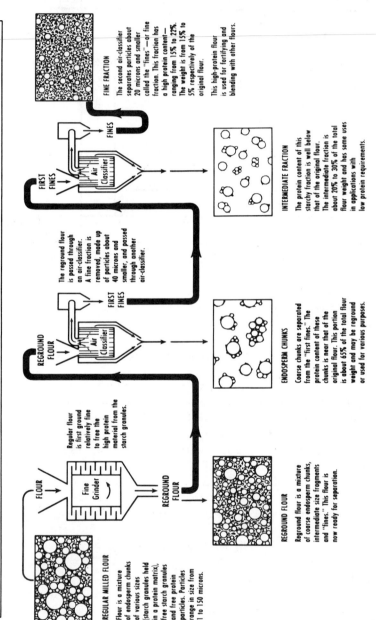

Figure 8-3. How flour is fractionated. (Source: Wheat Flour Institute, Chicago.)

and smaller, is removed and passed through a second air classifier. Particles of about 20 μm and smaller are separated; they comprise about 10% of the original flour and contain up to about twice the protein of the unfractionated flour. This high-protein flour is used to fortify low-protein bread flours or for enrichment in the production of specialty baked goods. A comparable fraction containing about half the protein content of the unfractionated flour is also obtainable.

Air classification has created considerable interest in the milling industry. Its advantages are numerous, such as manufacture of uniform flours from varying wheats; increase of protein content of bread flours and decrease of protein content in cake and cookie flours; controlled particle size and chemical composition; and production of special flours for specific uses. A number of equipment and process patents on fine grinding and separation have been issued. The technology of the process is well known, yet its benefits and potential have not been fully utilized mainly because of the availability of low- and high-protein wheats and the high energy cost involved in air classification. In recent years there has been interest in air-classified low-protein fractions as a replacement of chlorinated wheat flour in high-ratio cake production.

Soft Wheat Milling

Soft wheats are milled by the method of gradual reduction, similar to the method for milling hard bread wheats. Patent flours containing 7–9% protein, milled from soft red winter wheats, are especially suitable for chemically leavened biscuits and hot breads. Special mixtures of soft wheats are used to make cake flours for use in cookie and cake making; such flours usually contain 8% protein or less and are milled to very short patents (about 30%). Treatment with heavy dosages of chlorine lower the pH to about 5.1–5.3, weaken the gluten, and facilitate the production of short pastry. Cake flours are sieved through silk of finer mesh than that used for biscuit or bread flours (see also chapter 15).

Durum Wheat Milling

In durum milling, the objective is the production of a maximum yield of highly purified semolina. Although the same sequence of operations is employed in the production of flour and semolina, the milling systems differ in design. In semolina manufacture, impurities and the mill offals must be removed by the cleaning and purifying systems. Durum wheat milling involves cleaning and conditioning of the grain, light grinding, and extensive purification. The cleaning, breaking, sizing, and purifying systems are much more elaborate and extensive than in flour mills. On the other hand, the reduction system is shorter in durum mills, because the primary product is removed and finished in the granular condition. For maximum yield of large endosperm particles, break rolls with U-cut corrugations are employed. The break system is extensive to permit lighter and more gradual grinding than in flour mills. Durum wheat of good milling quality normally yields about 62% semolina, 16% clear flour, and 22% feeds. Particle size distribution and granulation of semolina are highly important in the production of macaroni (see also chapter 20).

Flour Bleaching and Maturing

Bleaching of flour was introduced as early as 1879 in Britain and around 1900 in America. In the earliest days flour was treated with nitrogen peroxide. Subsequently other methods came into use to make the flour whiter and simultaneously improve the dough handling and bread characteristics. The treated flours possess baking properties similar to those of flour that has been stored and naturally aged. Today, much bread and practically all cake flours in the United States are bleached. In addition, maturing agents are used to obtain maximum baking performance. Flour improvers are used in Great Britain, Canada, and many other countries. In West Germany only ascorbic acid may be used legally as a flour improver. In still other countries no flour improver is allowed. Agents that have maturing action but little or no bleaching action include bromates, iodates, peroxysulfates, peroxyborates, calcium peroxide, and ascorbic acid (which is enzymatically converted to dehydroascorbic acid, an oxidizing agent). Agents that have both bleaching and maturing effect include oxygen, ozone, chlorine, and chlorine dioxide. The improvers azodicarbonamide and acetone peroxide have been approved by the Food and Drug Administration for inclusion with the Standards of Identity for flour as bleaching and maturing agents. Acetone peroxide performs a dual function of bleaching and maturing. Azodicarbonamide $H_2NCON=NCONH_2$ is reduced to hydrazodicarbonamide (biurea), $H_2NCONHNHCONH_2$. It has maturing action only. Benzoyl peroxide is added primarily as a bleaching agent. Additional agents, used less commonly for bleaching, include nitrogen peroxide, fatty acid peroxides, and certain preparations (e.g., from untreated soyflour) containing the enzyme lipoxygenase. Chemical formulas of flour improvers and their reduction products are listed in Table 8-1.

Quantitative requirement for oxidation of flours depends on several factors. Generally, as the protein content increases, the requirement for oxidants increases. Mixing time and oxidation levels compensate each other to some extent, even though they are not completely interchangeable. As the degree of milling refinement or flour grade is lowered, oxidation requirements increase, because protein sulfhydryl groups susceptible to oxidation are found in higher concentrations in the aleurone layer and the germ than in the starchy endosperm. Low-grade flours have more of those tissues than highly refined flours.

Table 8-1. Chemical Formulas of Flour Improvers and Their Reduction Products

Oxidizing agent		Reduction product	
Oxygen	O_2	Water	H_2O
Potassium bromate	$KBrO_3$	Potassium bromide	KBr
Potassium iodate	KIO_3	Potassium iodide	KI
Chlorine dioxide	ClO_2	Chlorine and water	Cl_2 and $H_2O(?)$
Dehydro-L-ascorbic acid		L-ascorbic acid	

Acetone Peroxide (Monomeric) → Acetone

Azodicarbonamide → Hydrazodicarbonamide (Biurea)

The tolerance of a flour to overtreatment varies with the improvers. Flour can tolerate overtreatment with ascorbic acid much better than with other improvers. A small excess of potassium iodate may be highly detrimental. In general, the level of potassium bromate, iodate, acetone peroxide, or azodicarbonamide added at the mill varies from 5 to 20 ppm; ascorbic acid is added at about 25 ppm. Chlorine dioxide containing about 20% free chlorine is used at the rate of about 15 ppm. For the continuous bread-making process, the amount of improver used is up to five times higher than that used in conventional breadmaking. The difference in oxidant is supplemented by the baker. Generally, a combination of oxidants—e.g., of bromate and iodate (about 4:1)—is used. The combination provides for a complimentary action of slow and rapid agent. Oxidants vary in their oxidation potential and oxidation rate. Atmospheric oxygen is

Table 8-2. Proximate Chemical Composition[a] of a Commercial Mill Mix of Hard Red Spring Wheat and Its Principal Mill Products

Product	Proportion of wheat (%)	Protein[b] (%)	Fat (%)	Ash (%)	Starch (%)	Pentosans (%)	Total sugars[c] (%)	Undetermined (%)
Wheat	100.0	15.3	1.9	1.85	53.0	5.2	2.6	6.8
Patent flour	65.3	14.2	0.9	0.42	66.7	1.6	1.2	1.4
First-clear flour	5.2	15.2	1.4	0.65	63.1	2.0	1.4	2.8
Second-clear flour	3.2	18.1	2.4	1.41	56.3	2.6	2.1	3.6
Red dog flour	1.3	18.5	3.8	2.71	41.4	4.5	4.6	11.0
Shorts	8.4	18.5	5.2	5.00	19.3	13.8	6.7	18.0
Bran	16.4	16.7	4.6	6.50	11.7	18.1	5.5	23.5
Germ	0.2	30.9	12.6	4.30	10.0	3.7	16.6	8.4

Source: USDA mimeographed publication ACE-189 (1942).

[a]13.5% moisture basis.
[b]Nitrogen × 5.7.
[c]Expressed as glucose.

Table 8-3. Average Nutritional Value of 100-g Milled Products

Milled product and type	Main nutrients (g)			Energy		Minerals/mg				Vitamins		
	Proteins	Fat	Carbo-hydrates	Calories (kcal)	Joules (kJ)	K	Ca	P	Fe	B$_1$ (µg)	B$_2$ (µg)	Niacin (mg)
Wheat flour, 550	10.6	1.1	74.0	348	1480	126	16	95	1.1	110	80	0.5
Wheat flour, 1050	12.1	1.8	71.2	349	1485	203	14	232	2.8	330	100	2.0
Wheat meal, 1700	12.1	2.1	69.4	345	1465	290	41	372	3.3	360	170	5.0
Rye flour, 997	7.4	1.1	75.6	342	1453	240	31	180	2.3	190	110	0.8
Rye meal, 1800	10.8	1.5	70.1	337	1432	439	23	362	3.3	300	140	2.9

Reproduced with permission (courtesy, AID, Ministry of Nutrition, Agriculture, and Forestry, Bonn, 1977).

a slow oxidant; chlorine dioxide functions rapidly. Iodates, acetone peroxide, and azodicarbonamide act much more rapidly than bromates.

The U.S. Code of Regulations (21, Food and Drugs, Parts 100–199, Federal Register, Washington, D.C.) of April 1, 1979, defines and describes cereal flours and related products as follows: *Flour* is defined as a food prepared by grinding and bolting clean wheat other than durum and red durum wheat. It may be supplemented by up to 0.75% malt or by small amounts of harmless preparations of alpha-amylase from *Aspergillus oryzae*. Its moisture content should not exceed 15%, and its ash percentage should not exceed (on a moisture-free basis) (% protein/20) + 0.35. At least 98% should pass a No. 70 sieve of ASTM. It may contain a maximum of 200 ppm ascorbic acid and optimum amounts of the following bleaching and/or oxidizing (aging) agents (alone or combinations): oxides of nitrogen, chlorine, nitrosyl chloride, chlorine dioxide, benzoyl peroxide (with carrier), acetone peroxides, and up to 45 ppm azodicarbonamide. Up to 50 ppm potassium bromate may be added to flours whose baking qualities are improved by such additions.

Enriched flour contains (mg/lb) 2.9 thiamine, 1.8 riboflavin, 24 niacin, and 13.0–16.5 iron. Its total calcium content should not exceed 960 mg/lb, and it may contain up to 5% wheat germ or partly defatted wheat germ. *Instantized flours* are prepared by selective grinding or bolting, other milling procedures, or by agglomerating procedures. *Phosphated flour* contains 0.25–0.75% monocalcium phosphate. *Self-raising flour* contains a mixture of sodium bicarbonate and one or more acid-reacting substances added to a maximum level of 4.5 parts per 100 parts of flour to produce at least 0.5% of carbon dioxide. *Cracked wheat* is produced by cracking, *crushed wheat* by crushing, and *whole wheat flour* by grinding cleaned wheat, other than durum and red durum, to meet specified granulation requirements. The maximum of potassium bromate in whole wheat flour is 75 ppm.

The ash content of *farina* may not exceed 0.6% and of *semolina* 0.92%, on a moisture-free basis, for both. Farina may be enriched to contain (per pound) 2.0–2.5 mg thiamine, 1.2–1.5 mg riboflavin, 16.0–20.0 mg niacin, at least 13.0 mg iron, 250 USP units of the optional ingredient vitamin D, and 500 mg of the optional ingredient calcium.

Proximate chemical composition of a commericial mill mix of hard red spring wheat and its principal milled products is given in Table 8-2. Table 8-3 shows the average nutritional value of 100-g milled products in West Germany.

BIBLIOGRAPHY

Anon. "From Wheat to Flour." Wheat Flour Institute; Chicago; **1965**.

CIGI. "Grains and Oilseeds, Handling, Marketing, Processing," 2nd ed. Canadian Int. Grains Inst.; Winnipeg, Canada; **1975**.

Inglett, G. E. (Ed.) "Wheat: Production and Utilization." Avi Publ. Co.; Westport, CT; **1974**.

Kent-Jones, D. W.; Amos, A. J. "Modern Cereal Chemistry," 6th ed. Food Trade Press; London; **1967**.

Lockwood, J. F. "Flour Milling," 4th Ed. Northern Publ. Co.; Liverpool; **1960**.

Pomeranz, Y. (Ed.). "Wheat Chemistry and Technology," 2nd ed. Am. Assoc. Cereal Chemists; St. Paul, MN; **1971**.

Scott, J. H. "Flour Milling Processes," 2nd ed. Chapman and Hall; London; **1951**.

Smith, L. "Flour Milling Technology," 3rd ed. Northern Publ. Co.; Liverpool; **1944**.

Swanson, C. O. "Wheat and Flour Quality." Burgess Publ. Co.; Minneapolis, MN; **1938**.

Tsen, C. C.; Hlynka, I. *Bakers Digest* **1967** *41*(5), 58–62, 64.

Chapter 9

Wheat Flour Components in Bread Making

It has been known for many years that flours milled from certain wheat varieties grown in some locations consistently produce bold loaves with good internal crumb grain and color and with excellent shelf life. On the other hand, some wheat flours produce compact bread that has a crumbly interior and stales rapidly. Researchers surmise that those differences result from qualitative and quantitative compositional variations. The problems of relating chemical composition and structure of wheat flour components to functional properties in bread making is complicated by several factors: the large number of components; the high molecular weight, limited solubility, and difficulty of separating or isolating pure components without altering them; and the interaction of the components during dough mixing, fermentation, and baking. Some classical fractionation methods make it possible to separate flour components; some modern biochemical techniques enable the researcher to characterize those components and to follow interactions among macromolecules; and the combination of separation and characterization techniques can be used to elucidate the roles of wheat-flour components in breadmaking.

PROTEIN CONTENT

When bread is baked from flours milled from wheat varieties grown under widely different climatic and soil conditions, protein content is the major factor to account for variation in loaf volume within a single variety. Indeed, the relation between loaf volume (provided the bread was baked by the optimized test) and protein content is linear (Finney and Barmore, 1948). Because the protein content–loaf volume relation is linear *within a single variety*, the bread-making quality of new wheat varieties can be easily determined.

In practice, a plant breeder can determine the slope of regression line from two or three samples of a new variety at different protein levels. As

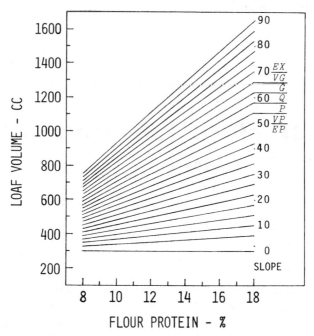

Figure 9-1. Loaf volume (100 g flour) versus protein content regression lines for correcting loaf volumes of wheat varieties to a constant protein basis. Slope is rate of change in loaf volume per 1% protein. EX = excellent; VG = very good; G = good; Q = questionable; P = poor; VP = very poor; EP = extremely poor. (From Finney, 1978).

shown in Figure 9-1 the regression of loaf volume with *protein quantity* (content) differs widely among *varieties*. This variation results in differences in the slopes of regression lines for individual varieties: the greater the slope, the greater the increase in loaf volume per 1% increase in protein. Those differences in slope reflect differences in bread-making quality of the protein in the different varieties of wheat (Finney and Yamazaki, 1967).

The significance of protein content as a factor that governs bread-making quality of wheat and wheat flour was confirmed in a study of the functional properties of dark hard and yellow hard red winter wheats (Pomeranz et al, 1976). In that study, functional properties of 14 commercial and six varietal hard red winter wheat samples, which varied widely in protein content and percentages of dark, hard, vitreous (DHV) kernels and of yellow hard kernels, were evaluated by laboratory milling and baking tests. Simple correlation coefficients were calculated for the relationships between protein content (of wheat or flour), DHV, baking absorption, mixing time, and loaf volume. The results are summarized in Table 9-1. The correlation coefficients were constantly higher for the commercial samples alone than for all samples. That difference probably resulted from relatively small effects of varietal differences in the commercial samples,

Table 9-1. Simple Correlation Coefficients for Protein and Baking Characteristics of Commercial and Commercial Plus Varietal Samples

	Simple correlation coefficients for all samples[a]					
	DHV[b]	Wheat protein	Flour protein	Baking absorption	Mixing time	Loaf volume
DVH		0.832**	0.829**	0.640**	−0.117	0.841**
Wheat protein	0.898**		0.995**	0.713**	−0.298	0.951**
Flour protein	0.916**	0.996		0.743**	−0.310	0.949**
Baking absorption	0.859**	0.841**	0.861**		0.092	0.625**
Mixing time	−0.270	0.420	−0.391	0.060		−0.353
Loaf volume	0.900**	0.981**	0.985**	0.842**	−0.425	

Simple correlation coefficients for commercial samples[c]

[a]N = 20.
[b]Dark, hard, and vitreous.
[c]N = 14.
**Denotes significance at the 1% level.
From Pomeranz et al, 1976.

which were mixtures of several varieties from several locations. Protein content explained about 90% (0.95^2) of the variability in loaf volume of all samples and about 96% (0.98^2) of the variability in loaf volume of the commercial samples; consequently, only 10% and 4% of loaf volume variability, respectively, was attributable to differences in protein quality and to experimental errors. The correlation was high between baking absorption and protein content but was not significant between mixing time and protein content.

Because the correlation between protein or DHV and bread making was so high, we calculated several partial correlation coefficients, keeping either protein content or DHV constant. The results are given in Table 9-2. Wheat protein and loaf volume were highly correlated ($r = 0.901$), irrespective of the DHV contents; the correlation between DHV and loaf volume ($r = 0.219$) was not significant when the protein content was constant.

In Table 9-1, the correlation coefficient between baking absorption and mixing time was not significant. However, when the protein content was kept constant (Table 9-2), the correlation between baking absorption and mixing time was highly significant (0.842). For flours with less than about 12% protein, mixing time increases with decreasing protein; above 12%, increases in protein content have little effect on mixing time. Consequently,

Table 9-2. Partial Correlation Coefficients for Wheat Protein and Baking Characteristics of Commercial Wheat Samples

	Partial correlation coefficients—DHV constant			
	DHV	Baking absorption	Mixing time	Loaf volume
Wheat protein		0.311	−0.419	0.901**
Baking absorption	0.434		0.591**	0.312
Mixing time	0.268	0.842**		−0.432
Loaf volume	0.219	0.161	−0.073	

Partial correlation coefficients—protein constant

**Denotes significance at the 1% level.
From Pomeranz et al, 1976.

when the effect of protein was eliminated, a relationship between baking absorption and mixing time was demonstrated.

Baking tests were made by the Rapid-Mix test on flours from two soft and three hard wheat cultivars from a single location and on 12 composites, six each, of soft and of hard wheat selections from many locations in Germany (Pomeranz et al, 1984). Maltose contents, amylose values, and starch damage were substantially and consistently lower in the soft than in the hard wheat flours. The difference was accompanied by higher water absorption in the hard wheat flours. Tested by the standard Rapid-Mix test, the soft wheat flours ranked lower in volume of baked rolls than in the hard wheat flours (Table 9-3). The soft wheat flours showed a larger volume response to increased sugar levels in the formula, to compensate for limited starch damage, than the hard wheat flours. When the sugar in the formula was optimized, the volume potential of the soft wheat flours was realized and was equal to that of hard wheat flours on a constant protein basis.

Two deductions can be made from this examination: (1) The relatively extensive starch damage in hard wheats provides adequate levels of fermentable sugars to bring out the loaf volume potential of the flour proteins. That potential cannot be realized, however, in flours from soft wheats that have limited starch damage. (2) If the increases in loaf volume per 1% protein are viewed as an index of functional bread making protein quality, there is no basic difference between the proteins in soft and hard wheats, although there were clearly varietal differences within both the soft and the hard wheats that were examined.

The loaf volume potential in soft wheats can be brought out by the addition of sugar. In a more general sense, however, this may not

Table 9-3. Loaf Volume Increases in Rapid-Mix Test of Composite Soft and Hard Wheat Flours

| Wheat type | Maximum volume increase (ml) per 1% protein | |
	Range	Mean
Soft	33.5–38.5	36.0
Hard	33.6–39.3	36.8

completely compensate for the effects of increased starch damage in hard wheats (eg, for the increase in water absorption). To compensate for the difference in water absorption, more extensive starch damage in soft wheats would be required. In addition, the inherent milling properties of hard wheats (response to conditioning, ease of sieving, and flour yield) are definitely desirable features of wheat with intermediate hardness. Other factors that need to be considered are increased excessive crust browning and impaired dough properties at high sugar levels.

Yet it is clear that soft wheats contain proteins that are equal, on a constant protein basis, in terms of functional properties to proteins of hard wheats and that, if there is a definite economic advantage in growing certain soft wheat varieties (eg, because of increased yield), those wheats can be made to perform more satisfactorily by increasing the sugar in the formula to compensate for the low starch damage.

Finally, the results of this study have implications for the manner in which the Rapid-Mix test is performed. Inasmuch as the formula is not optimized for the amount of fermentable sugars, soft wheats are penalized in testing. The adjustment of "falling number," by adding malt, does not correct for that deficiency. Increased malt levels, high in proteolytic enzymes, can be detrimental especially in low-protein wheat flour. In practice, increasing the amount of added sugar from 1% to about 2.5%, or alternatively increasing the starch damage from about 7% to 12%, would provide a more valid and meaningful comparison of the bread-making functional proteins in soft and hard wheats.

FRACTIONATION AND RECONSTITUTION

After the linear relation between protein content and loaf volume was established in the optimized baking test, wheat flours representing a wide range in quality (as indicated by loaf-volume response per 1% of flour protein) were fractionated into starch, gluten, and water-soluble components. Fractions from one variety were recombined in their original and in different proportions, and various fractions of different varieties were interchanged (Finney, 1943). Invariably, flours that were reconstituted in their original proportions gave bread that was identical to bread made with the original, unfractionated flour; this established that neither the fractionation nor reconstitution techniques impaired bread-making qualities.

When wheat gluten, starch, and water-soluble fractions from different varieties were interchanged and the flours so formed were baked, gluten proteins accounted for differences in bread quality of the varieties studied, as defined by loaf-volume increase per 1% protein. Although these results indicated that gluten proteins govern differences in bread-making potential of wheat varieties, bread of acceptable volume, texture, and freshness retention cannot be produced unless other flour components—namely starch, water solubles, and lipids—are also present in approximately the same quantities found in normal, unfractionated wheat flours (see also Mauritzen and Stewart, 1965).

GLUTEN PROTEINS

Once the key role of gluten proteins in governing varietal differences in breadmaking potential was established, the next logical step was to identify the gluten component(s) responsible for those differences.

Fractionation of gluten components on the basis of solubility damages the bread-making properties of those components. Reconstitution studies showed that separation by ultracentrifugation of the two main fractions of gluten into a centrifugate containing glutenin and a supernatant rich in gliadin did not impair the bread-making properties of the gluten components. The gliadin fraction controls loaf volume and varies in flours that differ in bread-making potential (Hoseney et al, 1969a,b). The factor responsible for mixing time and dough development is in the glutenin fraction. During dough mixing, the protein mass is converted from granular protein bodies into a homogenous network in which starch granules are imbedded. Proper dough development is essential to optimum performance in breadmaking.

More recently, Finney et al (1982) and Jones et al (1983) fractionated and characterized gluten proteins and studied their functional properties in bread-making. The fractionation scheme is described in Figure 9-2. Gluten proteins from hard winter wheat flour of good (RBS–76) and poor (76–412) bread-making quality were solubilized in dilute lactic acid and separated by ultracentrifugation into four protein fractions (Jones et al, 1982). The high-molecular-weight glutenin proteins sedimented at 100,000 g as a dense, relatively insoluble pellet (35 min for 76–412 flour; 2 h for RBS–76 flour). The low-molecular-weight glutenin proteins sedimented at 435,000 g as a gel (6 h for 76–412 flour; about 10 h for RBS–76 flour). The high-molecular-weight gliadin proteins also sedimented at 435,000 g as a clear viscous solution (6 h for 76–412 flour; about 10h for RBS–76 flour). The low-molecular-weight gliadins of both flours remained in the corresponding supernatants. Removing total free

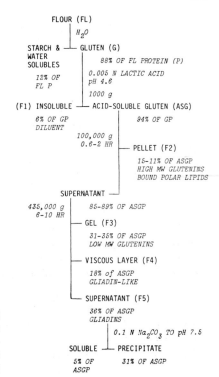

FLOUR (FL)

H_2O

STARCH &
WATER   GLUTEN (G)
SOLUBLES
                          88% OF FL PROTEIN (P)

12% OF   *0.005 N LACTIC ACID*
FL P   *pH 4.6*

                1000 g

(F1) INSOLUBLE   ACID-SOLUBLE GLUTEN (ASG)

6% OF GP   *94% OF GP*
DILUENT

            100,000 g
            0.6-2 HR   PELLET (F2)

                          15-11% OF ASGP
                          HIGH MW GLUTENINS
                          BOUND POLAR LIPIDS

SUPERNATANT

435,000 g   *85-89% OF ASGP*
6-10 HR
                GEL (F3)

                31-35% OF ASGP
                LOW MW GLUTENINS

                VISCOUS LAYER (F4)

                18% of ASGP
                GLIADIN-LIKE

                SUPERNATANT (F5)

                36% OF ASGP
                GLIADINS

                    0.1 N Na_2CO_3 TO pH 7.5

SOLUBLE   PRECIPITATE

5% OF   *31% OF ASGP*
ASGP

Figure 9-2. Scheme to fractionate wheat flour into crude protein and starch plus water solubles and to fractionate the acid-soluble gluten into two glutenin and two gliadin fractions. (From Finney et al, 1982.)

lipids from flour before washing out and solubilizing the gluten materially increased the sedimentation rate of the high-molecular-weight glutenin (pellet) proteins of the poor-quality bread flour 76–412.

The corresponding gel and viscous layer plus supernatant fractions of the good- and poor-quality flours were interchanged singly in reconstituted flours containing the starch plus water-soluble fraction and baked into bread (Finney et al, 1982). The gel glutenin proteins of the acid-soluble gluten proteins controlled mixing requirement and baking absorption, and the viscous layer and supernatant gliadin proteins controlled loaf volume and crumb grain. It was suggested that the relative ease with which the high-molecular-weight pellet glutenins sedimented already after 35–120 minutes at only 100,000 g indicates that they are relatively free compared to the low-molecular-weight gel glutenins that required an additional 6–10 hours at the high centrifugal force of 435,000 g to sediment. The gel glutenins appeared to interact tenaciously with the gliadin proteins. The extent to which the bound polar lipids interacted with the pellet proteins, both intra- and inter-molecularly to produce very high molecular weight

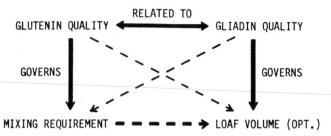

Figure 9-3. Diagram of likely direct and indirect relationships between glutenin quality, gliadin quality, and the functional properties mixing requirement and loaf volume (optimum). (From Finney et al, 1982.)

aggregates, may render them as relatively noninteractive with the gel glutenin and gliadin proteins, so that the pellet glutenin proteins are free to sediment at low centrifugal forces (Finney et al, 1982).

The ease of sedimenting the pellet glutenins indicates that they are not involved in formation of additional lipoprotein complexes and that the free lipids become bound probably by interacting with reactive gel glutenin and free gliadin proteins. Similarly, when the dough is formed, reactive gel glutenins probably interact with reactive gliadins. Thereby, the gel glutenin proteins become bound. When the centrifugation forces are greater than the protein interaction forces but less than the molecular forces that keep the relatively small gliadin proteins in solution, then the gel glutenins sediment (Figure 9.3).

It was postulated (Finney et al, 1982) that the tenacity with which gliadin proteins interact with free lipids and glutenin proteins may be the physical-chemical criterion of why poor quality bread wheats are poor and good ones are good.

AMINO ACIDS

Automated techniques to determine the amino acid composition of proteins have facilitated the assay of amino acids in gluten proteins. Gluten proteins are characterized by high concentrations of glutamic acid (about 32%) and proline (about 10%). The acidic amino acids, glutamic and aspartic acids, occur mainly as amides. Thus far, studies on the amino acid composition of cereals have been disappointing, because they have failed to explain the differences in breadmaking characteristics of wheat varieties.

According to Wall (1971, 1979 a,b), numerous studies that involved amino acid composition, molecular-weight peptide maps, and N-terminal sequences have shown that present gliadin nomenclature is inadequate to describe true relationships and that a more suitable system must be developed. Mifflin et al (1981) suggested that classification should be based on structural loci in the genome and their chemical structure, as determined by the base sequence of the genes. Based on those considerations, Mifflin et al (1981) suggested to classify prolamines into high-molecular-weight (glycine-rich), S-poor, and S-rich fractions. These three fractions differ in physical and chemical properties and in their ability to form aggregates. They postulated that differences in relative amounts of these prolamines account for different processing properties of wheat, rye, and barley.

In light of the postulates of Mifflin et al (1981), how much significance should be attached to the role of the glutenin:gliadin ratio in bread making (Lee, 1975)? The effects of gliadin and glutenin are not additive. Their effects are the result of interaction. One of the interesting pieces of evidence on their interaction stems from the fact that one cannot mix a dough to optimum consistency if all gliadin is replaced by glutenin. Similarly, how are those findings related to proposed structures of gluten proteins: aggregating linear molecules versus linear chains cross-linked by disulfide bonds that govern resistance of dough to elastic deformation? How do these structures relate, in turn, to various degrees of disulfide bond reactivity? And, finally, how is the picture modified—complicated by the possibility that some, or even most, storage proteins are actually glycoproteins?

CHEMICAL BONDS IN DOUGH

Bread dough is a highly complex chemical system. The wheat flour components include proteins; mono-, oligo-, and polysaccharides; lipids; minerals; and enzymes. The additives yeast, milk solids, and malt each contain several chemical moieties, but water, sugar, and salt are basically chemically pure. In addition, the dough contains a mixture of gases. The components interact and are constantly modified during mixing, fermentation, and baking.

To explain chemical and physical modifications in breadmaking it is essential to understand the nature of the chemical bonds among dough components. Wheat proteins are generally credited with attaining the final objectives of the baker—the production of a well-risen loaf of bread of satisfactory texture. In addition, wheat proteins largely govern the flour's water absorption, oxidation requirement, and mixing and fermentation tolerance. Yet it is well recognized that flour proteins are not the only wheat component uniquely suited for bread making and that practically all components are essential to produce an acceptable bread. Consequently, while the following discussion centers around the chemical bonds in wheat proteins, the involvement of other components is outlined or implied.

Three-Dimensional Structure

Proteins are generally considered to have at least three levels of structure—primary, secondary, and tertiary—although for some proteins no clear

Table 9-4. Main Chemical Bonds in Dough Proteins

Type of bond	Mechanism	Energy (kcal/mole)
Covalent	Two atoms are bound by a common electron pair	30–100
Ionic	Attraction between opposite charges	10–100
Hydrogen	Affinity of hydrogen for electronegative atom (ie, oxygen)	2–5
Van der Waals	Long-range interaction between nonpolar groups	up to 0.5

Source: Wehrli and Pomeranz, 1969a.

distinction can be made between the last two categories. The sequential arrangement of amino acids in a protein chain constitutes the first level of organization and is termed primary structure. The chains of amino acids are not straight but are folded to form characteristic three-dimensional structures, called secondary structure. To account for the folding of the chains into relatively compact and rigid globular molecules, we visualize another level of organization, the tertiary structure. Finally, if the protein consists of two or more loosely linked globular subunits we speak of its quarternary, or multimeric, structure. The native structures of most proteins are relatively compact and rigid; there is little internal space that is accessible to solvent. In general, the nonpolar amino acids (such as leucine and valine) are buried in the interior of the molecule, where they are shielded from contact with solvent. Finally, and most significantly, the complex molecular organization of a protein is essentially dictated by its amino acid sequence. That sequence governs the primary, secondary, and tertiary structures; and of the large number of theoretically possible structures, the native one is most favored.

Types of Bonds

It is possible to use four models to describe chemical bonds in proteins. They are summarized in Table 9-4 and illustrated schematically in Figure 9-4. There are several types of covalent bonds. They include bonds within the amino acid, between amino acids (peptide linkages), and disulfide links within or between peptide chains. The energy associated with (or required to break) those bonds is high. In addition we have three types of noncovalent bonds that generally are much weaker than covalent bonds. It must be emphasized that the four types of bonds are theoretical models that are useful in describing the natural bonds. The latter are never pure covalent, ionic, hydrogen, or Van der Waals bonds, but always a mixture of bonds.

In addition to four chemical bonds, there is the hydrophobic bond that is often confused with the Van der Waals bond. Hydrophobic and Van der Waals bonds have one thing in common; both are significant in interactions between nonpolar groups. However, the hydrophobic bond is not a

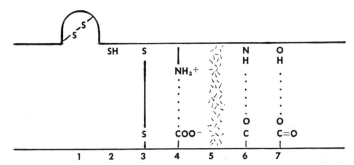

Figure 9-4. Schematic diagram of bonds within and between polypeptide chains; solid lines, covalent bonds; dotted lines, noncovalent bonds. (From Wehrli and Pomeranz, 1969a.)

chemical bond (ie, due to the attraction between protons and electrons) but a thermodynamic phenomenon or entrophy effect (Wehrli and Pomeranz, 1969a).

Covalent Bonds

The only covalent bonds that are known to be significant in dough structure are the disulfide linkages between proteins. Such bonds have an energy of 49 kcal/mole and probably are not broken at room temperature except by a chemical reaction.

About 1.4% of the amino acids in gluten are either cystine or cysteine (Wall, 1964). The disulfide bonds in cystine can link together portions of the same polypeptide chains and contribute to dough firmness. A network with permanent cross-links cannot show viscous flow, which characterizes bread dough. Viscous flow requires opening of the rigid cross-links. If the dough is not to lose cohesion, the cross-links must re-form at the same rate at other sites. By reacting with sulfhydryl groups (of a cysteine residue), disulfide groups can cause interchange reactions in rheological properties of dough, as first suggested by Goldstein in 1957.

The effects of oxidizing agents on the rheological properties of dough may be quantitatively explained by breaking disulfide cross-links, and their concurrent re-formation may be explained by exchange reactions with sulfhydryl groups (Bloksma, 1968). The velocity of exchange reactions would be expected to be proportional to the combined number of sulfhydryl and disulfide groups; viscosity would be inversely proportional to the sulfhydryl content. In practice, those effects are complicated by variations in sulfhydryl and disulfide reactivity. That reactivity depends on the size of the molecules of which the groups are a part (Villegas et al, 1963), on the three-dimensional availability of the groups, and probably on interaction with proteins and nonprotein components.

According to Belderok (1967), baking quality of wheat is governed by protein content and disulfide:sulfhydryl ratio (SS/SH). For a given protein level, optimum bread making results were obtained if the SS/SH ratio was above 7 in strong flours investigated by Kuchumova and Strelnikova (1967); the ratio increased during the flour storage.

During dough mixing only 1–2% of all gluten disulfide bonds can be broken by exchange with free sulfhydryl groups (Mauritzen, 1967), indicating that only a few reactive disulfide bonds are crucial to rheological properties. Total and reactive sulfhydryl groups and the ratio of reactive to total sulfhydryl groups increase with decreasing flour strength. Total disulfide content decreases slightly, whereas reactive (or the ratio or reactive to total) disulfide groups increase with decreasing mixing strength. Thus, mixing strength appears inversely related to reactive sulfhydryl and disulfide contents (Tsen and Bushuk, 1968).

In addition to the postulated interchange reactions, disulfide and sulfhydryl groups may react with each other by oxidation. Sulfhydryl groups can be added to double bonds of fatty acids to form a lipoprotein, but thus far such complexes have not been identified in doughs. Lipid peroxides, and their degradation products, may oxidize sulfhydryl groups or react with amino groups of proteins forming lipoproteins or polymers resulting from protein-protein interaction. Lipid peroxides or a mixture of highly unsaturated lipids in the presence of the enzyme lipoxidase significantly improve gas retention, loaf volume, and softness retention of bread (Auermann et al, 1965).

When the acidic moieties of amides in gluten, glutenin, or gliadin are methylated, solubility, intrinsic viscosity, and cohesion are changed significantly (Beckwith et al, 1963). Dough-handling properties can be modified, either beneficially or adversely, by the addition of minute amounts of reducing agents or sulfhydryl-blocking reagents. Also, the performance of a flour in bread making can be improved significantly by addition of an appropriate amount of oxidizing agent, such as 20–50 ppm potassium bromate.

The oxidation requirement (the amount of oxidant needed to produce the best loaf of bread in terms of volume, crumb texture, and freshness retention) is related to total protein content and to protein sulfhydryl groups and disulfide linkages. Continuous interchange reactions between sulfhydryl and disulfide groups and the reactivity of protein sulfhydryl groups affect oxidation requirements.

Ionic Bonds

The importance of ionic bonds is demonstrated by several effects of adding salt to the dough. Theoretically, ions influence their environment in the following ways: (1) Ions increase the dielectric constant of water and thereby reduce ionic as well as dipole-dipole (hydrogen bond) attraction and repulsion of dough components; (2) ions form aggregates with water

Table 9-5. Ionic Components in 100 g Wheat Flour[a]

Component	Positive charges (mval)	Negative charges (mval)
K^+	2.87	
Mg^{++}	1.16	
Ca^{++}	0.96	
Zn^{++}	0.04	
Fe^{+++}	0.075	
Na^+	0.091	
Cu^{++}	0.006	
Cl^-		1.28
Phytate		0.82
Phospholipids	1.50	1.93
Gliadin	2.10	1.62
Glutenin	2.97	2.36
Albumin + globulin	1.80	2.00

[a]14% moisture basis, 70% extraction, 15% protein.

molecules and decrease the amount of free water that is essential to both hydrophobic binding and dough mobility; (3) ions reduce the zeta potential (the thickness and charge of the ionic double layer) around the charged protein molecules and enhance their interaction; (4) ions decrease the entropy of polymerization, thereby favoring dissociation of proteins; and finally (5) ions in a dough may complex with ionic groups of lipids. Theoretically, ions may enhance both association and dissociation of dough components. In practice, the former prevails, since adding salts increases dough rigidity and reduces extensibility.

The ions occurring in wheat flour are summarized in Table 9.5 . About 7.3% of all amino acid residues in gliadin and 9.3% of those in glutenin are charged. In addition to those ions, 2% sodium chloride and 0.5% other mineral salts (in form of yeast food) are added in commercial baking.

Hydrogen Bonds

Hydrogen bonds result from the affinity of hydrogens of hydroxyl, amide, or carboxyl groups for the oxygen of carbonyl or carboxyl groups. Peptide hydrogen bonds between the amide hydrogen of one amino acid residue and the carbonyl oxygen of another amino acid residue are the only types of noncovalent bonds that can be found in large numbers (2 per amino acid) to produce regular folded structures. The energy of an $N\text{-}H \ldots O=C$ hydrogen bond was originally estimated at about 8 kcal/mole. Recent studies, however, show that the strength of interpeptide hydrogen bonds in an aqueous medium is extremely small, which has tended to discount the hydrogen bond as important to the stability of folded polypeptide chains in globular proteins.

However, conditions in the interior of a large protein molecule differ substantially from conditions in an aqueous environment. The interior of the molecule is likely to have a high concentration of hydrocarbon-like amino acid residues, is relatively free of water, and has regions of low dielectric constant. Under those conditions, the interpeptide bond is greatly strengthened. In addition, the high concentration of peptide bonds in the interior of protein molecules favors hydrogen bond formation. Thus, while hydrogen bonds apparently are not the dominant noncovalent force stabilizing folded protein structures, their role cannot be dismissed (Wehrli and Pomeranz, 1969a).

Indirect evidence of the importance of hydrogen bonds in dough is twofold: (1) the presence of the necessary elements for their formation, and (2) profound changes in rheological properties from the action of hydrogen bond–disrupting agents. About 90% of dough components are highly polar. Over one-third of the amino acids in gluten are glutamine. That amino acid has an amide group that can participate in hydrogen bonding. Studies with water-insoluble and -soluble synthetic polypeptides containing side-chain amide groups have indicated hydrogen-bonding phenomena (Krull and Wall, 1966; Krull et al, 1965). Deuterium bonds in most cases have somewhat higher bond energies than hydrogen bonds. When deuterium oxide is used instead of water in baking, gluten is strengthened and elasticity is increased (Vakar et al, 1965). Since deuterium oxide also enhances hydrophobic bonding, its effect does not prove conclusively the importance of hydrogen bonds. If hydroxyl protons of proteins and carbohydrates are allowed to exchange for deuterium, the strength of gluten is further increased, which is direct experimental proof that hydrogen bonds are involved in stabilizing dough structure, because hydroxyl groups do not participate in hydrophobic interaction.

Adding 3 M urea destroys dough structure (Jankiewicz and Pomeranz, 1965). Although urea has been used for many years in protein investigations, it is not known whether its disruptive effect is from the breakage of hydrogen or hydrophobic bonds. Urea forms weaker hydrogen bonds than water and cannot lower hydrogen bond energy of water. It has therefore been suggested (Whitney and Tanford, 1962) that urea breaks hydrophobic bonds.

The most effective hydrogen bond–forming components in wheat flour, in addition to water, are pentosans. The total (soluble and insoluble) pentosan content of flour is only about 1.5%, but pentosans can absorb 15 times their own weight of water (Bushuk, 1966), so up to 23% of the water dough could be bound to pentosans. Although most hydroxyl groups of the flour are in the starch, the starch granules are so tightly packed that only the groups at the surface are available for hydrogen bonding. The specific surface of starch is 0.1249 m^2/g (Gracza and Greenberg, 1963). In 100 g dough there is about 44 g starch that absorbs 8.7 g water, corresponding to about 3.5 monomolecular layers of water around the starch granules, which in baking swell and gelatinize and absorb most of the water. The role of glutamine, a strong hydrogen bond–forming residue of wheat proteins, was mentioned earlier. Naturally present or added sugars further increase the content of polar substances that may participate in hydrogen bonding. Finally, naturally occurring or added polar lipids may form hydrogen bonds with any of the previously mentioned components.

Hydrogen bonds between small molecules, such as water, significantly increase viscosity of a liquid. Mobility and much of the plasticity of doughs are affected by such bonds. In macromolecules, hydrogen bonds may lead to elastic (eg, DNA, certain proteins) or even rigid (eg, cellulose, starch) structures. The bonds are weak enough to be temporarily extended, exchanged, or broken. As soon as the stress is released, the molecule returns to its original stable form.

Van der Waals Bonds

Van der Waals bonds and dipole-induced dipole interactions provide very weak bonds (up to 0.5 kcal/mole). Although they occur between all atoms separated by at least four covalent bonds, they are not significant in the presence of stronger bonds or at distances longer than 5 Ångstrom. They may play a role in attraction between nonpolar amino acid residues or fatty acid side chains in systems in which hydrophobic bonds are impossible because of limited water. The starch-glyceride complex is the only type of complex that is stabilized by dipole-induced dipole interaction and which has been postulated to affect baking and bread properties. Such complexes occur naturally in starch granules or are formed between starch and artificial surfactants, or starch and sucroesters (Wehrli and Pomeranz, 1969a).

Hydrophobic Bonds

Many of the amino acids in proteins have nonpolar side chains. The contribution of those groups to secondary and tertiary structures was postulated in 1954 by Kauzmann. He proposed the term "hydrophobic bond" to account for the forces responsible for the tendency of the nonpolar residues to adhere to one another and to avoid contact with the aqueous surrounding. Klotz (1960), on the other hand, visualized the nonpolar side chains forming crystalline hydrates with water and hydration "icebergs" (Frank and Evans, 1945) coalescing to produce stable ice lattices over the protein surface.

Both Kauzmann (1954) and Klotz (1960) used simple model systems and extrapolated from the observed properties of small molecules the expected properties of large proteins. The conflict between the two concepts of hydrophobic bonds stems from disagreement as to which simple model is best suited to explain the behavior of complex proteins.

Schachman (1963) concluded on the basis of the available evidence that neither model can be adopted unequivocally to explain hydrophobic bonds. Regardless of the model we accept, it is clear that the interaction of nonpolar amino acid residues is a small stabilizing force in wheat proteins. However, because of the unique physical properties of water and the tendency of water molecules to exist as hydrogen-bonded clusters, the indirect effect of hydrophobic interactions may be quite significant.

Without free water no hydrophobic bonds can be produced. The question arises: Are there enough free water molecules in a dough to form hydrophobic bonds? According to Toledo et al (1968), only about 50% of the water in dough is bound. Consequently, hydrophobic bonds are possible. Experimental evidence for the importance of hydrophobic bonds was obtained by modifying dough properties by organic solvents (Ponte et al, 1967) and by hydrocarbons (Pomeranz et al, 1966). Existence of hydrophobic bonds is also indicated by nuclear magnetic resonance studies (Wehrli and Pomeranz, 1970a).

There is no doubt that lipids in dough can form hydrophobic bonds. Whether there are enough available nonpolar groups for hydrophobic interaction between proteins and lipids is another question. If the amino acid composition of a protein is known, its hydrophobicity can be calculated by the method of Bigelow (1967). Hydrophobicities of proteins range from 440 to 2,000 cal per average amino acid residue. Calculated hydrophobicities of glutenin and gliadin from data on amino acid composition reported by Woychik et al (1961) are 1,016 and 1,109, respectively. Thus, the hydrophobicity of both gluten proteins is theoretically high enough to make intra- and intermolecular hydrophobic bonds.

Hydrophobic bonds may contribute to both plasticity (as in forming biomolecular lipid layers) and elasticity (as in the alpha helix of proteins). The bond energies are low enough to allow rapid interchange at room temperature and to contribute to dough plasticity. On the other hand, they might stabilize conformations with small surfaces. Protein that has been stretched and its surface increased would tend to return to the original globular conformation as soon as the stress is released. Hydrophobic bonds could thus contribute to elasticity.

The hydrophobic bonds could also be important in early stages of baking, especially in oven-spring formation. All chemical bonds are weakened as the temperature increases. However, hydrophobic bond formation is an endothermic process favored by increasing temperature up to about 60°C.

Conclusions

The above discussion enumerates the theoretically feasible types of bonds between single dough components. In interactions between glycolipids and starch, or between glycolipids and gluten components, several types of bonds (hydrogen, hydrophobic, and Van der Waals) are involved. Different types of bonds may assume strategic significance, depending on

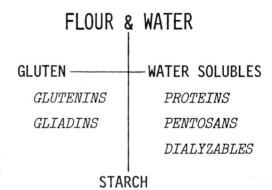

Figure 9-5. Scheme for fractionating wheat flour into gluten, water solubles, and starch.

whether single dough components interact or there is free competition between all components for interaction.

Studies with model components or individual flour components frequently tempt researchers to attribute the central role to a single type of bond and to discount the role of other bonds. Despite the appeal of such hypotheses, it seems that practically all types of bonds contribute to the structure of dough. While covalent and ionic bonds primarily increase cohesiveness of doughs, dipole, hydrogen, and hydrophobic bonds contribute to elasticity and plasticity. Van der Waals interactions are apparently of limited significance. Such a diversity of forces complicates interpreting factors that govern dough structure. At the same time it opens several avenues for improving bread-making potentialities of wheat flours.

WATER SOLUBLES

The three fractions obtained when wheat flour dough is washed under a gentle stream of water are gluten, starch, and water-soluble components (Figure 9-5). The water-soluble components of flour can be separated, in turn, by dialysis and centrifugation into four fractions: (1) membrane-permeable dialyzate (a mixture of soluble carbohydrates, amino acids, peptides, minerals, and growth factors essential for yeast fermentation); (2) globulins, which precipitate in the dialysis bag in the absence of minerals and can be removed after centrifugation; (3) heat-coagulable albumins in the supernatant; and (4) residual supernatant (Figure 9-6).

Baking experiments showed that the role of water-soluble components in bread making is twofold. The membrane-permeable dialyzate is essential to panary fermentation because it contributes to gas formation of yeast; however, that contribution can be replaced by synthetic yeast food. Neither

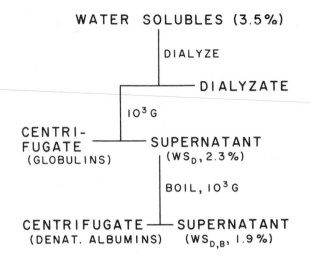

Figure 9-6. Fractionation scheme for certain water-soluble fractions of flour. Percentages are based on total flour weight. Abbreviations: WS, water solubles; D, dialyzed; B, boiled. (From Hoseney et al, 1969c.)

the globulins nor the albumins are essential to produce a normal loaf of bread. The residual supernatant is rich in water-soluble pentosans and glycoproteins and contributes to gluten extensibility and to retention of gas in a fermented dough.

Fincher and Stone (1974) and Fincher et al (1974) showed that the water-soluble pentosan can be fractionated into two fractions with saturated ammonium sulfate. The precipitated fraction (Figure 9-7) is an arabinoxylan contaminated with free protein. The soluble fraction (Figure 9-8) is an arabinogalactan covalently linked to a peptide. The water-soluble pentosans contain a small amount of ferulic acid associated with the largest part of the arabinoxylan fraction.

Wheat flour contains water-soluble and water-insoluble pentosans (Hoseney, 1984). Water-insoluble pentosans amount to about 2.4% in wheat endosperm. When total starch is centrifuged, it divides into two fractions—the prime starch and a gelatinous fraction referred to as "squeegee," "sludge," or "tailings." The tailings fraction contains damaged and small-granule starch along with proteins and pentosans. The polysaccharides in this fraction are made up of L-arabinose, D-xylose, and D-glucose. Reports vary on the presence of galactose and mannose.

The water-insoluble pentosans are similar to water-soluble pentosans except for a greater degree of branching. The main chain is D-xylopyranosyl units linked beta-1,4, with L-arabinofuranosyl side chains attached to the 3 position (if single) or 2 and 3 positions of the xylose chain. About 60% of the xylose residues are branched, and about one-half of those are branched at positions 2 and 3.

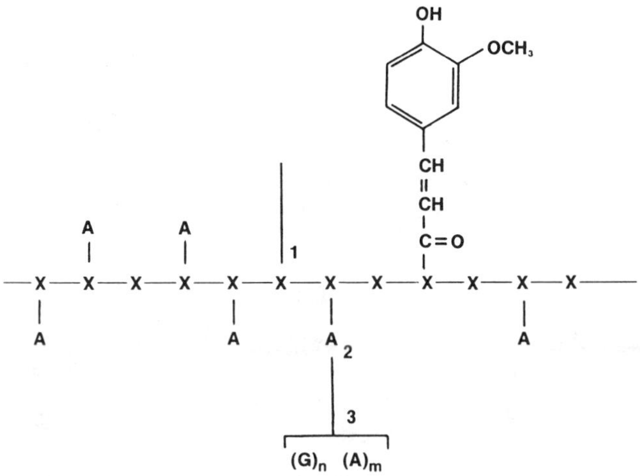

Figure 9-7. Hypothetical structure of glycoprotein from wheat flour. Vertical line at 1 indicates polypeptide chain; X, β-D-xylopyranose units; A, α-L-arab-inofuranose units; G, Galactose units; 1, 2, 3, possible linkages between carbohy-drates and protein. (From Neukom et al, 1967; Fincher et al, 1974.)

Water-Soluble Pentosans

Cold-water extracts of flour give 1.0–1.5% water-soluble pentosans. In addition to the pentoses xylose and arabinose, the polymers contain galactose and protein. The predominant water-soluble pentosan from wheat flour is a straight chain of anhydro-D-xylopyranosyl residue at the 2 or 3 position. Removal of the side chains leaves a xylan chain that is not soluble. The oxidative gelation of pentosans appears to involve the addition of a protein thiyl radical to the activated double bond of ferulic acid. The ferulic acid is esterified to the arabinoxylan. The covalent binding of protein and polysaccharide chains creates high-molecular-weight entities. These would be reflected by the viscosity increases. Such cross-linking may change the rheological properties of a dough (Hoseney 1984).

Hoseney et al (1969c) showed that removal of the water-soluble fraction of wheat flour reduced the loaf volume of the resultant bread. The fraction responsible for that effect was identified as a water-soluble pentosan fraction.

Water-Insoluble Pentosans

Casier and co-workers (Casier and Soenen, 1967; Casier et al, 1969; 1973) have studied the influence of water-insoluble pentosans in both wheat and rye on baking properties. Adding 2% water-insoluble endosperm pentosans increased the bread volume by 30–45%. In addition, other properties of the bread such as cell uniformity, crumb characteristics, and

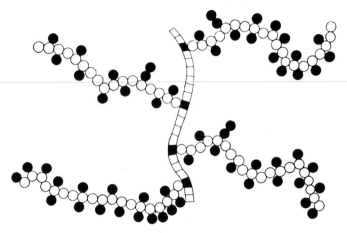

Figure 9-8. Proposed model of arabinogalactan-peptide. ○, galactose; ●, arabinose; ■, hydroxyproline; □, other amino acids. Arabinose is present in the α-L-arabinofuranyl configuration. The D-galactopyranose residues are linked to hydroxyproline by glycosidic linkages in the β-anomeric configuration.

elasticity were improved. Water-soluble pentosans from rye and wheat flour also improved quality of bread baked from nonwheat flours. No beneficial effects of water-insoluble pentosans were found by Hoseney et al (1971) for strong, high-protein U.S. wheat flours.

Water-insoluble pentosans impair quality of cookies (Yamazaki, 1955; Sollars, 1956) and influence the rate of starch retrogradation and bread staling (Kim and D'Appolonia, 1977).

LIPIDS

Wheat flour lipid composition is shown schematically in Figure 9-9 (adapted from data reported by MacMurray and Morrison, 1970). Total wheat flour lipids consist of about equal amounts of nonpolar and polar components. Triglycerides (TG) are a major component of nonpolar lipids, digalactosyldiglycerides (DGDG) are a major component of glycolipids; and lysophosphatidylcholines (LPC) and phosphatidylcholines (PC) are major components of phospholipids. Differences in solubility provide a convenient and useful means of separating wheat flour lipids into major categories: free and bound (Figure 9-10). Free lipids can be extracted with nonpolar solvents such as ether or petroleum ether (PE). For extraction of bound (mainly to protein) lipids, polar solvents, such as water-saturated butanol (WBS) or a mixture of chloroform-methanol-water, are required. Lipids extracted by PE are arbitrarily defined as free, and those extracted by WBS, following PE extraction, are defined as bound lipids.

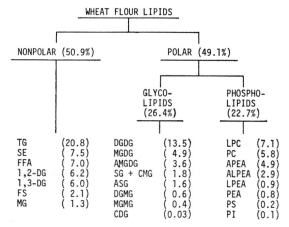

Figure 9-9. Composition of total wheat flour lipids (extracted with water-saturated butanol). (Adapted from MacMurray and Morrison, 1970.)

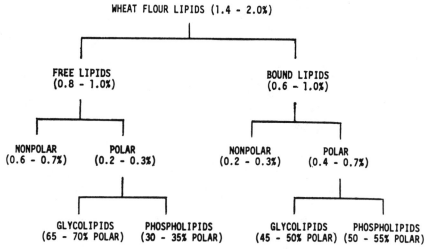

Figure 9-10. Classification and distribution of the main lipids in wheat flour. (From Pomeranz and Chung, 1978.)

The free lipids can be fractionated according to their elution from a silicic acid column. About 70% of the total free lipids can be eluted with chloroform; and they form what is arbitrarily called the nonpolar fraction, containing TG as a major component. The residual 30% free lipids can be eluted from the column with a more polar solvent, such as methanol, and comprise a mixture of free polar lipids. Among the free polar lipids, about two-thirds are glycolipids, containing DGDG as a major component, and one third are phospholipids, with PC as a major component.

About 0.6–1.0% bound lipids can be extracted from flour with WSB after PE extraction. Bound lipids contain about 30% nonpolar and 70% polar lipids. Bound polar lipids are rich in phospholipids, with LPC as a major component. Because glycolipids are important in bread making, it is important to distinguish clearly whether they are free or bound. Although percentagewise the free polar lipids are richer in glycolipids than the bound polar lipids, the actual amounts of both glycolipids and phospholipids are higher in the bound polar than in free polar lipids (Figure 9-10).

Morrison et al (1975) divided flour lipids into those inside starch granules, which are true starch lipids, and all other lipids outside the starch granules, which are called nonstarch lipids. Even polar solvents consisting of alcohol-water mixtures, such as WSB, extract mainly nonstarch lipids at room temperature. Starch lipids can only be extracted efficiently with hot N-butanol-water (65:35) or hot WSB. Starch lipids are almost exclusively monoacyl lipids, and 86–94% of the total starch lipids are lyso-phospholipids, with LPC as the chief constituent (Morrison, 1978).

Lipids in Bread Making

In tests to demonstrate the role of lipids in bread making, gluten was isolated from both untreated and defatted flours and then remixed with starch and water-soluble components in different proportions. Loaves were then baked from these doughs. At each gluten level, loaf volume was higher in bread baked from untreated (ie, lipid-containing) than from defatted flour (Figure 9-11) (Chiu et al, 1968). Studies with defatted flours have shown that nonpolar lipids are deleterious and that polar lipids are beneficial in bread making (Figure 9-12). The deleterious effects of nonpolar lipids can be counteracted by polar lipids, and the effect on loaf volume has been shown to depend on the levels of nonpolar and polar lipids as well as on their ratio. The effects of lipids demonstrated that they are minor components that have major importance in bread making.

The significance of protein-lipid interactions was examined by Chung and Pomeranz (1978). Using 0.05 N acetic acid, they determined the extractability of proteins from untreated flours of good and poor baking quality and from the same flours after removal of fat by acetone and isopropanol. Next, they dialyzed, lypophilized, and fractionated the protein extracts by gel filtration on a Sephadex G-100 column. The whole extracts and their fractions were characterized by starch and sodium dodecyl sulfate polyacrylamide gel electrophoresis. Both the nondefatted and the defatted poor (Chiefkan/Tenmarq) flours showed higher protein extractability than did the good flours (Shawnee). More proteins were extracted from the untreated good and poor control flours than from the corresponding defatted flours. In elution profiles from the gel column, all protein extracts showed similar retention times for the glutenin, gliadin, and water-soluble fractions. Quantitatively, the glutenins were the most

Figure 9-11. Microloaves (10 g flour) baked from untreated (top) and defatted (bottom) flours. From left to right: original flour containing 13% protein; mixture of starch and water solubles to give 10, 13 and 16% protein respectively. (From Chiu et al, 1968.)

variable among the gel-separated fractions in the acetic acid extracts from the flours. Whereas concentrations of gliadins were similar in all extracts, concentrations of glutenins depended on flour samples and the defatting solvent. Defatting decreased glutenin/gliadin ratios. The glutenin and gliadin fractions of strong and weak flours and of nondefatted and defatted flours differed in number and intensity of electrophoretic bonds.

In a subsequent study, Chung and Pomeranz (1979) examined the binding of acid-soluble proteins from untreated and defatted good (Shawnee, C.I. 14157) and poor (Chiefkan/Tenmarq, KS501097) baking quality flours to phenyl Sepharose CL-4B by bath and column elution techniques. Batchwise elution of the proteins from the hydrophobic gel with five solvents showed that total absorbance at 280 nm of the eluates was higher for the poor than for the good baking quality flour. Acid-soluble proteins of defatted good and poor baking flours differed little from those of nondefatted flours in hydrophobic binding capacity. Glutenins from isopropanol-defatted poor baking quality flour (KS501097) were less hydrophobic, and gliadins were more hydrophobic, than glutenins and gliadins from the good baking quality flour. The difference in apparent hydrophobic interaction was more pronounced for glutenin than for gliadin. The relationship between the amount of protein eluted and the amount of protein absorbed by the hydrophobic gel varied with source and concentration of protein. The results indicate that acid-soluble proteins

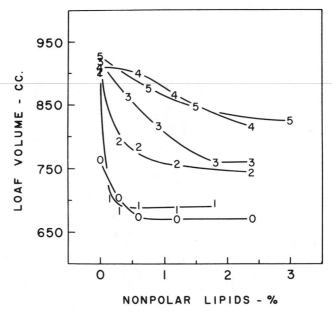

Figure 9-12. Effects of nonpolar and polar lipids on loaf volume of bread from petroleum-ether-extracted flour baked with shortening and with various combinations of nonpolar and polar lipids (numbers 1–5 denote 0.1–0.5 g polar lipids per 100 g flour). (From Daftary et al, 1968.)

from good and poor bread-making flours differ in apparent hydrophobic properties from those of nondefatted flours.

Glycolipids

Olcott and Mecham postulated in 1947 that a gluten-lipid complex is formed during dough mixing or gluten preparation. However, the precise structure of that complex and the manner in which it contributes to bread making have been elucidated only recently as a result of intensive investigations of wheat flour glycolipids, which, as the name implies, are complexes of carbohydrates and lipids. Structural formulas of glycolipids are given in Figure 9-13. Because glycolipids combine the polar features of polyols (carbohydrates) with the lipophilic behavior of long aliphatic chains (fatty acids), some of them show considerable solubility in lipid solvents, but they also form aqueous solutions. The combination of polar and nonpolar properties suggests that glycolipids might be associated with lipid-aqueous interfaces. Noncharged surface-active glycolipids apparently interact with structural proteins by hydrophobic bonding and, together with other

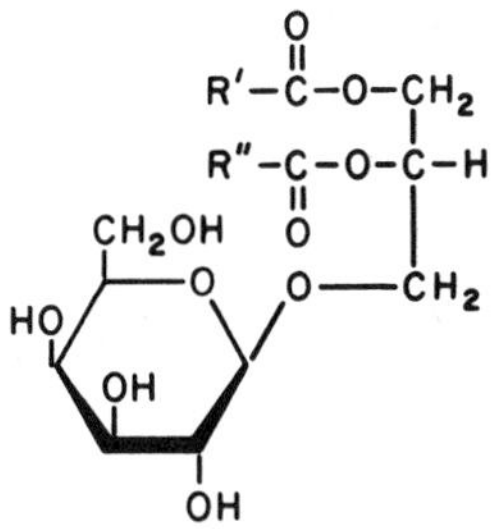

I,2-Diacyl-3-β-D-galactosyl-   I,2-Diacyl-3-[6-ᴬ-D-galactosyl-
sn-glycerol   β-D-galactosyl]-sn-glycerol
(Monogalactosyldiglyceride)   (Digalactosyldiglyceride)

(I)   (II)

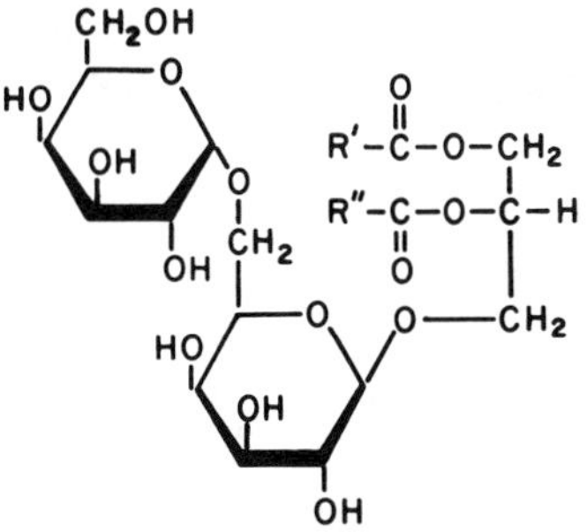

Figure 9-13. Structural formulas of glycolipids.

complex lipids, may be responsible for interactions between lipid micelles and proteins in chloroplasts.

Polar Lipids and Bread Making

Chung et al (1982) extracted lipids (with petroleum ether) from 21 samples of hard red winter wheats and 23 samples of experimentally milled straight-grade flours that varied in bread-making potential. Wheat protein content varied from 11.5% to 15.7%, flour mixing time from 7/8 to 9 minutes, and loaf volume (LV) per 100 g of flour from 523 to 1,053 cc. The total lipids from 10 g of flour dry basis (db) were fractionated into polar lipids (PL) and nonpolar lipids (NL); total lipids were analyzed colori-metrically for carbohydrates, mainly galactose (GAL). PL content varied from 14.8 to 28.1 mg per 10 g of wheat and from 10.6 to 27.3 mg per 10 g of flour; NL/PL ratios were 6.31–11.32 for wheat and 2.47–6.91 for flour; lipid GAL ranged from 1.61 to 5.49 mg and from 2.64 to 5.61 mg in 10 g of wheat and flour, respectively. Significant linear correlations were found between LV and the following variables; PL content (r = 0.877 for wheat and 0.888 for flour), NL/PL ratios (r = 0.902 for wheat and 0.907 for flour), and lipid GAL (r = 0.745 for wheat and 0.905 for flour). PL, NL/PL ratio, and lipid GAL were curvilinearly related to mixing time requirement. The correlation coefficients of LV with PL, NL/PL ratio, and lipid GAL generally were somewhat improved when LV and lipid contents were corrected to a constant protein basis. The data indicate that the quantity of PL or galactolipids occurring naturally in wheat is related to bread-making (functional) properties and may govern or be closely related to other factors that govern functional properties of good and poor varieties of wheat. The highly significant correlations point to the potential usefulness of PL, NL/PL ratio, and lipid GAL for estimating LV potential of hard red winter wheat flours.

INTERACTION BETWEEN LIPIDS AND OTHER FLOUR COMPONENTS

In washing out gluten or in dough making, one-third to two-thirds of the wheat flour free lipids (some free nonpolar and practically all polar components) are bound and subsequently unextractable with petroleum ether (Chiu and Pomeranz, 1966; Mecham, 1971). Extractability of lipids depends on several factors including particle size, composition and age of the wheat flour, dough composition (including amounts and types of flour and added lipids), water content of extractants and flour lipids, presence of shortening and/or surfactants, work input and atmosphere in dough mixing, and stage in bread production (Pomeranz, 1980a–d; Chung and Pomeranz, 1981).

Starch lipids can be excluded from consideration of binding during dough mixing, because starch lipids are bound very tightly as inclusion compounds with amylose and consequently are not available for reacting with gluten proteins (Acker et al, 1968).

Available information indicates that in dough, polar lipids (glycolipids and phospholipids) interact mainly with gluten rather than with the soluble wheat flour proteins. This is of significance, as gluten is the skeleton or framework of wheat flour dough and is responsible for gas retention, which is required for the production of light, yeast-leavened products. In the baked bread, much of the interaction is with starch. This is of significance, as the starch governs, to a large extent, freshness retention of the baked bread.

MODEL SYSTEMS

Methods of study, bonds, and mechanisms of interaction between lipids and proteins and starch are summarized in Table 9-6; some of the proposed models are shown in Figure 9-14. Almost a quarter century ago, Hess (1954) proposed, on the basis of x-ray, electron microscope, and optical measurements, a structural relationship of proteins, lipids, and starch in wheat flour, in which wedge protein deposits are surrounded by a lipid layer, beyond which lie adhesive layers and still further starch granules. Hess (1954) proposed a model (Figure 9-14) in which the adhesive protein is bound to starch through a lecithin layer.

Table 9-6. Bonds in Glycolipids and Wheat Flour Macromolecule Complexes

Method of study	Type of bond between glycolipid and		
	Starch	Gliadin	Glutenin
Solvent extraction of gluten proteins	—	Hydrogen	Hydrophobic
Lipid binding in starch dough	Hydrogen	—	—
Infrared	Hydrogen	Van der Waals, hydrogen	Van der Waals, hydrogen
NMR	Hydrogen, some induced dipole interaction	—	Hydrophobic and hydrogen
Autoradiography	Strong interaction in bread	—	Interaction in dough
Baking test	Hydrophobic and hydrogen bonds are essential for improvement in bread making		

From Wehrli and Pomeranz, 1969.

In 1957 Traub et al reported that a 46-Å spacing observed in x-ray studies of wheat and flour was due to phospholipids associated with protein fibers in the form of bimolecular leaflets. Grosskreutz (1961) studied the structure of wheat gluten by electron microscope and x-ray techniques and proposed a lipoprotein model involving a bimolecular lipid layer structure (the Danielli and Dawson type of membrane) as shown in Figure 9-14. Grosskreutz showed that the proteins in moist gluten consist of folded, polypeptide chains in the alpha-helix configuration, arranged into flat platelets of the order of 70 Å thick. Extraction of the phospholipids did not affect the basic platelets but seriously impaired their ability to bond into sheets capable of sustaining large plastic deformation. From x-ray evidence of the phospholipid structure in gluten, Grosskreutz (1961) deduced (1) that proteins formed well-oriented bimolecular leaflets of the type found in myelin, (2) that lipoprotein occupies about 2–5% of the elastic gluten structure, and (3) that protein chains are bound to the outer edge of a phospholipid bimolecular leaflet array (probably via salt-type linkage between acidic groups of the phospholipid and the basic protein groups).

Hoseney et al (1970) found that free polar lipids (principally glycolipids) are bound to the gliadin proteins by hydrophilic bonds and to the glutenin proteins by hydrophobic bonds, as shown in Figure 9-14. In unfractionated gluten, the lipid is apparently bound to both protein groups at the same time. The simultaneous binding of polar lipids to both gliadin and glutenin may contribute to the gas-retaining ability of gluten.

Wehrli and Pomeranz (1970a) supported the model proposed by

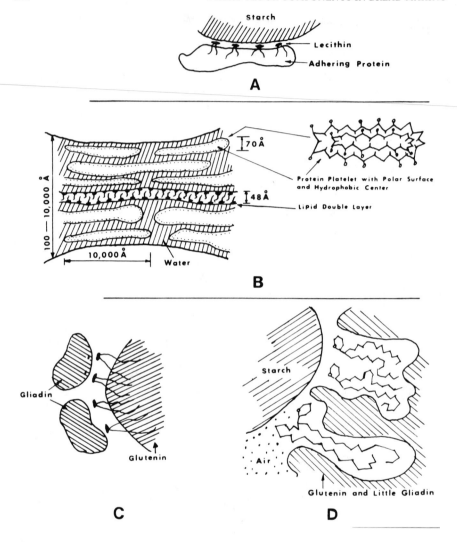

Figure 9-14. Proposed models of the complex formed in breadmaking: (A) starch-lipid-adhesive protein complex in flour, (Hess, 1954); (B) lipoprotein model by Grosskreutz (1951); (C) gliadin-glycolipid-glutenin complex by Hoseney et al (1970); (D) starch-glycolipid-gluten complex by Wehrli (1969).

wheat flour macromolecules. They investigated by infrared and nuclear magnetic resonance (NMR) spectroscopy complexes between galactolipids and raw starch, gelatinized starch, gliadin, and glutenin. Infrared spectroscopy indicated hydrogen bonds between glycolipids and gelatinized starch or gluten components and Van der Waals bonds between glycolipids and gluten components. The NMR spectra showed an inhibition of the methylene signal of glycolipids by glutenin, indicating hydrophobic bonding.

In studying the distribution of lipids between gliadin and glutenin of wheat proteins, Olcott and Mecham (1947) found that more than 80% of the lipids were associated with the glutenins fractionated from gluten by acid precipitation. Ponte et al (1967), however, found that the major part of lipids was associated with the gliadins fractionated from gluten by 70% alcohol. Later, Hoseney et al (1969b) found that association of lipids with gliadin or glutenin depended on the condition used to fractionate glutenins and gliadins from gluten.

Simmonds and Wrigley (1972) reported that much less protein was extractable with 6 M urea from gluten than from "storage protein" that had been prepared with organic solvents and was depleted in lipid. So named because it exists between the starch granules in the mature grain, "storage protein" forms gluten by interacting with wheat flour components in the presence of water. Reconstitution experiments, which involved the wetting of storage proteins in the presence of restored flour lipids, suggested that the reduced protein solubility in gluten was due to lipid-protein association during dough formation. Charbonnier (1973) showed that solubility of proteins in 55% ethanol depended on interaction with lipids. The proteins were rendered insoluble after treatment of the flour with ethanol-ether-water (2:2:1, by volume). Based on the studies of fractions from Sephadex G-100 column chromatography, electrophoretic characterization, and amino acid analyses of fractionated column chromatography eluates, it was deduced that the decrease in protein extractability was mainly due to glutenins.

Chung et al (1977) also found that the decrease in protein extractability in 3 M urea in a pyrophosphate buffer, pH 7.0, depended on amounts of lipids removed from a poor bread-making quality flour (Chiefkan/ Tenmarq Cfk/Tm, KS 501097) and from a good-quality flour (Shawnee); protein extractability was determined in flours that were previously defatted with Skelly B, benzene, acetone, and 2-propanol. The amount of lipids extracted during defatting ranged from 0.88% to 1.41% for Cfk/Tm and from 1.05% to 1.47% for Shawnee flour (least with Skelly B, next with benzene, and most with 2-propanol). The increase in total extractable lipids was primarily due to an increase in extracted polar lipids. Regardless of lipid contents in flour, more proteins were extracted from the poor- than from the good-quality flour. Similarly, the absorbance of protein extracts of Shawnee flour in 0.01 M sodium pyrophosphate (Na-pp) and 0.005 N and 0.05 N acetic acid (AcOH) decreased curvilinearly with increasing amounts of lipids removed from flour by Skelly B, benzene, acetone, and 2-propanol.

Reconstituting the defatted flours with the lipids extracted by the various solvents restored protein solubility, but to different extents depending on the lipid extractant (Skelly B, benzene, acetone, and 2-propanol) and protein extractant (0.001 M Na-pp buffer, 0.005 N and 0.05 N AcOH). The solubility of nongluten proteins (soluble in Na-pp) was completely restored by lipid reconstitution except for the 2-propanol treatment. Protein solubility in 0.005 N AcOH was restored completely by lipid addition for Skelly B and benzene treatment but only partially for the acetone and 2-propanol treatments. For 0.05 N AcOH-soluble proteins, presumably gluten proteins, restoration of protein solubility was complete only for Skelly treatment, which involved free lipids only. The restoration was almost irreversible with 2-propanol treatment, which involved bound as well as free lipids. Only nonpolar lipids showed solubilizing effects on proteins of both poor- and good-quality flours; polar lipids had no solubilizing effects or a slight insolubilizing effect (protein aggregation) for good flour when they were separately reconstituted with defatted flours by blending in the dry state.

LIPID-PROTEIN-STARCH INTERACTIONS

Wehrli and Pomeranz (1970b) studied interactions that take place between starch and gluten proteins in dough and bread. For that purpose tritium-labeled galactosyldidecanoyl-glycerol was synthesized (Wehrli and Pomeranz, 1969a,b). Sections prepared from dough and bread containing the labeled galactolipid were studied by autoradiography. In the dough, the galactolipid was distributed in the gluten and, to a limited extent, in the starch; in the bread, most of the galactolipid was in gelatinized (by oven heat) starch granules and formed a complex, which seemed to be responsible for the improved retention of freshness in bread baked with glycolipids. Wehrli (1969) proposed that glycolipids improved gas retention and, hence, loaf volume, by sealing the gas cells, presumably by complexing between the swelling starch and the coagulating proteins, as shown in Figure 9-14.

DeStefanis et al (1977) drew a conclusion similar to that drawn by Wehrli and Pomeranz (1970b)—that the lipids bound to proteins during dough mixing are translocated and bound to starch during the baking. They found that at the sponge stage, binding between added surfactants and the major flour components was not extensive. The additives were firmly bound to the gluten proteins during dough mixing and strongly bound to the starch by complexing with both amylose and amylopectin fractions in bread. They concluded, after studying model systems, that two concurrent phenomena occurred during baking: (1) The bonds between the gluten proteins and the additives weakened progressively (protein denaturation) as the dough temperature increased; and (2) as starch gelatinized above 50°C, the additives, weakly bonded to proteins, readily formed a strong complex with starch and could thus be translocated from proteins to starch.

In addition to the additives, triglycerides, free fatty acids, and lysophosphatidyl choline were also bound to the starch.

WHEAT FLOUR LIPIDS AND THE SHORTENING RESPONSE

Interaction between lipids and wheat flour macromolecules was followed in studies on the mechanism of the "shortening effect"—ie, the increase in loaf volume and improvement of crumb grain from the addition of 1–3% shortening or hardened vegetable fat. The shortening effect was reviewed by Bell et al (1977), who visualized two mechanisms of shortening effects: chemical and physical. The chemical effect would involve lipid oxidation; the mechanism was considered to be inoperative, or at least insignificant, in bread making. The following physical effects were reviewed: lubrication, sealing, foam formation, involvement of hydrogen and hydrophobic bonds, and delayed carbon dioxide release. Bell et al (1977) showed that the rate of carbon dioxide release was faster in doughs baked without than in doughs baked with shortening. Fat in dough increased gas retention in the initial stage of rapid expansion. The final loaf volume depends on the permeability of the dough to carbon dioxide in the earliest stages of baking. Bell et al suggested that the difference in carbon dioxide release might explain the shortening response. They postulated that physical mechanisms account for the increased loaf volume owing to shortening in the dough. Loaf volume increases when sufficient solid shortening components remain free in the dough. The free components are especially important in the initial stage of rapid dough expansion—the so-called oven spring. The free shortening components presumably facilitate the production of oriented structures in dough. The structures persist even when the temperature exceeds the melting point of the fat, and they seem to favor gas retention in the early stages of baking. The concept of delayed carbon dioxide release mechanism agrees with several well-established facts, such as the significance of free lipids and the critical oven spring. It seems, however, that attributing the differences in shortening effect among wheat flours to delayed carbon dioxide release alone is an oversimplification.

Wheat flour lipids and their role in bread making have been the subject of several comprehensive reviews (for a detailed list, see Chung and Pomeranz, 1977, and Chung et al, 1978). It is well established that in PE-defatted flours, nonpolar lipids are detrimental and that polar lipids, especially glycolipids, are effective improvers. The roles of polar and nonpolar lipids are modified, however, by the addition of shortening and/or surfactants. MacRitchie and Gras (1973) emphasized that if baking formulations include shortening or other lipid additives, the net effect of natural flour lipids may be obscured. On the other hand, it is difficult to determine the effects of adding lipids to an untreated flour because of the presence of natural lipids. Studies in the author's laboratories indicate that interactions among wheat flour components are important in the shortening response (Pomeranz and Chung, 1978).

Those studies showed that the shortening response is affected to a large extent by wheat flour quality, which is governed by the inherent wheat flour components (presumably gluten proteins), and is modified by the removal of wheat flour lipids. According to Wehrli (1969), since shortening in the absence of glycolipids is detrimental to loaf volume, it probably interferes with the formation of a stable membrane between starch and proteins. Such a membrane, presumably, can form only if the starch surface is covered with glycolipids. The shortening response is further affected by the nature of wheat flour lipids that are removed (Chung et al, 1979).

Shortening had a detrimental effect on loaf volume and crumb grain in the absence of native flour lipids, and severity of the detrimental effects was linearly related to the amount of polar lipids removed from the defatted flour. It was concluded that in flour containing no nonpolar lipids, loaf volume was governed by the quantity of polar lipids, shortening, and interaction of shortening and polar lipids.

PROTEIN-ENRICHED BREAD

Experiments with glycolipids synthesized in the author's laboratory (Wehrli and Pomeranz, 1969b) indicated that a certain hydrophilic-hydrophobic balance is required for best effects in bread making (Pomeranz, 1971, 1973). The significance of those findings with synthetic compounds was threefold: They confirmed the role of natural flour glycolipids in bread making; they opened the possibility of synthesizing glycolipids superior to those found in flour; and they provided the basis for the production of protein-enriched bread through the addition of small amount of glycolipids (Pomeranz et al, 1969a,b). (See also Chapter 14)

Wheat flour contains fairly large concentrations of proteins, but the proteins are not well balanced in amino acids and are particularly low in lysine. The quality of protein poses practically no problem if bread is part of a diet rich in protein foods of high biological value, such as milk, eggs, meat, fish, and legumes. In countries where high-protein foods are not generally available, however, wheat flour enriched with plant and animal proteins, including defatted oilseed flours, edible yeasts, and fish flour, might be used to improve nutrition in low-priced, protein-enriched breads and other baked products.

Without the addition of extra glycolipids even relatively low levels, of about 5%, of lysine-rich protein supplements (ie, soy flour) decrease loaf volume, impair crumb texture, and reduce shelf life. If small amounts of glycolipids are added, however, at least 10% soy flour, containing about 50% protein, can be included in the formula without impairing consumer acceptance of the bread. Soy protein contains about 6% lysine (compared to only 2% in wheat flour), so the addition of 10% soy flour easily doubles the amount of nutritionally limiting lysine that would be supplied by flour. The use of glycolipids in the production of acceptable, nutritionally improved bread is particularly promising, because little or no changes would be required in bread-making processes, formulations, schedules, or equipment. The processes for production of protein-enriched bread and for the synthesis of glycolipids have been granted public patents (Pomeranz and Finney, 1972; Pomeranz and Wehrli, 1973); and the patents are assigned to the U.S. Department of Agriculture, which is authorized to grant royalty-free licenses for their use.

The functional (bread-making) properties of commercially available sucrose esters were improved by Chung et al (1976). As much as 4% sucrose tallowate previously was required to restore bread-making quality of high-protein breads (Pomeranz et al, 1969a,b); but now, acceptable high-protein breads can be made with as little as 0.5% sucrose esters. The effect on loaf volume of as little as 0.25–0.50% of the improved sucrose esters, added to wheat flour alone, is eqivalent to the effects of 3% shortening. On the other hand, addition of unimproved sucrose tallowate, even to wheat flour alone, is deleterious to loaf volume and crumb grain. Effectiveness of sucrose esters is proportional to their monoester contents (Chung et al, 1976). Their effectiveness can be further enhanced by removal of undesirable nonpolar and other components together with, presumably, di- and triester fractions.

An "ideal" nutritionally improved high-protein bread should be low priced; require no special formulations, schedules, and equipment to produce; and be acceptable to the consumer with regard to overall eating qualities, including freshness retention. Although it is possible to produce a high-protein bread that is superior to the one that was produced about 15 years ago, it has not been possible to meet satisfactorily all the above requirements for ideal bread. Studies from the author's laboratories (Pomeranz et al, 1977) showed that the use of flour from germinated soybeans (in the presence of small amounts of surfactants such as sucrose esters or sodium stearoyl lactylate) overcomes some of the difficulties. Scanning electron micrographs indicated differences in the size and shape of particles constituting soy milk and flours from chemically treated, high-temperature-treated, and germinated heat-treated soybeans. The last-mentioned treatment modified the particles the most. Of the three flours, nitrogen solubility index was highest for the flour milled from germinated soybeans. When bread was baked with each of the four soy products in a no-sugar formula, soy milk and high-temperature-treated soy flour produced unacceptable bread with regard to loaf volume and crumb grain. Bread baked from 90 g wheat flour and 10 g flour from the chemically treated or germinated product, in the presence of 0.50 g sucrose palmitate or 0.50 g sodium stearoyl lactylate plus 3 g vegetable shortening, was consumer-acceptable with regard to loaf volume, crumb grain, crumb color, freshness retention, taste, and flavor. Wheat flour enriched with flour from germinated soybeans could well be the solution to the problems associated with producing low-priced, nutritionally improved, protein-enriched bread.

BIBLIOGRAPHY

Acker, L.; Schmitz, H. J.; Hamza, Y. *Getreide Mehl Brot* **1968**, *18*, 45.

Auermann, L. Y. A.; Kretovich, V. L.; Polandova, R. D. *Appl. Biochem. Microbiol.* **1965**, *1*, 44–50.

Beckwith, A. C.; Wall, J. S.; Dimler, R. J. *Arch. Biochem. Biophys.* **1963**, *103*, 319.

Belderok, B. *Getreide Mehl* **1967**, *17*, 20–23.

Bell, B. M.; Daniels, D. G. H.; Fisher, N. *Food Chem.* **1977**, *2*, 57–60.

Bigelow, C. C. *J. Theor. Biol.* **1967**, *16*, 187–211.

Bloksma, A. H. *Soc. Chem. Ind. Lond. Monogr.* **1968**, *27*, 135–166.

Bushuk, W. *Bakers Digest* **1966**, *40*(5), 38–40.

Casier, J. P. J.; Soenen, M. *Getreide Mehl* **1967**, *41*, 46.

Casier, J. P. J.; Stephan, H.; De Paepe, G.; Nicora, L. *Brot Gebaeck* **1969**, *12*, 231.

Casier, J. P. J.; De Paepe, G.; Bruemmer, J. M. *Getreide Mehl Brot* **1973**, *27*, 36.

Charbonnier, L. *Biochemie* **1973**, *55*, 1217.

Chiu, C. M.; Pomeranz,Y. *J. Food Sci.* **1966**, *31*, 753.

Chiu, C. M.; Pomeranz, Y.; Shogren, M. D.; Finney, K. F. *Food Technol.* **1968**, *22*, 1157.

Chung, O. K.; Pomeranz, Y. *Bakers Digest,* **1977** *51*(5), 32–44, 153.

Chung, K. H.; Pomeranz, Y. *Cereal Chem.* **1978**, *55*, 230.

Chung, K. H.; Pomeranz, Y. *Cereal Chem.* **1979**, *56*, 196.

Chung, O. K.; Pomeranz, Y. *Bakers Digest* **1981** 55(5), 38–50, 55, 56, 97.

Chung, O. K.; Pomeranz, Y.; Goforth, D. R.; Shogren, M. D.; Finney, K. F. *Cereal Chem.* **1976**, *53, 615–626*.

Chung, O. K.; Pomeranz, Y.; Finney, K. F.; Shogren, M. D. *J. Am. Oil Chem. Soc.* **1978a**, *55*, 635–641.

Chung, O. K.; Pomeranz, Y.; Finney, K. F. *Cereal Chem.* **1978b**, *55*, 598–618.

Chung, O. K.; Pomeranz, Y.; Finney, K. F. *Cereal Chem.* **1982**, *59*, 14–20.

Daftary, R. D.; Pomeranz, Y.; Shogren, M. D.; Finney, K. F. *Food Technol.* **1968**, *22*, 322–330.

DeStephanis, V. A.; Ponte, J. G., Jr.; Chung, F. H.; Ruzza, N. A. *Cereal Chem.* **1977**, *54*, 13–24.

Fincher, D. B.; Stone, B. A. *Aust. J. Biol. Sci.* **1974**, *27*, 117.

Fincher, G. B.; Sawyer, W. H.; Stone, B. A. *Biochem. J.* **1974**, *139*, 535–545.

Finney, K. F. *Cereal Chem.* **1943**, *20*, 381–396.

Finney, K. F. In "Cereals 78' Better Nutrition for the World's Millions." Pomeranz, Y. Ed. Am. Assoc. Cereal Chem.; St. Paul, MN. **1978**.

Finney, K. F.; Barmore, M. A. *Cereal Chem.* **1948**, *25*, 291–312.

Finney, K. F.; Yamazaki, W. T. In "Wheat and Wheat Improvement." Quisenberry, K. S.; Reitz, L. P., Eds. Am. Soc. Agron.: Madison, WI; **1967**.

Finney, K. F.; Jones, B. L.; Shogren, M. D. *Cereal Chem.* **1982**, *59*, 449–453.

Frank, H. S.; Evans, M. W. *J. Chem. Phys.* **1945**, *13*, 507–532.

Goldstein, S. *Mitt. Gebiete Lebensm. Hyg. (Bern)* **1957**, *48*, 87–93.

Gracza, R.; Greenberg, S. I. *Cereal Chem.* **1963**, *40*, 51–61.

Grosskreutz, J. C. *Cereal Chem.* **1961**, *38*, 336–349.

Hess, K. *Kolloid Z.* **1954**, *136*, 84.

Hoseney, R. C. *Food Technol.* **1984**, *38*(10), 114–117.
Hoseney, R. C.; Finney, K. F.; Shogren, M. D.; Pomeranz, Y. *Cereal Chem.* **1969a**, *46*, 126–135.
Hoseney, R. C.; Finney, K. F.; Pomeranz, Y.; Shogren, M. D. *Cereal Chem.* **1969b**, *46*, 495–502.
Hoseney, R. C.; Finney, K. F.; Shogren, M. D.; Pomeranz, Y. *Cereal Chem.* **1969c**, *46*, 117–125.
Hoseney, R. C.; Finney, K. F.; Pomeranz, Y. *Cereal Chem.* **1970**, *47*, 135–140.
Hoseney, R. C.; Finney, K. F.; Shogren, M. D.; Pomeranz, Y. *Cereal Chem.* **1971**, *48*, 191–201.
Jankiewicz, M.; Pomeranz, Y. *Cereal Chem.* **1965**, *42*, 37–43.
Jones, B. L.; Finney, K. F.; Lookhart, G. L. *Cereal Chem.* **1983**, *60*, 276–280.
Kauzmann, W. In "The Mechanism of Enzyme Action." John Hopkins University Press; Baltimore; **1954**.
Kim, S. K.; D'Appolonia, B. L. *Cereal Chem.* **1977**, *54*, 225–229.
Klotz, I. M. *Biol.* **1960**, *13*, 25–48.
Krull, L. H.; Wall, J. S. *Biochemistry* **1966**, *5*, 1521–1527.
Krull, L. H.; Wall, J. S.; Zobel, H.; Dimler, R. J. *Biochemistry* **1965**, *4*, 626–633.
Kuchumova, L.; Strelnikova, M. M. *Prikl. Biokhim. Mikrobiol.* **1967**, *3*(5), 592–596.
Lee, J. W. *Proc. R. Aust. Chem. Inst.* **1975**, *42*(2), 33.
MacMurray, T. A.; Morrison, W. R. *J. Sci. Food Agric.* **1970**, *21*, 520.
MacRitchie, F.; Gras, P. W. *Cereal Chem.* **1973**, *50*, 292–302.
Mauritzen, C. M. *Cereal Chem.* **1967**, *44*, 170–182.
Mauritzen, C. M.; Stewart, P. R. *Aust. J. Biol. Sci.* **1965**, *18*, 173.
Mecham, D. K. In "Wheat Chemistry and Technology." Pomeranz, Y. Ed. Am. Assoc. Cereal Chem.: St. Paul, MN. **1971**, pp. 393–451.
Mifflin, B. J.; Field, J. M.; Shewry, P. R. In "Seed Proteins," Proc. Phytochem. Soc. Symp., Versailles, France; **1981**.
Morrison, W. R. *Adv. Cereal Sci. Technol.* **1978**, *2*, 221–348.
Morrison, W. R.; Mann, D. L.; Wong, S.; Coventry, A. M. *J. Sci. Food Agric.* **1975**, *26*, 507.
Neukom, H.; Providoli, H.; Gremli, H.; Hui, P. A. *Cereal Chem.* **1967**, *44*, 238–244.
Olcott, H. S.; Mecham, D. K. *Cereal Chem.* **1947**, *24*, 407–414.
Pomeranz, Y. In "Wheat Chemistry and Technology." Pomeranz, Y., Ed. Am. Assoc. Cereal Chem.: St. Paul, MN, **1971**, pp. 585–674.
Pomeranz, Y. *Adv. Food Res.* **1973**, *20*, 153–188.
Pomeranz, Y. *Bakers Digest* **1980a**, *54*(1), 20; (2), 12–25.
Pomeranz, Y. *Getreide Mehl Brot.* **1980b**, *34*, 11.
Pomeranz, Y. *Cereal Foods World* **1980c**, *25*, 656–662.
Pomeranz, Y. In "Cereals for Food and Beverages." Inglett, G. E.; Munck, L., Eds. Academic Press, Inc.: New York, **1980d**, pp. 201–232.
Pomeranz, Y.; Chung, O. K. *J. Am. Oil Chem. Soc.* **1978**, *55*, 285–289.
Pomeranz, Y.; Finney, K. F. *U.S. Patent Office, July 25*, **1972**, No. 3, 679, 433.
Pomeranz, Y.; Wehrli, H. P. U.S. Patent Office, April 24, **1973**, No. 3, 729, 461.
Pomeranz, Y.; Rubenthaler, G. L.; Finney, K. F. *Food Technol.* **1966**, *20*, 105–108.
Pomeranz, Y.; Shogren, M. D.; Finney, K. F. *Cereal Chem.* **1969a**, *46*, 503–511.
Pomeranz, Y.; Shogren, M. D.; Finney, K. F. *Cereal Chem.* **1969b**, *46*, 512–518.
Pomeranz, Y.; Shogren, M. D.; Bolte, L. C.; Finney, K. F. *Bakers Digest* **1976**, *50*(1), 35–40.
Pomeranz, Y.; Shogren, M. D.; Finney, K. F. *J. Food Sci.* **1977**, *42*, 824–842.
Pomeranz, Y.; Bolling, H.; Zwingelberg, H. *J. Cereal Sci.* **1984**, *2*, 137–143.
Ponte, J. G., Jr.; De Stephanis, V. A.; Cotton, R. H. *Cereal Chem.* **1967**, *44*, 427–435.
Ponte, J. G.; De Stephanis, V. A.; Titcomb, S. T.; Cotton, R. H. *Cereal Chem.* **1967**, *44*, 211–220.
Schachman, H. K. *Cold Spring Harbor Symp. Quant. Biol.* **1963**, *28*, 409–430.
Simmonds, D. H.; Wrigley, C. W. *Cereal Chem.* **1972**, *49*, 317–323.
Sollars, W. F. *Cereal Chem.* **1956**, *33*, 121–128.
Toledo, R.; Steinberg, M. P.; Nelson, A. I. *J. Food Sci.* **1968**, *33*, 315–316.
Traub, W.; Hutchinson, J. B.; Daniels, D. G. H. *Nature* **1957**, *179*, 769.
Tsen, C. C.; Bushuk, W. *Cereal Chem.* **1968**, *45*, 58–62.
Vakar, A. B.; Pumpyanski, A.; Semenova, L. V. *Appl. Biokhim. Mikrobiol.* **1965**, *1*, 1–13.
Villegas, E.; Pomeranz, Y.; Shellenberger, J. A. *Cereal Chem.* **1963**, *40*, 694–703.
Wall, J. S. In "Proteins and Their Reactions." Avi Publ. Co.; Westport, CT; **1964**, pp. 315–341.
Wall, J. S. *Cereal Sci. Today* **1971**, *16*, 412–417, 429.

Wall, J. S. *Cereal Foods World* **1979a**, *24*, 288–292, 313.
Wall, J. S. In "Recent Advances in the Biochemistry of Cereals." Laidman, D. L.; Wynn, R. G.,
    Eds. Academic Press; London; **1979b**, pp. 275–311.
Wehrli, H. P. Ph.D. dissertation. Kansas State University; **1969**.
Wehrli, H. P.; Pomeranz, Y. *Chem. Phys. Lipids* **1969c**, *3*, 357–370.
Wehrli, H. P.; Pomeranz, Y. *Cereal Chem.* **1970a**, *47*, 160–166.
Wehrli, H. P.; Pomeranz, Y. *Cereal Chem.* **1970b**, *47*, 221–224.
Whitney, P. L.; Tanford, C. *J. Biol. Chem.* **1962**, *237*, PC 1737.
Woychick, J. H.; Boundy, J. A.; Dimler, R. J. *J. Agric. Food Chem.* **1961**, *9*, 307–310.
Yamazaki, W. T. *Cereal Chem.* **1955**, *32*, 26–37.
Yasunaga, T.; Bushuk, W.; Irvine, G. N. *Cereal Chem.* **1968**, *45*, 269–279.

Chapter 10

Dough and Bread Structure

The structure of wheat-flour doughs and the development of doughs into bread have been studied by the use of transmission electron microscopy (TEM) by Simmonds (1975) and Khoo et al (1975). Sandstedt et al (1959) pointed out difficulties in the preparation of thin sections for examination by light microscopy (LM); preparation of dough and very thin sections for TEM is much more difficult. Simmonds (1975) described some changes that occurred during the conversion of flour to dough and suggested that two types of inclusions occurred in the protein phase of the dough: Type I inclusions were irregular, stained densely, and presumably had been formed from the endoplasmic reticulum; type II inclusions were spherical, had not been formed in doughs from defatted flours, and were therefore thought to be lipid-rich. Khoo et al (1975) briefly described various stages of bread making—freshly mixed dough, fermented and proofed dough, and fully baked bread. During baking, the protein fraction changed little (in microscopically visible structures), but the starch granules, particularly the large ones, became gelatinized. The unique gluten-forming properties of proteins are known to contribute to the excellent baking quality of wheat flour in production of white bread. The gluten forms a film containing starch granules and gas bubbles. Starch participates in this process as the granules take on various shapes that fit around vacuoles and give a desired crumb grain and texture. During dough mixing, fully hydrated and developed gluten proteins form a web of fibrils with numerous microscopic vacuoles. After fermentation, this protein lattice structure shows larger air cells. Many of the small air cells enmesh minute starch granules with them. The veil-like protein coating on the surface of the starch granules, mainly because of increase in the size of air cells, stretches and rolls up into fibrils. After kneading and proofing, these fibrils aggregate and form longer and larger fibrils. Bread crumb is characterized by thin walls and large gas cells. The strands are swollen and fused, and the veil-like protein and starch form a cohesive mass.

LIGHT AND TRANSMISSION ELECTRON MICROSCOPY

Bechtel et al (1978) used LM and TEM (1) to elucidate microscopic changes associated with undermixed, optimally mixed, and overmixed flour-water doughs; (2) to determine differences among doughs formed from poor- and good-quality flours; and (3) to examine the ultrastructural changes that take place in a dough containing all bread ingredients during mixing, fermentation, proofing, and baking. An optimally mixed dough formed from a good bread-making composite flour results in even and continuous distribution of protein around starch granules and even distribution of inclusions within the protein. The undermixed composite-flour dough failed to meet those conditions. In the overmixed dough, distribution of inclusions within the protein was even, but the protein was not continuous. Protein strands were broken, and the protein contained many large vacuoles. The vacuoles apparently weakened the protein by producing localized thin protein strands that were disrupted easily. Those results agree with the findings of Baker and Mize (1946), who measured (by changes in dough density) vacuole formation in doughs and effects of vacuole formation on bread quality. Doughs that are inadvertently overmixed can be relaxed and then remixed to form an optimally mixed dough (Bloksma, 1971). When an overmixed dough was allowed to relax, the protein vacuoles decreased in size and number. Possibly the relaxation and gentle remixing "mends" the broken protein strands so that an optimally mixed dough is restored.

Although both flour lipids and shortening contribute to the size and overall quality of the bread, the mechanisms of that contribution differ (Pomeranz, 1971). It is possible to differentiate between wheat-flour lipids and shortening, but only the flour lipids are consistently associated with the protein. The shortening is not consistently associated with either protein or starch. A poor-quality flour dough differed strikingly from a good-quality flour dough. The protein strand of the poor flour dough broke easily, even before the protein inclusions were uniformly distributed. In good-quality flour, the tensile strength of the hydrated proteins apparently governs the structure of the dough and the bread. That was quite apparent when a good-quality flour was grossly overmixed: the protein strands remained continuous, with relatively few being broken. This "stability" to overmixing (and even undermixing) is one of the most important and desirable characteristics of good-quality flours.

The results point to the significance of starch and starch-protein interaction in the baked bread. They are in agreement with the findings of Yasunaga et al (1968) on factors that govern gelatinization of starch during baking, of Derby et al (1975) on the significance of limiting amounts of water in gelatinization of starch in baked or cooked products, and of

Wehrli and Pomeranz (1970) on the significance of starch-protein interactions in dough and bread.

FREEZE-FRACTURE INVESTIGATIONS

Fretzdorff et al (1982) conducted freeze-fracture investigations (1) to study the ultrastructure of dough ingredients, (2) to follow structural interactions of the dough components in dough and bread, and (3) to evaluate changes in water distribution and overall structure in doughs containing several combinations of components and in fermented doughs and bread crumb.

Dehydrating or freeze-drying specimens during sample preparation may produce artifacts or mask surface details. Moreover, exposure to buffers, fixatives, and dehydrating agents before drying may alter the protein matrix and liberate starch granules from the matrix. To examine the relationship between the structures of starch and those of baked goods, the components of the system should be practically undisturbed, and the best treatment is no treatment (Chabot, 1979).

For those reasons and because water, next to starch, is the main quantitative ingredient of dough and bread, dough and bread should be studied with minimal, or preferably without, chemical fixation and dehydration. The freeze-fracture technique, therefore, is a promising method to investigate water distribution in dough and bread. In the freeze-fracture technique, a replica of the sample is made by rapid freezing, fracturing, and finally shadowing with platinum. The cleaned replicas of the frozen specimens can be evaluated with high magnification and resolution in an electron microscope. Even though the freeze-fracture method overcomes artifacts resulting from fixation and dehydration, artifacts may still occur during freezing, fracturing, and shadowing. Contamination and heat damage must be minimized, and adequate contrasting platinum shadowing must be used to demonstrate details of the fine structure of biological systems.

The freeze-fractured isolated gluten had a sheetlike appearance. The TEM structure of whole gluten proteins described by Crozet et al (1974) was a smooth, compact matrix that had fibrillar and granular zones associated with numerous lipid inclusions. In the freeze-fracture study of Fretzdorff et al (1982), the sheetlike structures of isolated gluten and those of doughs and bread differed. But the small lipid inclusions in gluten were similar to those in the doughs and bread containing sugar. Similarly, the structure of the isolated water-soluble fraction had no equivalent in the dough; however, in flour-water doughs (containing yeast and soy flour), the "dry clay soil" appearance in the water droplets and large water areas partially resembled isolated water solubles.

Although TEM studies (Bechtel et al, 1978; Khoo et al, 1975; Simmonds, 1972) showed a protein network composed of fibrils that interact with starch granules, SEM studies (Aranyi and Hawrylewicz, 1969; Chabot et al, 1979; Evans et al, 1977; Varriano-Marston, 1977) demonstrated a veil-like protein sheet enveloping the starch granules. In the freeze-fracture study of Fretzdorff et al (1982), a pattern of protein development as additional ingredients were added to the basic flour-water dough was observed (Table 10-1). The protein network, including water droplets, was transformed to a sheetlike structure after all ingredients were added. After sugar was added, a transition stage appeared between the two structures. At the same time, lipid inclusions became increasingly obvious in the sheetlike protein. The sheetlike structure in the mixed complete dough was converted during fermentation into a network. This agrees with SEM results of Khoo et al (1975), who found that the veil-like protein in a mixed dough stretches and rolls up into fibrils during dough fermentation and proofing. In the bread crumb, dense protein sheets showed protein denaturation and water uptake by starch during gelatinization.

In several samples starch-protein interaction was observed. In the doughs containing salt and shortening, the connection was formed by thin "pearl chains." The extent of interaction decreased after sugar was added and thin protein strands adhered to the starch granules. After fermentation, however, the interaction intensified in the flour-water-yeast-salt dough. In the bread crumb, a tight connection formed between protein and starch.

Water distribution in dough and bread could be visualized by the freeze-fracture technique but not by TEM and SEM. The starch granules and yeast cells were coated by water, which is comparable to a definite separating space between starch granules and protein. After sugar was added, the broad water layer around the starch granules narrowed and was accompanied by transformation of protein from a network to a sheetlike structure. The protein network enclosed water droplets, which became smaller and more diffuse after sugar was added and almost disappeared after oxidant was added (complete dough). Large water areas were visible in all doughs; in flour-water doughs, they contained a soluble material and showed a weblike structure. After salt was added, the areas became smoother and the web seemed to be occupied by "pearl chains." After sugar was added, the large water areas were smooth. No change was detected in the large water areas after fermentation. In the bread, the water was taken up by the gelatinized starch, and only the water surrounding the yeast cells was still visible. No vacuoles could be detected by freeze-fracture.

Several salient, novel, and (in part) unexpected findings resulted from the investigation: (1) The space between the starch granules was filled by gluten protein and large water areas. (2) Added ingredients caused protein development in mixed doughs from a network to a sheetlike structure. No change was detectable in protein development after fermentation in the flour-water-yeast-salt dough, but a dramatic change occurred in the complete dough, from sheetlike protein to a network. (3) Water distribu-

Table 10-1. Summary of Freeze-Fracture Observations[a]

Dough composition and/or processing stage	Starch granule water coat	Water distribution in protein network	Large-water area	Protein distribution	Starch-protein interaction
Mixed dough					
FW or FWY	Broad	Droplets	"Dry clay soil"	Strands in coarse network	"Pearl chains"
FWSo	Broad	Droplets	"Dry clay soil"	Strands in network are stretched	"Pearl chains"
FWYS	Broad	Droplets	Contains "pearl chains"	Strands in network are stretched	"Pearl chains"
FWYSSh	Broad	Droplets	Contains "pearl chains"	Strands in network are stretched	"Pearl chains"
FWYSShSu	Narrow	Diffuse	Smooth	Transition between network and sheet	Little, with adhering strands
FWYSShSuM	Narrow	Diffuse	Smooth	Transition between network and sheet	Little, with adhering strands
Complete or complete + soy	Narrow	Almost no free water	Smooth	Sheet	Not detected
Fermented dough					
FWYS	Broad	Droplets	Contains "pearl chains"	Network	Fairly intensive, adhering protein strands
Complete	Broad	Droplets	Smooth	Network	Adhering protein strands
Bread crumb	—	—	—	Sheet, dense particle distribution	Tight connection

[a]F = flour; W = water; Y = yeast; S = salt; M = malt; So = soy; Sh = shortening; Su = sugar.

tion changed with protein development. The protein network included water droplets; the sheetlike protein contained irregularly distributed water. At the same time, the ultrastructure of the large water areas changed (from "dry clay soil," to "pearl chains," to a smooth surface). (4) Two types of protein-starch interactions were observed ("pearl chains" and thin strands), depending on the composition of the dough. The interaction was intensified after fermentation.

THE ROLE OF STARCH AND ITS MODIFICATION

Chabot et al (1979) reported that air cells in bread were about 20 μm thick. Starch granules were embedded in a matrix but in most cases were disguised by the protein covering them. Small vacuoles were covered in the protein layer covering the granules. The protein covering may be so complete that it may prevent iodine staining of starch granules. During baking, starch granules gelatinize and are flexible enough to fit around the air cells. The granules remain intact and are identifiable, in part at least, because limited water is available during gelatinization. Under conditions of limited water availability, a strong bond between starch and proteins may be formed.

Modifications in starch during baking are of particular interest in relation to the role of starch in wheat and rye bread. According to Kulp and Lorenz (1981), functionality of wheat starch as a bread ingredient remains elusive. Integrity of wheat starch granule is essential for optimal performance in both bread and cake systems. Mechanical, chemical, or biochemical disruptions of the native granule organization have reverse effects on bread-forming properties of wheat starch. The effect of excessive mechanical starch damage is further aggravated by abnormally high levels of cereal alpha-amylase from malt or sprouted grain.

Yasunaga et al (1968) determined gelatinization of starch in white bread crumb by pasting a slurry of the crumb in the amylograph. The degree of gelatinization depended mainly on the available moisture but also on the temperature during baking. Starch in the outer layers of the crumb was more gelatinized than starch in the center. It was concluded that starch was only partly gelatinized during baking. Higher baking absorption, higher baking temperature, and longer baking time each produced more extensive gelatinization. Derby et al (1975) stated that under baking conditions, changes in starch are a function of temperature and amount of available water. If the temperature is maintained around 100°C, changes in starch depend primarily on the amount of available water. The availability of water is determined by the formula or recipe used and by the presence of ingredients or components such as proteins, pentosans, or sugars, which compete with starch for water. The amount of moisture available for gelatinization is also affected by the degree of protection against water absorption that fat provides to the starch particles. Starch in bread was at an intermediate stage of gelatinization, in which birefringence was retained in a relatively small number of granules.

More recently, Lineback and Wongsrikasem (1980) measured starch gelatinization in baked products by microscopic and enzymatic methods. Starch isolated from white bread had collapsed granules, but folding and deformation were not complete, indicating that pasting was incomplete before water availability became a limiting factor. The starch had lost all birefringence and was 96% susceptible to the action of glucoamylase. Those results must be interpreted with caution, as the isolated starch was only 54% of the total in the bread.

Vassileva et al (1981) followed, by SEM, changes in starch structure in the crumb and crust of bread. Starch modification increased as crumb temperatures during baking increased from 93/94°C to 97/98°C. Maintaining the crumb temperature at 97/98°C caused relatively small changes in starch appearance but improved bread quality. Light-microscopic studies of starch isolated from bread crumbs heated during baking at 93/94°C and 100°C showed little birefringence. The results were used to explain why crumb of bread heated for a short time during baking at 97/98°C is sticky and balls during chewing. It is known that there is a transfer of water from denatured gluten to gelatinized starch during baking. That transfer may be suboptimal if the gelatinization is very limited. Starch granules beneath 200–300 μm of the outer crust surface were more gelatinized than granules in the outer crust layer. This may be attributed to higher water availability for gelatinization beneath the outer crust surface.

THE EFFECTS OF ADDITIVES AND SUPPLEMENTS

Changes in structure of white bread resulting from addition of single cell protein were studied by Evans et al (1977). When flour and water were mixed to an optimally developed dough, the starch granules were enveloped by a continuous sheet of gluten protein. As a result of protein-starch interaction during dough development, a smooth, veil-like network stretched over the starch granules. That network was disrupted after adding 6% single-cell protein (SCP). Supplementation with SCP thickened gluten sheets, contours of starch granules were masked, and rupture of gluten increased. There was no translucency to the poorly formed gluten sheets; the gluten mass was opaque, rough-textured, and thick, and it lacked cohesiveness. Adding sodium stearoyl-2-lactylate (SSL) to dough improved gluten sheeting and adding SSL to the SCP dough partially counteracted the adverse effects of SCP on the gluten layer.

Pomeranz et al (1977) studied the effects of celluloses, wheat brans, and oat hulls on functional bread-making properties of panned white bread. Adding up to 5% fiber materials decreased loaf volume to an extent expected from dilution of functional gluten proteins. At levels above 7%, fiber materials decreased loaf volume much more than expected from dilution of gluten. Light microscopy and SEM revealed a major difference between the crumb structure of the control and experimental breads. The control bread had a fine crumb structure composed of thin sheets and filaments. Such a structure was essentially absent in fiber-containing breads.

RYE BREADS

Information on the structure of rye breads is limited. Wasserman and Dorfner (1974) compared by SEM structures of white, mixed, and rye bread crumbs. It was calculated that the surfaces of starch and gluten structures in baked wheat bread are 1,000 and 2,000 dm^2/g, respectively. It was postulated that there are no basic differences between structures of wheat and rye breads; in both, crumb cell walls are permeated by small micropores. In Knacke bread, however, a mass of gelatinized starch-protein forms a continuous wall free of micropores.

According to Drews and Seibel (1974) and Stephan (1982), water-imbibing and swelling substances (proteins and pentosans), as well as amounts and properties of starch are of major significance in rye bread making. Wheat starch, which gelatinizes at higher temperatures (above 70°C) than rye starch (above 50°C), is less susceptible to enzymic attack. Consequently, wheat products can be baked in relatively neutral media and may require light acidification, only if produced from highly sprouted grain. Similarly, the significance of starch in white wheat bread is much smaller than in rye bread. Most of the attention is directed to the contribution of gluten quality and quantity. Gluten can exert its unique viscoelastic properties more distinctly in the relatively neutral wheat dough than in the pronounced acidic rye dough. In addition, wheat breads are predominantly baked from white flours, which contain relatively small amounts of pentosans that are concentrated in the outer kernel layers. Those pentosans absorb water and swell but produce only with difficulty porous and well-leavened baked products. Those pentosans may be a strain on the whole system.

In wheat bread, the well-leavened gluten system is reinforced during baking by the gelatinized starch. The protein components cannot realize their full potential in the acidic, highly viscous rye dough. Consequently, the degree of leavening of rye bread is reduced. Starch is a major structural component, rather than a reinforcing contributor, in rye bread. The role exterted by the starch is affected by the pH of the system, dough composition and formulation, enzymic activities, and extent of starch degradation (mechanical or enzymic).

Use of a coarse meal, rather than a flour, is especially important in the production of bread from highly sprouted rye. In the meal, surface area, starch damage, and effective enzymic activities are reduced, dough acidification is enhanced, and excessively high, stress-effecting levels of water are eliminated. For all practical purposes, in some parts of central Europe, much ripe grain has some degree of sprout damage. Starch damage in rye bread making, unlike in wheat bread making, is highly undesirable. Similarly, the tendency to use finely milled flour in wheat bread production (to accelerate processes, produce more uniform products, and increase water absorption and starch damage) is a disadvantage in rye processing. In addition to the above reasons, practice has shown that some deficiencies in rye processing caused by raw material of somewhat inferior quality can be mitigated by the use of meals rather than flours.

WHEAT, WHEAT-RYE, AND RYE COMPARED

Pomeranz et al (1984) followed changes that occur during dough mixing, flour production, fermentation, and baking and compared the structures of wheat, mixed wheat-rye, and rye doughs and bread. To this end, three types of bread (100% white wheat flour, 60% wheat–40% rye flours, and 90% rye meal–10% rye flour) were examined by SEM along with appropriate samples drawn during dough mixing, various stages of sour dough production and fermentation, and baking. It was found that in white wheat bread, dough structure is formed by a protein matrix that becomes thinner, finer, and better distributed by fermentation and formation of small vacuoles. In the bread crumb, there is considerable interaction between proteins and swollen-modified starch. In dough formation, both immediately after mixing and after fermentation, some structure formation involves "stringing" of small starch granules. In light of the high numbers of small starch granules, this contribution of starch to dough structure may be quite significant (Meuser and Klingler, 1977). Evers and Lindley (1977) have shown that starch granules below 10 µm in diameter accounted for one-third to one-half of the total weight of wheat endosperm starch (Brocklehurst and Evers, 1977). The large starch granules in the dough of white wheat bread probably contribute little to structure formation. As shown by SEM, the contribution of small starch granules in the baked bread is relatively smaller than that of large granules, which swell and interact with protein. Thus, in white wheat bread, the structure of the dough involves the gluten protein matrix and the strings of small starch granules glued together by gluten proteins. The structure of the baked wheat bread involves primarily interaction of denatured gluten–swollen starch (mainly large granules) and small starch granules that are strung together. Much of the starch in the crumb and crust is modified, but substantial amounts of starch granules, especially small ones, seem to retain their size and shape in "protected" areas inside the vacuoles. Starch in the area immediately beneath the outside crust is less modified than starch in the area next to the outside crumb. The difference seems to be related to water availability. The coherence and continuity of the protein matrix is weakened by small and disrupted by large amounts of bran particles.

The mixed wheat-rye bread resembles the wheat bread except that there is less organization, and some pieces appear to be glued together rather than spread uniformly over the starch to provide a uniform structure. Some contribution to dough structure seems derived from gumlike materials and protein and to a small extent from starch granules that are damaged mechanically or enzymatically. The higher bran-aleurone content (than in white wheat flour) may be disruptive, but those particles may be "anchored" by the various contributors to dough and bread structure. Whether dough acidification (sour) contributes to incorporation of bran aleurone particles in the dough and bread depends on several factors. Extent of starch-granule modification varied widely, probably reflecting differences in gelatinization of small and large granules, higher temperature of wheat starch than of rye starch gelatinization under baking conditions, and conditions in fractured areas or inside vacuoles. The crumb may be fine-textured and porous but relatively fragile and crumbly in part.

Rye meal samples (from sour through dough to bread) were much denser than white wheat flour or mixed wheat-rye samples. The rye meal dough is held together in part by gums and in part by a dense mass. The small starch granules are strung out and clustered (glued) to the large starch granules. Additional structural dough components include finely textured gum components and rye meal pieces, thin-threaded in three dimensions. Whereas pericarp-aleurone particles are part of the problem in white wheat or mixed wheat-rye, in rye meal the chunks containing particles from the outer grain layers help to provide a coherent and continuous dough or bread structure. The major contributor to the structure of the rye bread is expanded starch. The structure of the rye meal bread crust is coarse, has large vacuoles, and is affected by bran and varied contributions of starch. Despite the long baking time, high water content, and lower gelatinization temperature of rye than of wheat starch, some small granules seem to retain much of their size and shape both in the crumb and the crust of rye meal bread. Starch has a much greater role in rye than in wheat bread, especially wheat bread baked from relatively high-protein flours. In the rye bread, the starch forms the sheetlike component. The gums, probably in cooperation with protein, form fibrils. The extensively expanded starch holds large pericarp-aleurone chunks together, binds large amounts of water, and makes it possible to produce a delectable bread of fine taste and flavor and good shelf life. These findings

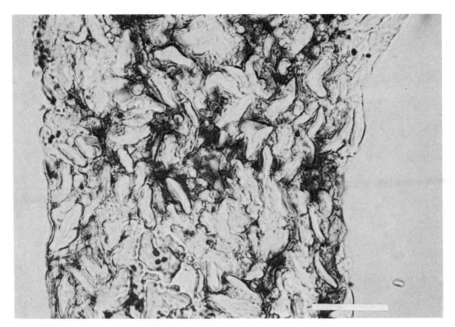

Figure 10-1. LM of a cross section through the wall of a vacuole of wheat bread crumb. Bar = 100 μm.

were extended in a subsequent study that compared structures of crumb and crust layers of the breads (Pomeranz et al, 1984b).

In the course of those investigations it became clear that some of the interpretations of the results depended on the manner in which the specimen was prepared. To resolve this problem, additional methods were used to examine the structure of bread crumb (Pomeranz and Meyer, 1984).

Examination under the light microscope (LM) makes it possible, through staining, to determine semiquantitatively the distribution of protein and starch. Figures 10-1 and 10-2 indicate that distribution of protein in the crumb of wheat and rye bread is not uniform. There were considerably more areas that stained for protein in the wheat bread crumb than in the rye bread crumb. In addition, the wheat bread crumb had several areas of high protein concentration distributed at random throughout the crumb. The general impression is that whereas in the wheat bread crumb the protein matrix holds the crumb together, in the rye bread crumb no such coherent matrix is present. The rye starch granules are more modified and expanded than the wheat starch granules, probably because of the lower gelatinization temperature of rye starch under baking conditions. The proximity of highly expanded starch, even in the micrograph of the rye bread crumb, does not prevent the

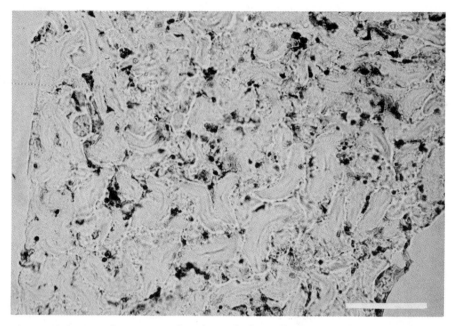

Figure 10-2. LM of a cross section through the wall of a vacuole of bread crumb from rye flour. Bar = 100 μm.

recognition of well-delineated, individual starch granules. To the extent that protein is present, it surrounds the starch granules in the rye bread crumb and forms a matrix in which rye starch granules are embedded. A similar conclusion was reached during observation under the LM of iodine-stained preparations. Protein comprised the major matrix in the bread crumb of wheat bread and starch in the crumb of rye bread. Scanning electron micrographs of the wheat and rye bread crumb are shown in Figures 10.3–10.10. The greater depth in the field observed in scanning electron micrographs present a good overview of the structure of the bread crumb in wheat (Figure 10-3) and rye (Figure 10-7) bread. In samples prepared by cutting the surface before freeze-drying (method A), one can see in areas surrounding the vacuoles the arrangement and distribution as well as modification-expansion of individual starch granules and compare them with apparently less modified starch granules (Figures 10-3, 10-5, 10-7, 10-9). In agreement with the micrographs from LM (Figures 10-1, 10-2), the starch is substantially more modified-expanded in rye than in wheat bread crumb. In wheat (Figure 10-5—high magnification), the starch granules are clearly embedded in a protein matrix. On the other hand, it is more difficult to discern the outline of the individual starch granules in the inner structure of the vacuole in rye bread (Figure 10-9—high magnification). This is due to the more

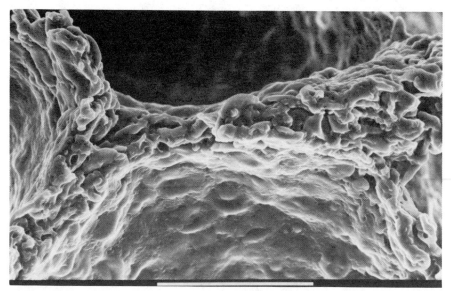

Figure 10-3. SEM of crumb of wheat bread, method A. Bar = 100 μm.

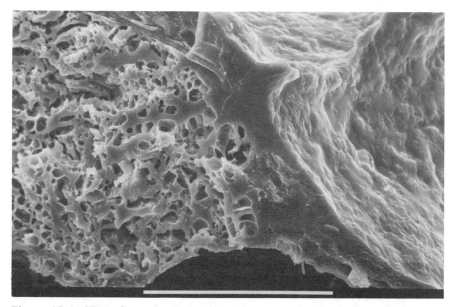

Figure 10-4. SEM of crumb of wheat bread, method B. Bar = 100 μm.

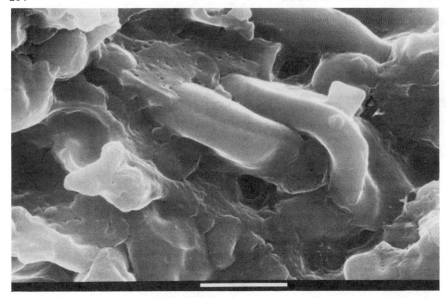

Figure 10-5. SEM of crumb of wheat bread, method A. Bar = 10 μm.

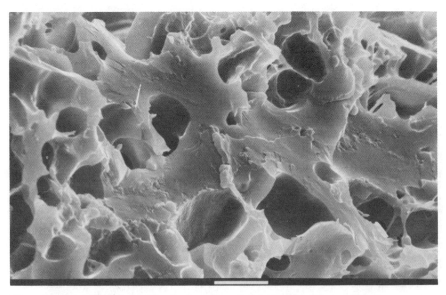

Figure 10-6. SEM of crumb of wheat bread, method B. Bar = 100 μ.

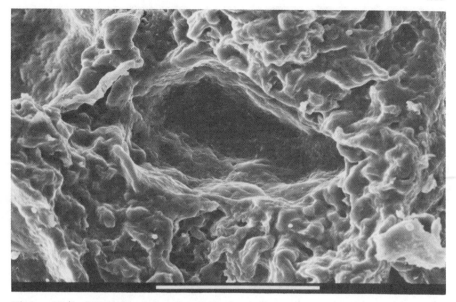

Figure 10-7. SEM of crumb of rye bread, method A. Bar = 100 μm.

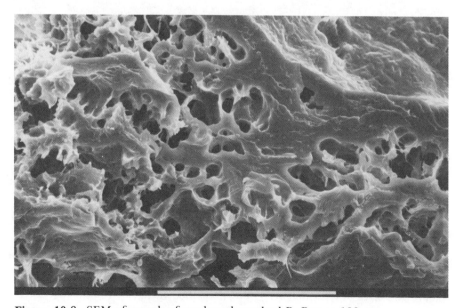

Figure 10-8. SEM of crumb of rye bread, method B. Bar = 100 μm.

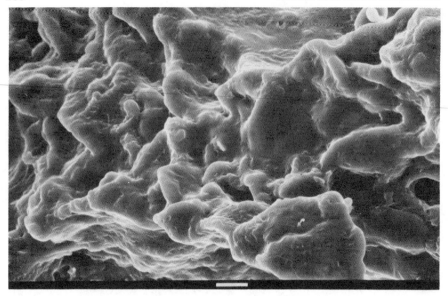

Figure 10-9. SEM of crumb of rye bread, method A. Bar = 10 μm.

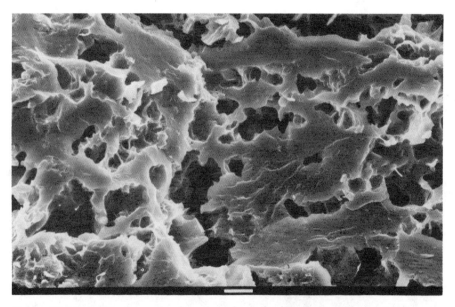

Figure 10-10. SEM of crumb of rye bread, method B. Bar = 10 μm.

extensive expansion-modification of the rye starch granules but also to the fact that they may be covered by a layer of gums, proteins, and soluble starch.

In samples prepared by cutting the surfaces after freeze-drying (method B), the outer surfaces of the vacuoles remain relatively intact, and considerable damage was caused to the starch, proteins, and other components inside the vacuole walls (Figures 10-4, 10-6, 10-8, 10-10). A comparison of the micrographs of the crumb prepared by method A (Figures 10-3, 10-5, 10-7, 10-9) with the micrographs of crumb prepared by method B (Figures 10-4, 10-6, 10-8, 10-10) indicates that the former show little and the latter show a considerable amount of micropores. Those micropores are the result of minute vacuoles in the walls surrounding the larger vacuoles. They can be seen only after the walls of the vacuoles are fractured. They can be seen also in the preparations examined under LM. The possibility that some of the micropores are the result of shrinkage during freeze-drying cannot be excluded.

Thus, the information obtained by each method of sample preparation complements that obtained by the other. Useful information can be obtained by examination of micrographs at high magnification. Method B shows the folding and layerlike stratification of starch in the walls of the vacuoles as well as a partial cover of a protein matrix. The results of micrographs for method B, as expected, parallel those for LM of their sections from frozen material. This information is particularly useful in examining the vacuolar wall structure of rye bread. Whereas it is difficult to conclude from Figure 10-9 about the manner in which the starch granules are held together, Figure 10-10 demonstrates the "spot welding" interaction at fairly regular intervals that is responsible for the structure of the crumb of rye bread.

The findings were also confirmed by scanning electron micrographs of bread crumb macerated with water to remove the soluble starch prior to mounting. In the case of wheat flour bread (Figure 10-11), even after the soluble starch was washed out there remained a residual coherent matrix of the denatured gluten. In the case of rye bread, a less coherent structure was left as the main components were washed out (Figure 10-12).

In summary, whereas in the wheat bread the crumb is held together by a matrix of denatured protein, in the rye bread crumb highly expanded starch granules fulfill that role. Fracturing freeze-dried crumb provided different information than sectioning prior to freeze-drying. In the first case, little damage was caused to components of outer surfaces of vacuoles. In the second case, the protein matrix and starch granules were broken. At the same time, the presence of micropores in the material surrounding the vacuole was observed and confirmed the findings from LM of sections of the bread crumb. Examination by SEM of residues of bread crumb macerated to wash out soluble starch demonstrated the presence of a

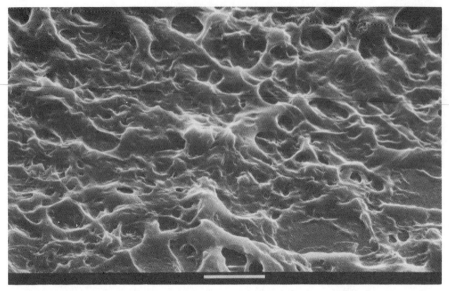

Figure 10-11. SEM of water-macerated crumb of wheat bread. Bar = 10 μm.

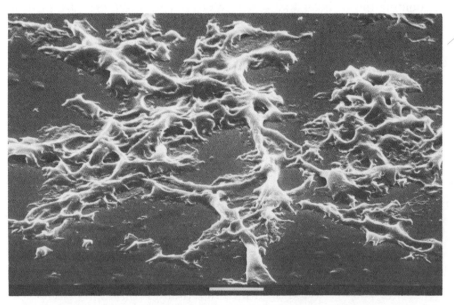

Figure 10-12. SEM of water-macerated crumb of rye bread. Bar = 10 μm.

residual coherent structure of apparently denatured gluten proteins in wheat bread. In rye bread there were only few similar, less coherent structures.

BIBLIOGRAPHY

Aranyi, C.; Hawrylewicz, E. J. *Cereal Sci. Today* **1969**, *14*, 230.

Baker, J. C.; Mize, M. D. *Cereal Chem.* **1946**, *23*, 39.

Bechtel, D. B.; Pomeranz, Y.; De Francisco, A. *Cereal Chem.* **1978**, *55*, 392–401.

Brocklehurst, P. A.; Evers, A. D. *J. Sci. Food Agric.* **1977**, *28*, 1084–1089.

Bloksma, A. H. In "Wheat Chemistry and Technology," Pomeranz, Y., Ed. Am. Assoc. Cereal Chemists; St. Paul, MN; **1971**.

Chabot, J. F. *Scan. Electron Microsc.* **1979**, *3*, 279–296, 298.

Chabot, J. F.; Allen, J. E.; Hood, L. F. *J. Food Sci.* **1978**, *43*, 727.

Chabot, J. F.; Hood, L. F.; Liboff, M. *Cereal Chem.* **1979**, *56*, 462–464.

Crozet, N.; Godon, B.; Petit, L.; Guilbot, A. *Cereal Chem.* **1974**, *51*, 288.

Derby, R. I.; Miller, B. S.; Miller, B. F.; Trimbo, H. B. *Cereal Chem.* **1975**, *52*, 702–713.

Drews, E.; Seibel, W. *Getreide Mehl Brot* **1974**, *28*, 307–312.

Evans, L. G.; Volpe, T.; Zabik, M. E. *J. Food Sci.* **1977**, *42*, 70–74.

Evers, A. D.; Lindley, J. *J. Sci. Food. Agric.* **1977**, *28*, 98–102.

Fretzdorff, B.; Bechtel, D. B.; Pomeranz, Y. *Cereal Chem. 59, 113–120.*

Khoo, U.; Christianson, D. D.; Inglett, G. E. *Bakers Digest* **1975**, *49*(4), 24–26.

Kulp, K.; Lorenz, K. *Bakers Digest* **1981**, *55*(5), 24–28, 36.

Lineback, D. R.; Wongsrikasem, F. *J. Food Sci.* **1980**, *14*, 71–74.

Meuser, F.; Klingler, R. N. *Getreide Mehl Brot* **1977**, *33*, 57–62.

Pomeranz, Y. In "Wheat Chemistry and Technology." Pomeranz, Y., Ed. Am. Assoc. Cereal Chem.; St. Paul, MN; **1971**.

Pomeranz, Y.; Meyer, D. *Food Microstruct.* **1984**, *3*, 159–164.

Pomeranz, Y.; Shogren, M.D.; Finney, K. F.; Bechtel, D. B. *Cereal Chem.* **1977**, *54*, 25–41.

Pomeranz, Y.; Meyer, D.; Seibel, W. *Cereal Chem.* **1984a**, *61*, 53–59.

Pomeranz, Y.; Meyer, D.; Seibel, W. *Getreide Mehl Brot* **1984b**, *28*, 138–146.

Sandstedt, R. M.; Schaumburg, L.; Fleming, J. *Cereal Chem.* **1954**, *31*, 43–49.

Simmonds, D. H. *Cereal Chem.* **1972**, *49*, 324–335.

Stephan, H. *Getreide Mehl Brot* **1982**, *36*, 16–20.

Varriano-Marston, E. *Food Technol.* **1977**, *31*(10), 32–36.

Vassileva, R.; Seibel, W.; Meyer, D. *Getreide Mehl Brot* **1981**, *35*, 303.

Wassermann, L.; Dorfner, H. H. *Getreide Mehl Brot* **1974**, *28*, 324–328.

Wehrli, H. P.; Pomeranz, Y. *Cereal Chem.* **1970**, *47*, 221.

Yasunaga, T.; Bushuk, W.; Irvine, G. N. *Cereal Chem.* **1968**, *45*, 269–279.

Chapter 11

The Art and Science of Bread Making

BREAD MAKING IN ANCIENT TIMES

There is no record of when or where bread originated. It was first eaten in some crude form long before man learned to scribble hieroglyphics. It is certain, however, that the known history of bread is longer than the history of any other food and that its history runs parallel with the known history of man. Religions have long used bread as a sacred symbol. To Christians at Communion it is the body of Christ. To Jews the unleavened bread (matzah or matzoth) symbolizes the release of the Children of Israel from slavery in Egypt. The Greeks built a temple in honor of the goddess of bread grains, Demeter; in Rome the mother of the church was Ceres. The Egyptians used bread as part of religious rituals, offered it as a sacrifice, and cast it on the Nile as a tribute to their gods. A shortage of bread has been suggested as causing the fall of Rome, the French Revolution, and the Bolshevik Revolution in Russia. In the history of Great Britain, bread has been linked with politics at two vital times: during the years that led to the repeal of the Corn Laws in 1846, and a century later when a Labour Government rationed bread in peacetime (Sheppard and Newton 1975).

Although the history of bread is nearly as old as the history of mankind, the basic formulation has changed little. The history of bread is to a large extent the history of the baking oven and of the raw materials used in the preparation of bread—mainly the flour and the leaven.

Bread has faithfully accompanied humanity for many centuries. In very early times, grain was pounded and consumed as a watery paste. The grain was later ground in primitive mills, and the ground product was sieved. The flour was made into dough and shaped into flat cakes that were baked on hot stones or directly in a fireplace. The next step was the baking of fermented doughs.

The Egyptians and Babylonians knew baking and brewing more than 100 years before the Christian era (Figure 11-1). They attributed the art of

Figure 11-1. Diagrammatic description of a bakery in Egypt (Ramses III, about 1200 BC). From left to right: Kneading the dough. Transfer in jars to the bakery. The dough is shaped. Two bakers work by the heated oven, one of them preparing a spiral piece of dough. The baker at extreme right prepares the oven. The bread is pasted on the inside of the oven for baking. (Courtesy, Max Wahren, Bern, Switzerland.)

fermentation to the wisdom and grace of the god Osiris. Bread was made by grinding the wheat and forming it into a dough which was permitted to ferment. The fermented dough was then shaped into loaves and placed in ovens to bake. Two kinds of bread were produced: a nutritious, coarse, dark variety for the poor, and a refined, white loaf for the rich.

The ancient Hebrews distinguished between leavened and unleavened bread, the latter being prepared even today for Passover. Knowledge of fermentation was carried over to the Greeks and Romans, who particularly developed the art of wine making. Ancient Greece grew barley and malt. In the seventh century BC, grain was the most important imported item. Homer mentions bread in his writings, though it is not known whether it was flat or leavened bread. Herodotus mentions an oven in the seventh to sixth century BC. At the time of Demosthenes, Athens imported 400,000 midimnos (about 31,000 tons) of wheat mainly from Egypt, Syria, Libya, and Sicily. Solon (640–540 BC), one of the "seven wise men of Greece," prohibited export of cereal grains. The weight of bread was controlled and the price was fixed.

We know that in 168 BC the Roman baker (*pistor*) used a stone or clay mixer (*alveus*) to prepare his bread (*panis*) from a fermented dough (*fermentum*). The bread was baked in a stone or clay oven (*furnus*), and the bakery (*officium pistoria*) was part of a relatively large complex that included a mill and brewery. Roman bakers, at the time of Caesar, were members of trade unions that included a special union (Siliginarus) of high-quality bread makers. When Aurelius decided to distribute bread (instead of grain), bakers became government employees. In the year 312, there were 254 bakeries in Rome. The weight and price were regulated. There were several types of bread. Some were apparently of poor quality, as Marcjalis recommends not to hit a slave but rather feed him bread from Rhodos so he will "loose" his teeth.

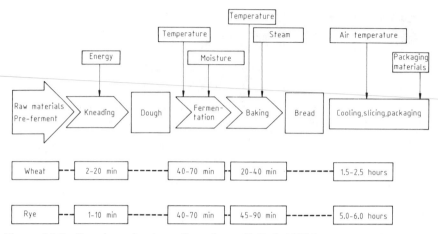

Figure 11-2. Bread production—flow sheet. (Seibel, 1979.)

The Middle Ages saw development of bakeries in cities; the bakeries were combined with mills and breweries. The poor farmer or worker in town seldom enjoyed the luxury of bread. Wars, drought, and pestilence brought hunger and decimated populations. The situation was somewhat improved in the 12th century with the formation of the Hansa, and in the 14th century with the development of bakers' guilds. In England, King Henry III established a bread price in 1226 that was changed only in the middle of the 18th century. Industrial cities of north Italy enticed workers by low prices of bread.

BREAD MAKING IN MODERN TIMES

Today, baking in most countries is mechanized. The production of baked goods comprises the following five steps (Figures 11-2, 11-3): (1) preparation of raw materials (selection, preparation, and scaling of ingredients); (2) dough formation and development; (3) dough processing (fermentation and leavening, dividing, molding, and shaping); (4) baking, and (5) manufacture to finish products (including measures to retain quality, slicing, packaging, sterilization, etc).

FLOUR

Flour is the single most important and basic ingredient in bread making. Although bread has been produced from meals and flours milled from most cereal grains, the type of bread accepted by the customer in the Western world is normally prepared from wheat meal or flour or from wheat-rye meals or flours (see also Chapters 7 and 9). Flour used in bread

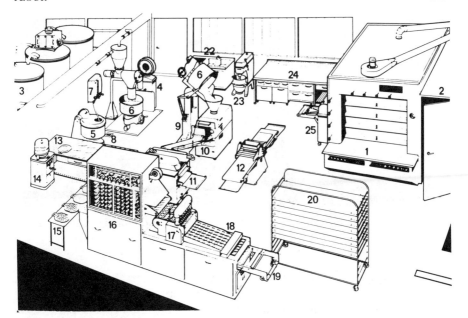

Figure 11-3. Production diagram of a modern handicraft bakery. (1) Baking oven for rolls and bread, (2) Final proof, (3) Flour storage, (4) Flour scaling, (5), (8) Dough mixing, (6), (9), (10) Dough dividing, for bread, (7) Water measurement, (11) Dough moulding, for bread, (12) Sheeting, (13), (14), (15) Dough dividing for rolls, (16) Fermentation, (17), (18), (19) Surface, treatment, (20) Final proof, (21), (22), (23), (24), (25) Dough handling of other baked goods. (Courtesy, Werner and Pfleiderer, Stuttgart, West Germany.)

making is generally milled from common (or so-called vulgare) wheat. Flour from durum wheat is used in some parts of the world (mainly the Middle East) to make flat bread. In the Western world, durum is used mainly to make semolina for macaroni production. Bread making quality of a flour depends on the quality and quantity of the flour proteins. Proteins of flours milled from common wheat possess the unique and distinctive property of forming gluten when wetted and mixed with water. Wheat gluten imparts to dough physical properties that differ from those of doughs made from other cereal grains. It is gluten formation, rather than any distinctive nutritive property, that gives wheat its prominence in the diet.

When water is added to wheat flour and mixed, the water-insoluble proteins hydrate and form gluten, a complex and coherent mass in which starch, added yeast, and other dough components are imbedded. Thus, the gluten is, in reality, the skeleton or framework of wheat-flour dough and is responsible for gas retention, which makes production of light leavened products possible. Gluten is not present, as such, in wheat flour. Gluten is formed when water is added and the dough is mechanically handled. Handling "develops" the gluten or dough and involves hydration, modifi-

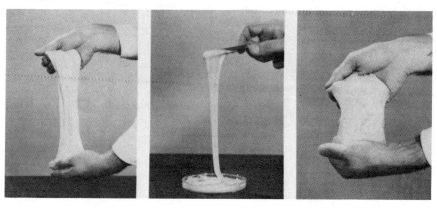

Figure 11-4. Hydrated glutenin (left), gliadin (center), and gluten (right). (Dimler, 1965.)

cation of gluten proteins, and interaction with other flour components (mainly lipids). The gluten can be separated from other flour components by washing under a gentle stream of water. The gluten thus separated is composed of two-thirds water and one-third dry matter. The dry matter is about 90% proteins and lipids, and small amounts of other components. The gluten proteins can be further separated into gliadin (a prolamin soluble in 70% alcohol) and into glutenin (a glutelin insoluble in alcohol but soluble in acids and alkali). The two gluten fractions have diametrically different properties (Figure 11-4). Only a combination of gliadin and glutenin (about 1:1) gives a gluten of desirable properties in bread making. During maturation, profound changes take place in composition of wheat proteins. Those changes dramatically affect the bread making performance of flours milled from wheat harvested at various stages of maturity.

On the basis of their suitability to produce yeast-leavened bread, common wheats and the flours milled therefrom are classified broadly into two groups, strong and weak. Strong flours contain a relatively high percentage of proteins that form a tenacious, elastic gluten of good gas-retaining properties and are capable of being baked into well-risen, shapely loaves that possess good crumb grain and texture. They require considerable water to make a dough of proper consistency to give a high yield of bread. The doughs have excellent handling properties and are not critical in their mixing and fermentation requirements. For this reason they yield good bread over a wide range of baking conditions and have good fermentation tolerance.

In contrast, weak flours have a relatively low protein content and form weak gluten of low elasticity and poor gas-retaining properties. They have relatively low water-absorbing capacity, yield doughs of inferior handling quality, and have mixing and fermentation requirements that render them more likely to fail in baking. Weak flours require less mixing and fermentation than strong flours to give optimum baking results. The protein test, while not included as an official grading factor for wheat, is an accepted marketing factor. Customarily, data on protein content are made available to buyers of wheat and flour.

Various degrees of strength are required by the bakery trade. Thus, the strengths of flours sold for family trade, commercial pan bread, and commercial hearth bread are increasingly greater in the order named. Home baking involves rather mild treatment (hand or low-speed mixing and gentle fermentation) so that good results are obtained with flour of low protein content (about 10.5%) and more easily conditioned gluten than would be required for mechanized commercial bakeries. For the manufacture of pan bread, a medium-strong flour with about 11.5% protein is required to withstand high-speed mixing and produce a dough possessing the physical characteristics that will permit machine manipulation. Bread baked on the hearth of the oven requires a flour of still higher protein content (about 12% or more) to yield a strong dough that will not flatten unduly under its own weight. Other types of bread flours must be supplied by the flour millers to suit different markets.

OTHER INGREDIENTS

The amount of water added during dough mixing depends on water absorption of the flour, the method and equipment used to make and process the dough, and the characteristics desired in the baked bread. Generally, water absorption increases with increase in protein content and with increase in flour extraction. Basically, the baker aims at an optimum water absorption for a specified product. Excessive amounts of water make the dough sticky and difficult to handle and make the bread wet, soggy, and susceptible to microbial damage; in addition, the moisture content may exceed maximum levels permitted by regulatory agencies. On the other hand, doughs that are too dry do not develop satisfactorily during mixing, fermentation by yeast is reduced, and the bread stales and crumbles rapidly. In such bread the bakers' return per unit flour is reduced. Water used is generally drinking water of intermediate hardness. Water is added to bind flour and other dough ingredients into a coherent mass, to dissolve certain ingredients (ie, sugars, soluble proteins, and pentosans), for development of yeast and/or sour dough microorganisms, and for leavening action at the baking stage. Mineral components in the water toughen the gluten, improve the stand of the dough during fermentation, increase gas retention, and improve crumb grain.

Salt is added (about 1.5% of flour weight) for taste and to improve dough handling. Salt slows down water imbibition and swelling of flour proteins, shortens the gluten, reduces dough extensibility, and improves gas retention, bread crumb grain, and slicing properties. The amount of yeast is

Table 11-1. Use of Compressed Yeast in Various Types of Baked Goods

Bread or baked goods	Yeast usage (%)	
	Average value	Range
Rye bread	1	0.5–1.5
Rye mixed grain bread	1.5	1.0–2.0
Wheat mixed grain bread	2.2	2.0–2.4
Wheat bread	2	1.5–2.5
Rolls	4	3.0–4.0
Zwieback	8	6.0–10.0
Sweet bread	5	4.0–6.0
Berlin filled doughnuts	7	6.0–8.0

Source: vom Stein, 1952.

about 2% (flour basis) in regular white bread; for rolls, larger amounts of yeast are used. The amounts of compressed yeast in various types of baked goods are listed in Table 11-1. (from vom Stein 1952).

Bakers' yeast ferments the available sugars (in flour or added) to yield carbon dioxide (and alcohol) to provide light, porous, yeast-leavened products. Leavened bread was prepared for many years by mixing a portion of leftover dough with a new dough batch and permitting the mix to ferment. A more advanced practice was to mix residual yeast from a brewery or distillery with a bread dough. Today it is recognized that there are many types and strains of yeast, each possessing distinctive characteristics, properties, and uses. Bakers' yeast is carefully cultured to bring about definite and desirable changes in structure and flavor of yeast-leavened doughs.

Bread can be produced from flour, water, salt, and yeast and/or sour dough. Optional ingredients include fats or oils, sugars, milk powder or mixtures of vegetable (i e soy flour) proteins and whey proteins, oxidants, enzymic supplements, dough conditioners, dough softeners, and others. Some of the additives and their roles are listed in Table 11-2.

Figure 11-5 compares the relative loaf volumes and crumb grains of breads made from flour and water plus yeast, sugar, salt, malt, oxidant (ascorbic acid), shortening, and soy flour. The effects of optimum mixing and shortening (in an otherwise optimized formula) are compared in Figure 11-6. Loaf volume of breads made from good- and poor-quality flours (100 g) and very lean to optimized formulations are compared in Figure 11-7.

Fats (or oils) mellow baked products, make the crumb finer and silkier, and in small amounts (up to about 3%, flour weight) increase loaf volume and freshness retention. The crust is more pliable and soft. Surface active agents (emulsifiers) enhance the mellowing effect of fats and have a fat-saving effect. In the United States for many years the only fats used were lard and various types of shortening. Application of fluid shortenings

Table 11-2. Dough Conditioner Ingredients Used in Standardized Bakery Products in The USA

Compound	Principal functions	Proprietary compositions
Emulsifiers		
Lecithin, hydroxylated lecithin	Dispersing agent	Component of emulsified blends
Monoglycerides and diglycerides	Crumb softener	Single and multicomponent emulsifier blends in beaded and plastic forms
Propylene glycol mono- and diglycerides	Crumb softener	
Diacetyl tartaric acid esters of mono-glycerides and diglycerides	Crumb softener, volume improver, dough strengthener	
Polysorbate 60	Volume improver, dough strengthener	Combinations with monoglycerides and diglycerides in beaded and plastic forms
Calcium stearoyl-2-lactylate, lactylic stearate, sodium stearoyl-2-lactylate	Dough strengthener, volume improver, crumb softener, processing tolerance	Primarily as single components in beaded form
Ethoxylated monoglycerides and diglycerides	Dough strengthener, volume improver	Combinations with monoglycerides and diglycerides in beaded and plastic forms
Oxidizing agents		
Potassium bromate, calcium bromate, potassium iodate, calcium iodate	Oxidation of gluten mercapto groups to improve gas retention of dough	Concentrated tablets containing bromate or bromate and iodate. Also as components of yeast foods
Azodicarbonamide	Fast-acting oxidizer that may replace iodate	Combinations with bromate
Calcium peroxide	Strengthens gluten, increases absorption	Proprietary compositions as single active component or in combination with enzyme-active soy flour, fats, emulsifiers and buffer salts
Reducing agents		
L-cysteine	Dispersal of gluten proteins to reduce mixing requirements	Combinations with whey, whey and bromates
Ascorbic acid		Proprietary short-time dough conditioner compositions
Enzymes		
Diastatically active malt preparations	Provides fermentable carbohydrate; stimulates gas production in dough	Malt syrups and powders

Table 11-2. (Continued)

Compound	Principal functions	Proprietary composition
Enzymes (Continued)		
Enzyme preparations from:		
Aspergillus oryzae	Source of amylase and protease. Stimulates gas production and improves dough machinability and extensibility	Concentrated enzyme tablets, bulk powders, component of some short-time dough conditioners
Bromelain	Source of vegetable protease	
Papaya-papain	Source of vegetable protease	
Soybean lipoxidase	Oxidizes carotene pigments of wheat, whitens crumb	Proprietary combinations with calcium peroxide
Fermentation accelerators		
Ammonium phosphate, ammonium sulfate, ammonium chloride	Nitrogen source for yeast metabolism	Proprietary yeast food formulations
Acidulants and buffers		
Monocalcium hydrogen-phosphate, vinegar, sodium diacetate, lactic acid	Acid salts and acids to aid in lowering dough pH. Flavor additives	Monocalcium hydrogenphosphate from yeast foods or as pure compound. Other acid salts or acids added singly. Lactic acid as component of sour flavors
Calcium sulfate, calcium lactate, calcium carbonate, dicalcium dihydrogenphosphate	Buffer salts to control pH. Source of calcium ion to strengthen gluten	Primarily in yeast foods

Source: Cole, 1973.

in baking started in the early 1960s; at first mainly in cake making and more recently in bread making. At present, a third generation of fluid shortenings is available (Table 11-3). Fluid shortenings are pumpable, liquid at room temperature, less expensive, highly effective in combination with surface-active agents, and relatively rich in unsaturated fatty acids.

Sugar enhances both fermentation and browning; improves dough stability, elasticity, and shortness; and makes the baked product somewhat mellow. As the amounts of sugar and fat in the formula increase, the amount of liquid required to attain a specific dough consistency decreases. Formula balance in yeast-leavened doughs is summarized in Table 11-4.

Over the past 25 years there have been changes in bread formulations in the United States. Those changes are listed in Table 11-5, which compares average old (1961) and revised (1981) formulations.

Figure 11-5. Relative loaf volumes and crumb grains of breads made from flour and water (1) plus yeast (2); yeast and sugar (3); yeast, sugar, and salt (4); yeast, sugar, salt, malt, ascorbic acid, and shortening (5); and yeast, sugar, salt, malt, ascorbic acid, shortening, and soy flour (6). (From Finney, 1978.)

Malt preparations have been largely replaced by microbial alpha-amylase preparations. Alpha-amylase preparations are listed in Table 11-6. Thermostability of alpha-amylases from various sources are compared in Table 11-7. The effects of various sources of alpha-amylase supplements on baking results are listed in Table 11-8.

FERMENTATION

The word fermentation is derived from the Latin verb *fermentare*, meaning to cause to rise. Whereas the English wort yeast is related to brewing, the French *levure* is derived from the Latin verb *levare* and the German *Hefe* from the verb *heben*, both words denoting rise.

According to Fowler and Priestley (1980), bread-making processes can be divided into those that depend on biological and those that depend on mechanical dough development. In processes that employ biological dough development in production of wheat breads, the effect of yeast development is critical. The biological processes include the following.

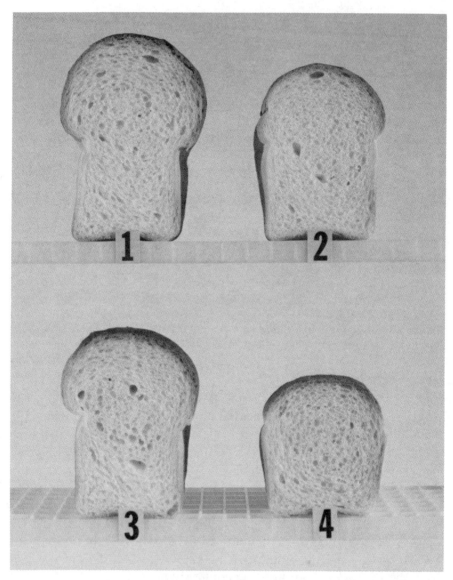

Figure 11-6. Bread baked from 100 g of flour in an otherwise optimized formula and optimum mixed (4⅜ min; loaf volume 998 ml; No. 1); (undermixed, 2 min; 748 ml; No. 2); optimum mixed, no shortening (875 ml; No. 3); and undermixed, no shortening (575 ml; No. 4). (Courtesy, K.F. Finney.)

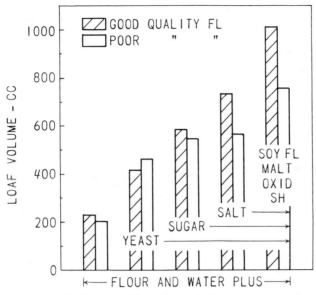

Figure 11-7. Loaf volumes of breads made from good- and poor-quality wheat flours (100 g) and very lean to optimized formulations. (From Finney, 1978.)

Table 11-3. Progressive Changes in Fluid Bread Shortening Composition and Functionality[a]

Fluid bread shortenings	Development period (year)	Fluid bread shortening composition (%)[b]				Additional functional properties
		MG+ DG	EMG+ EDG	SSL	Soybean oil	
Type I	Early 1960s	2.2–2.8	0	0	97.2–97.8	Lipophilic as crumb softener
Type II	Early 1970s	5.0–6.0	4.80–5.50	0	88.5–90.2	More hydrophilic than type I. Crumb softener, dough strengthener, loaf volume improver, dough tolerance increased
Type III	1977	3.4–4.7	4.25–5.75	4.25–5.75	88.1–83.8	Improvement of overall functionality of type II. Water absorption increased, shelf life and crumb softness extended, loaf volume further expanded

[a]Adapted from Smith (1979).
[b]MG+DG = mono- and diglycerides; EMG+EDG = ethoxylated mono- and diglycerides; SSL = sodium stearoyl-2-lactylate; soybean oil = partially hydrogenated.

Table 11-4. Formula Balance in Yeast-Leavened Doughs[a]

| Ingredient | Breads | | | | Rolls[c,d] | Sweet goods[c] |
	White pan[b]	White premium[b]	Whole wheat[b,c]	Pumper-nickel[b]		
Flour	100	100	100	100[e]	100	100
Water	64	30	66–68	71	64	60
Sweeteners						
Sugar	6–8	4	4–6	2	8	14
Honey	—[f]	3	0–2	—	—	—
Fats						
Shortening	2–3	4	3–5	1	8–10	12
Butter	—	4	—	—	—	—
Surfactant	0.2–0.5	—	0.2–0.5	—	0.2–0.5	0–0.25
Milk						
Nonfat dry	0.5–3	—	3	—	6	6
Condensed	—	28	—	—	—	—
Yeast	3	3	3	2	3–4	4
Salt	2.25	2	2.25	2.5	2–2.25	2
Vital gluten	—	—	2	—	—	—
Eggs (whole)	—	—	—	—	—	10

[a]Parts by weight based on 100 parts flour. Rolls include dinner rolls, hamburger buns, etc. Sweet goods include sweet rolls, coffee cakes, etc.
[b]From Ponte (1981).
[c]From Desrosier (1977).
[d]From Tressler and Sultan (1975).
[e]A mixture of clear wheat flour (24 parts), dark rye flour (38 parts), and rye meal (38 parts).
[f]Not included.

Straight dough. The dough is prepared by incorporating all ingredients in a single stage, and fermentation is carried out in bulk.

Delayed salt method. As above, except that the salt is added after completing two-thirds of bulk fermentation to reduce the effect of salt, which slows down gluten development and yeast action.

Delayed fat method. The fat is added at the knock-back stage of dough development to improve machinability.

Sponge and dough process. Bread dough is prepared in two stages, as described in detail later in this chapter.

Liquid ferment process. A liquid ferment contains the essential ingredients for yeast growth (with or without wheat flour) and after fermentation constitutes all or part of the dough liquor.

In processes employing mechanical dough development, traditional bulk fermentation can be replaced by intense mechanical energy input to a dough. The inclusion of an oxidizing agent (such as bromate) led to the development of the "no-time" system of bread making. Those developments were the basis of continuous bread-making systems as the John C. Baker Do-Maker and Amflow in the United States. In the Chorleywood Process in England, a no-time batch system involves mechanical development. For best results, the flour (treated in the mill with 15 ppm bromate

Table 11-5. Comparison of White Pan Bread Formulae[a]

Ingredient	Formula	
	Old	Revised
Flour	100	100
Water	variable	65.5
Sweeteners		
Sucrose (granulated)	9.2	0
High-fructose corn syrup	0	6.2
Corn syrup	0	1.2
Fats		
Lard	2.6	0.6
Shortening (vegetable)	0.7	0
Soybean oil	0	1.7
Surfactants		
Emulsifier/dough strengthener	0.35	0.75
Miscellaneous dough conditioner	0.35	0.50
Enzymes (protease)	0	0.25
Dairy-type products		
Nonfat dry milk	2.50	0
Soy-whey blend	0	2.20
Yeast	2.70	2.75
Salt	2.20	2.10
Yeast food	0.60	0.50
Mold inhibitor	0.20	0.20
Malt	Trace	0
Bread produced from formula	158.40	160.79

[a]Parts by weight based on 100 parts flour (Schnake, 1981).

Table 11-6. Alpha-Amylase Preparations

Source	Form	Potency (alpha-amylase units)	Point of usage
Barley malt	flour	about 50/g	at mill and at bakery
Wheat malt	flour	about 50/g	at mill and at bakery
Malt syrup	viscous liquid	about 5–30/g	added to doughs
Dried malt extract	sl. hygroscopic powder	about 5–30/g	added to doughs
Fungal	tablet	5000/tablet	added to doughs
Fungal	powder	50–200/g	added to doughs

and 15 ppm chlorine dioxide) is supplemented in the bakery with 30 ppm bromate and 30 ppm ascorbic acid plus 0.7% (flour basis) of fat with a slip point of 35°C. In the activated dough development (ADD), about 35–40 ppm of L-cysteine hydrochloride (or equivalent amount of sodium metabisulfite) is added at the mixing stage and dough maturity is completed by adding a relatively high level of an oxidizing agent. For additional reviews of mechanical dough processes, see Smerak (1973), Tsen (1973),

Table 11-7. Thermostability of Alpha-Amylases from Various Sources

Temperature		Percent of enzyme retention		
°C	°F	Fungal	Wheat malt	Bacterial
65	149	100	100	100
70	158	52	100	100
75	167	3	58	100
80	176	1	25	92
85	185	—	1	58
90	194	—	—	22
95	203	—	—	8

Source: Pyler, 1969.

Table 11-8. Effect of Various Sources of Amylase Supplements on Baking Results

Supplement type	Supplement amount (mg/100 g flour)	Loaf volume (cc)	Grain %	Texture %	External appearance	Remarks
Flours adjusted to the same gassing power						
Malted wheat flour	250.0	3,075	96	95	Very good	Excellent dough properties
Fungal	6.3	3,188	90-o[a]	90	Good	Soft dough
Bacterial	63.2	2,968	70-o[a]	70	Very good	Sticky, gummy bread crumb
Flours adjusted to same maximum viscosity						
Malted wheat flour	250.0	3,075	95	95	Very good	Excellent dough properties
Fungal	224.0	Sponge dough liquefied, impossible to handle				
Bacterial	29.5	3,075	80	75	Very good	Sticky, gummy bread crumb

[a]o denotes open.
Source: Pyler, 1969.

Olsen (1974), Tipples and Kilborn (1974), Moss et al (1979), Stenvert et al (1979), Kammann (1979), Pyler (1982), and French and Kemp (1985).

As stated previously, we distinguish between straight and sponge fermentation. Straight fermentation is popular in some countries, as it is the simplest and best suited for the processing of weak, low-protein flours. Its simplicity makes it attractive to use in processing doughs that are made from relatively strong flours. In a straight fermentation all ingredients—flour, liquid, yeast, malt, sugar, and other optional ingredients—are mixed in one process. Generally, a suspension of yeast, malt, sugar, and salt solutions is added to the sieved flour and mixed by hand or in a mechanical dough mixer. The dough is mixed to optimum consistency and allowed to rest for about 2–2.5 hours. The fermented dough is punched three times at about 50-minute intervals and scaled. Leavening of the dough depends on the amount of yeast; regular doughs contain about 30–40 g of yeast per

Table 11-9. Straight-Dough and Sponge Dough Formulations

Ingredients[a]	Straight dough	Sponge dough	
		Sponge	Dough
Flour	100.0	65.0	35.0
Water (variable)	65.00	40.0	25.0
Yeast	3.0	2.5	—
Yeast food	0.2–0.5	0.2–0.5	—
Salt	2.25	—	2.25
Sweetener (solid)	8–10	—	8–10
Fat	3.0	—	3.0
Nonfat milk solids	3.0	—	3.0
Softener	0.2–0.5	—	0.2–0.5
Rope and mold inhibitor	0.125	—	0.125
Dough improver	0–0.5	—	0–0.5
Enrichment	as needed	—	as needed

[a]Ingredients based on 100 parts flour.
Source: Ponte, 1971.

liter of water, or 1.5–2% on flour basis. Too little yeast causes dragging of fermentation, sticky dough, and poor crumb grain and texture. Too much yeast results in an overly porous dough and a bread that stales rapidly.

There are many modifications of sponge dough fermentation; a typical formulation used in the United States is described in Table 11-9. In early times, when bakers' yeast was unavailable, dough was fermented and leavened by a starter from spontaneous fermentation. The starter was prepared by dispersing a piece of fermenting dough with cold water and flour to produce a very soft dough. The conditions were selected to minimize formation of organic acids. In the Berlin process, the sponge is prepared from one-third of the total fluid and all of the yeast. The soft dough also includes flour (1:1 ratio of water:flour) and malt; it is mixed and fermented for 20–30 minutes. The sponge is remixed with two-thirds of the fluid and appropriate amounts of flour, salt, and other ingredients. The dough is fermented for 1½ hours and punched twice. The Berlin sponge differs little from the direct dough. The classical sponge method was developed because brewers' yeast, which was used before compressed bakers' yeast was available, was not suited for making straight dough bread and failed, particularly, in the early stages of baking. Bread baked from doughs containing brewers' yeast had little "oven spring." If, however, a sponge is used, fermentation proceeds in a satisfactory manner and the indigenous dough yeasts, which resemble bakers' yeast, can multiply and act in the oven. Although bakers' yeast of high quality is readily available, the sponge method is still useful. The sponge gives better results with strong flours that are modified and mellowed in the sponge stage. Many bakers feel that the sponge method yields better and more uniform bread than the straight method. Saving in yeast from the two-stage fermentation method is apparently small. For general reviews on fermentation see Alwes and Jolly (1974), Thompson (1980, 1982), Peppler (1981), and Dubois (1981). Formulations of liquid preferments, liquid sponge technology, technology of brew systems, and nontraditional fermentations in the production of baked goods are discussed by Ulrich (1975), Thompson (1983), Kulp (1983), and Sugihara (1977).

STAGES IN BREAD MAKING

In yeast- and bacteria-leavened doughs, the products of microbial metabolism mellow and ripen the dough and are essential for production of light, well-aerated, and appetizing bread. The two main changes occurring during fermentation involve (1) fermentation of carbohydrates into carbon dioxide, alcohol, and small amounts of other compounds that act as flavor precursors, and (2) modification of the proteinaceous matrix for optimum dough development and gas retention during the baking stage. At the same time, in yeast-leavened bread the number of yeast cells increases.

The properly fermented dough is cut and scaled, and the pieces are worked mechanically and shaped. The mechanical action removes large gas bubbles and homogenizes the dough. The dough pieces are allowed a final fermentation (proof) after panning or in free form. The final fermentation lasts about 30 minutes for small baked goods and up to 1 hour for large loaves, but it also depends on flour quality (its strength—ie, capacity to maintain dough shape), yeast level, dough composition, and temperature. Generally, the proof temperature increases with flour darkness (extraction) from about 28°C in white bread to 30–32°C in whole meal bread.

CONTINUOUS AND CONVENTIONAL BREAD MAKING

In 1953, the My Bread Baking Company of New Bedford, Massachusetts, installed a continuous bread-making process for preparing white pan bread at a rate of 60 loaves per minute. By the end of 1959, 59 units were in operation; by the end of 1969 well over 50% of the bread in the United States was made by a continuous process. Today, relatively little bread in the United States is produced by the original continuous method. The two main reasons are the inferior taste, flavor, and texture and the more rapid staling of bread produced by the continuous process.

The continuous process is designed to feed dough ingredients continuously to a mixer. The homogeneous mass is then passed to a dough pump and to a developing apparatus. In the developer, the dough is kneaded until optimum structure and gas retention properties are attained. The developed dough is transferred to a machine that continuously divides and extrudes the shaped dough pieces of proper weight into pans that move continuously. The pan movement is synchronized with the cutting knives of the divider so that each piece is properly placed in the pan and is ready for proof (Matz, 1960).

The chemical and biochemical differences between conventional and continuous bread making are mainly in the fermentation stage. In the conventional sponge, 60–70% of the flour is exposed for up to 5 hours in the sponge stage and 1 hour during proof. In the Am-Flow liquid sponge method, a small amount of flour is added to the ferment. In the Do-Maker system, generally, no flour is exposed to fermentation, although a maximum of 10% can be added to a broth. Basically, continuous processes use fermentable sugars in preferments that constitute the main liquid ingredients. The fermented preferment produces the necessary leavening power.

The continuous process has several advantages over conventional bread making. It is more compact and requires less space; the closed system is easier to maintain in good sanitary condition; the product is uniform, and the bread has a fine and silky crumb and soft texture (which is considered unsatisfactory by some consumers); it is economical from the standpoint of labor and power requirements; and processing is easy to control. On the other hand, continuous bread making is not economical if less than 1 million pounds is processed per week. It is best suited for white pan bread, although it has been adapted to other types. It requires continuous attendance and allows for little flexibility and the personnel must have good mechanical-technical training.

In Europe and Australia several continuous mixing and processing systems have been developed. In the Soviet Union there are continuous mixing and processing large-scale plants for the production of rye breads and other goods. The Iverson process in Sweden is useful for the production of special Scandinavian products. The Strahman process in Germany seems adapted to hearth breads with relatively coarse and dense grain and texture. More than half of British bread is made by the Chorleywood Bread Process. Although the process can be adapted to continuous mixers (Elton et al, 1966), the batch process is more widely used. The process has found application in Israel and South Africa.

Bread production by the conventional and continuous processes is depicted in Figures 11-8 through 11-11. Figure 11-12 shows equipment for continuous mixing of doughs. Table 11-10 shows the basic differences between sponge dough and liquid pre-ferment dough processes. Typical dough formulations using sponges and liquid preferments are compared in Table 11-11. The formula of the English Chorleywood Bread Process Formula is given in Table 11-12, and that of the Australian Brimec Process is given in Table 11-13.

Figure 11-13 compares processing stages and times for two conventional and two no-bulk fermentation bread making methods. Examples of

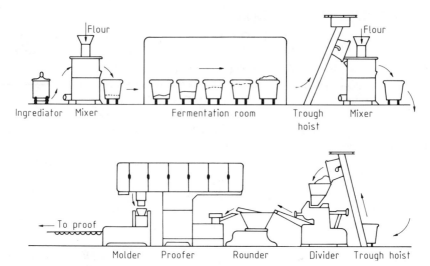

Figure 11-8. Equipment arrangement for conventional sponge and dough system. (From Seiling, 1969.)

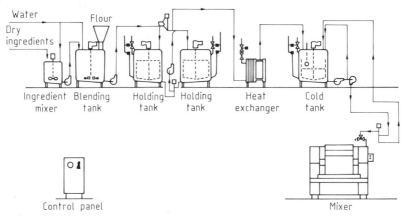

Figure 11-9. Flow sheet for the liquid ferment process. (From Seiling, 1969.)

equipment in a modern bakery in the United States are shown in Figure 11-14.

In the production of rye bread consisting entirely or partly of rye flour, the so-called sourdough method is used (Figure 11-15). A sourdough is one that is acidified and leavened by its microorganisms and that can be perpetuated ("refreshed") by adding flour and water to provide continuous acidification and fermentation (Figure 11-16). Sourdough performs three functions: leavening, acidification, and flavor development. Leavening of a sourdough by yeast in the sourdough is frequently augmented by adding bakers' yeast to the final dough. Acidification of the dough results from the action of lactic acid bacteria. Such acidification is essential for best results in

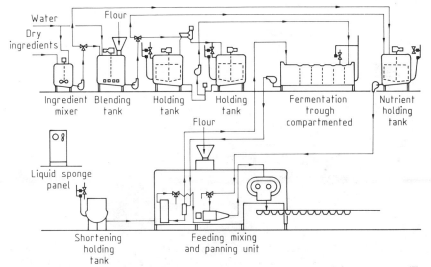

Figure 11-10. Arrangement for the Amflow continuous mixing system. (From Seiling, 1969.)

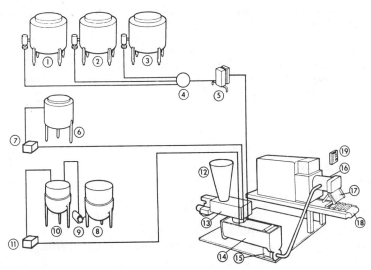

Figure 11-11. Continuous mixing process flow sheet for the Do-Maker process. (1) Broth tank; (2) Broth tank; (3) Broth tank; (4) Broth selector valve; (5) Broth heat exchanger; (6) Oxidation solution tank; (7) Oxidation solution feeder; (8) Shortening blending kettle; (9) Shortening transfer pump; (10) Shortening holding kettle; (11) Shortening feeder; (12) Flour hopper; (13) Flour feeder; (14) Premixer; (15) Dough pump; (16) Developer; (17) Divider; (18) Panner; (19) Control panel. (From Snyder, 1963.)

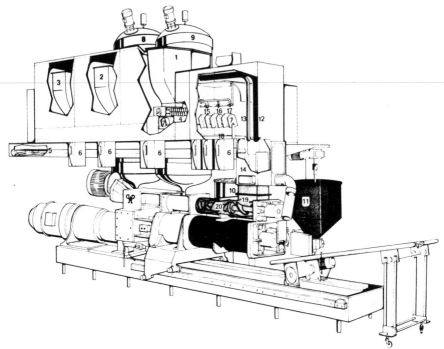

Figure 11-12. Equipment for continuous mixing of doughs. 1 Feeding device for flour; 2 Feeding device for other ingredients; 3 Feeding device for other powdered materials; 4 Feed auger; 5 Conveyor belt; 6 Sampling device; 7 Valves; 8,9 Ingrediator tanks for yeast, sugar, or salt solutions; 10 Mixer for viscous masses (e .g., fat); 11 Mixer for pre-dough; 12, 13 piping for viscous materials; 14 Feed tunnel; 15 Water feed pipe; 16 Yeast slurry feed pipe; 17 Feed pipe for other liquids (i. e., sugar solutions); 18 Collecting trough; 19 Feed cone for mixer; 20 Continuous mixer (From Spicher, 1983.)

baking rye bread. A high ratio of lactic acid to acetic acid is desirable. The indigenous yeasts of sourdough are more acid-resistant than are compressed bakers' yeast and can thrive in an acid medium. Extent of acidification, ratio of lactic acid to acetic acid, and yeast multiplication are affected by dough temperature, dough consistency, and duration of individual stages in sourdough production. These factors make it possible to control and regulate the quality of the final product (acidity, leavening, overall character) (Table 11-14).

There are two basic sourdough methods—the multiple-stage method and the so-called Berlin short, sour method (Table 11-15). Maximum acid formation occurs at around 38°C. In a soft dough, bacterial growth and acid formation are enhanced; in a stiff dough, acidification is reduced. In soft doughs fermented at about 30°C and containing up to 2% yeast,

Table 11-10. Basic Differences between Sponge Dough Process and Liquid Preferment Dough Process

Sponge side	Sponge	Liquid preferments			Concentrated brew
Flour levels(%)	50–100	40–70	15–40	0–15	none
Water (%)	30–42	59–64	60–65	61–66	25.0
Yeast (%)	2.5	2.5	3.0	3.5	3.5
Yeast nutrients (%)	0.5	0.5	0.55	0.60	0.60
Buffers (%)	none	0.05–0.10	0.10–0.125	0.1875	0.175
Sugar (%)	none	0.5	2.0	2.5	2.5
Salt (%)	none	none	0.25	0.25	0.50
Enrichment	Normal enrichment levels				
Weight of ferment (lbs)	variable	107–132	85–105	73–83	32
Set temperature (°C)	23–27	24–27	27–30	28–30	
Fermentation time (h)	3–5	2–3	2 or less		
Yeast-nutrient-oxidation (ppm)	15	20–50	26–60	35–75	
Mix reducing time					
Dough mixing time	normal	+10%	+40%	+60%	
Dough temperatures	normal (27°C)	Gradual increase to 35°C			
Dough floor time	normal (20 min)	Gradual decrease in time			

Source: Ulrich, 1975.

Table 11-11. Typical Dough Formulations Using Sponges and Liquid Preferments

Dough side	Sponge	Liquid preferments			Concentrated brew
Liquid preferment (kg)	Sponge	46.7–59.0	37.6–46.7	32.2–36.7	13.6
Flour (wt %)	0–50	30–60	60–85	85–100	100
Water (wt %)	22–34	0.5	0–6	0–4	41
Sugar (wt %)	8–10	8–10	8–10	8–10	8–10
Shortening (wt %)	3.0	3.0	3.0	3.0	3.0
Hard flakes (wt %)	none	Some may be required in continuous mixing			
Nonfat dry milk (wt %)	3–6	3–6	3–6	3–6	3–6
Salt (wt %)	2.25	2.25	2.0	2.0	1.75
Softeners (wt %)	0.50	0.50	0.50	0.50	0.50
Dough conditioners (wt %)	none	to 0.50	to 0.50	to 0.60	to 0.60
Enzymatic gluten mellowing agents	none	to 0.25	0.50	1.00	1.00
Additional oxidation (mg/kg)	none	10–25	15–35	20–50	50

Source: Ulrich, 1975.

acidification is markedly reduced. Acidification is considerably retarded by adding salt to rye sours. At high levels of salt (up to 5% on flour basis), fermentation can be prolonged.

The commonest method of rye bread production in Europe is the

Table 11-12. Chorleywood Bread Process Formula

Ingredients[a]	Formula
Flour	100
Water	61
Yeast	1.8–2.0
Salt	1.8–2.0
Fat	0.7
Ascorbic acid	75 mg/kg

[a]Ingredients based on 100 parts flour.

Table 11-13. Recipe for Production of White Bread According to the Brimec Process

Ingredients	No-time dough (ascorbic acid)	Traditional process (3-h fermentation)
Flour (kg)	50.0	50.0
Water (kg)	28.5	28.0
Compressed yeast (kg)	1.0	0.75
Salt (kg)	1.0	1.0
Fat (kg)	1.0	1.0
Malted flour (kg)	0.25	0.25
Ascorbic acid (mg/kg)	100	—
Potassium bromate (mg/kg)	10–30	7–10
Ammonium chloride (mg/kg)	300	300

Source: Bond, 1967.

so-called multiple-stage or progressive method. It involves producing several sours before the final dough is mixed (see Figure 11-16). Main acidification occurs in the basic sour that is held overnight. The need to "refresh" the sour, the large amount of labor, and the difficulty in controlling the individual stages have drastically reduced the manufacture of bread by that method. A simplified short sour (Berlin) method is more acceptable. In the short sour method, one aims primarily at a single-step acidification coupled with a strong aroma development. In the multiple-stage method, about 24 hours is required for a complete cycle. The short sour method requires 3 hours; it takes place at 35°C. The dough is soft; and compressed bakers' yeast is added at the final dough stage.

All sour-type doughs use a starter in the initial stage. The starter is taken from a ripe sour. It can also be produced by "spontaneous fermentation"— a process that is initiated by bacteria indigenous to the flour and by airborne organisms infecting the flour. When a dough undergoes spontaneous fermentation and is maintained at 35°C for several days, frequent "refreshing" by adding flour and water is required. Also available are bacterial cultures, which may be used as pure culture sours in the production of rye bread.

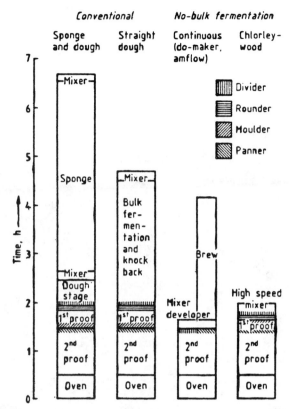

Figure 11-13. Comparison of processing stages and times for two conventional and two no-bulk-fermentation bread-making methods. (From Tipples, 1967.)

OVENS

There are many kinds of baking ovens. Their structure, mode of heating, and fuel vary widely throughout the world and are of interest in following the history and development of many nations. The ovens can be divided according to fuel (wood, coals, oil, gas, electricity, infrared, high frequency), form of oven (stationary and many forms of mobile hearths), heat (direct or indirect), and loading of oven (batch, continuous, automatic). The kind of oven determines, in many instances, the quality and characteristics of the bread (See Chapter 12). The oven temperature, presence of steam, and availability of upper and lower heat govern the texture of the bread crumb, the quality of the crust, the taste and flavor, and overall consumer appeal and acceptance. The baking temperature of modern ovens ranges from 100–180°C in baking pumpernickel to about 350°C in baking rye crisp or waffles; most white and medium-dark bread is baked around 230–250°C. The baking time is generally inversely related to the baking temperature and lasts up to 35 hours in genuine pumpernickel but

Figure 11-14. Examples of equipment in a modern bakery. Oven (top left); divider, rounder, and overhead proofer (top right); bread conveying and wrapping system (bottom left); and ingredient storage and conveying system (bottom right). (Courtesy, Baker Perkins, Inc.)

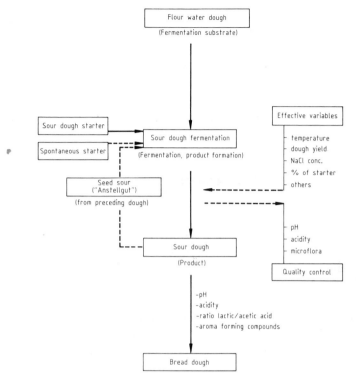

Figure 11-15. Schematic presentation of the rye dough process. (Spicher and Stephan, 1982.)

is only a few minutes for flat breads baked in the Middle East or for rye crisp (Table 11-16).

The oven is the most important part of the baker's equipment. It significantly affects the character of the baked bread and often the success or failure to produce an acceptable product. Vienna bakeries are famous throughout the world for the many types of bread they produce. All Viennese breads are characterized by a highly glazed, brightly colored crust. They are generally made in a special oven with a sole that slopes upward from the oven door, which makes it possible to load the bread under a ceiling of steam. Some of the steam condenses on the cool surface of the dough and keeps it moist during the early stages of baking. This allows maximum expansion of the dough piece as a result of oven spring. The steam also facilitates starch gelatinization on the outside of the loaf. When the steam is discontinued and the baking is completed in dry heat, the starch has a high glaze (Sheppard and Newton, 1957).

The first known method of baking by primitive people was on a flat stone previously heated by burning a fire on it, a simple method still used by the nomadic Bedouin Arabs. The first oven was a little "house" of large flat

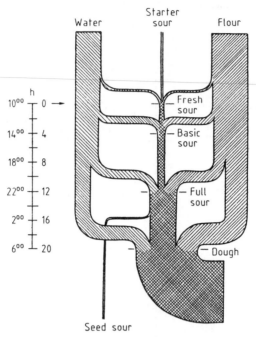

Figure 11-16. Multistage sourdough process flow sheet. (Spicher, 1983.)

stones assembled with enough free space in the middle for a loaf or two of bread. The heat was provided by a fire around the stones. The first portable oven was built by the Egyptians. It was a beehive- or barrel-shaped container of baked clay, usually divided by a central horizontal partition. The lower portion was heated; the upper part was accessible from the top and served as the baking chamber. The early Jews made a similar oven of hollowed stones. The dough pieces, instead of being placed for baking at the bottom of the oven or sole of the baking chamber, were plastered to the inside. As long as the pieces were wet and sticky, they adhered to the inside wall; after they were baked, they fell to the bottom of the oven. The oven in the ancient Roman public bakery was built in rock. The oven was heated by burning wood in the baking chamber, raking out the ashes, and putting the dough in to bake. The oven opening was closed with a large stone, sometimes sealed with clay. Ovens working on this principle, but constructed of bricks or small stones, may still be seen in the excavations of Pompeii. In fact, ovens based on this design formed most of those used in Europe until about two centuries ago. Although some of the Roman ovens had chimneys to improve the draft and carry away the steam, the use of chimneys and more efficient heat control and airflow were incorporated into ovens only many years later.

The predecessor of the modern ovens is the side flue. It was usually built of bricks with a thick sole and crown. Coal or coke (more recently gas or oil)

Table 11-14. Required pH and Acidity in Breads Made with Rye Flour
or Cracked Grain Rye

Bread type	Recommended specification	
	pH	Acidity[a]
Rye	4.20–4.30	8.0–10.0
Rye mixed grain (50–89% rye)	4.30–4.40	7.0–9.0
Wheat mixed grain (10–49% rye)	4.65–4.75	6.0–8.0
Rye cracked grain	4.00–4.60	8.0–14.0
Whole rye	4.00–4.60	8.0–14.0

[a]ml 0.1 N NaOH/10g.
Source: Spicher and Stephan, 1982.

Table 11-15. Percentage of Sourdough Used with Rye Flour (Type 997)

Sourdough process	Total time of process (h)	Fermentation time of full sour (h)	Degree of acidity of mature full sour[a]	pH of mature full sour	Sour-dough used (%)
Berlin short sour	3–4	3–4	9.5	4.1	45–50
2- and 3-stage sour process (basic sour overnight)	15–20	3	10.5	4.0	40–45
3-Stage sour dough process (full sour overnight)	20	8	13	3.9	35–40
Detmold single stage	15–20	17	12	3.8	30—35
Monheim salt sour dough process	18–48	48	20	3.6	25—30

[a]ml 0.1 N NaOH/10 g.
Source: Spicher and Stephan, 1982.

Table 11-16. Baking Times and Temperatures for Various Baked Goods

	Time	Temp (°C)
Waffles	2–9 min	200–400
Cookies	8 min	350
Crisp (flat) bread, Knäckebrot	8 min	340
Wheat bread and wheat mixed grain bread (1 kg)	35–40 min	220–230
Wheat bread (2 kg)	50 min	220–230
Rye bread (1 kg)	40–60 min	220–230
Rye bread (2 kg)	75 min	250–270
Rye bread (3 kg)	90 min	250–270
Rhenish black bread	2–4 h	210–230
Hamburg black bread	4–5 h	200–210
Westphalian black bread	8–10 h	180–200
Pumpernickel	16–35 h	100–180

was burned in a front corner of the oven, and the main chamber was heated by a fan-shaped flame across the interior. When the interior was very hot, burning was discontinued.

The steam-tube oven was developed by the American engineer A.M. Perkins, who first patented it in England in 1851. In that oven heat was provided by a series of sealed tubes containing distilled water converted to superheated steam.

The introduction of gas-heated ovens increased only after several improvements took place, including development of the Bunsen burner. Old ovens were nearly all peel-oven type. Bread pieces were loaded manually by long-handled, wooden peels shaped like a flat oar. In the draw-plate oven the baking chamber was loaded by a bottom plate mounted on a wheeled trolley. The reel oven was developed in the United States late in the last century. Its baking chamber is roughly cubical. The bread is placed on trays attached to a rotating reel. The more recent traveling oven is a long heated tunnel, through which the bread passes slowly on a moving belt until it is baked. A big traveling oven can hold as many as 2,000 medium-size loaves. Baking a medium-size 2-pound loaf in such ovens takes about 40 minutes. In high-frequency ovens, bread can be baked within a few minutes. In those ovens, heat is generated inside the dough and not applied externally. Consequently, the baked bread has no appetizing crust associated with its flavor. That can be remedied, in part, by exposing the bread piece to infrared radiation.

CHANGES DURING BAKING

The water content of the baked products is below 5% in freshly baked rye crisp and over 40% in large, dark bread. The water content in the whole loaf increases with the size of the baked product and darkness (extraction) of the flour used, and it decreases with increase in crust:crumb ratio. The water content of the crust in a freshly baked bread is practically nil, but after some time moisture migrates from the crumb to the crust and some is absorbed from the surroundings.

Several changes take place during baking both in the crust and in the crumb. The so-called browning reaction involves both caramelization of sugars and interaction between sugars and proteinaceous materials and imparts a deep color to the crust; thermal decomposition of starch and formation of dextrins contribute to crust luster. All this is accompanied by formation of flavor and taste components. At the same time profound changes take place inside the loaf of bread (Figure 11-17). At early stages the increase in temperature enhances enzymatic activity and growth of yeast and bacteria. At about 50–60°C the yeast and bacteria are killed. Above that temperature starch gelatinization, protein coagulation, and enzyme inactivation take place. Steam is formed around 100°C, at which

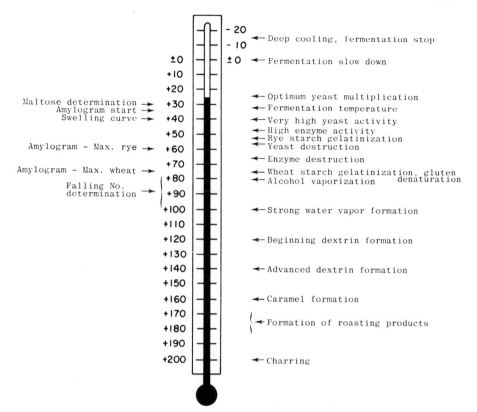

Figure 11-17 Indirect and direct baking changes as affected by temperature (°C). (Courtesy, J.M. Bruemmer.)

the final volume and crumb texture of the bread are set. The inside of the loaf does not exceed 100°C; in the crust, however, much higher temperatures are attained. In the range 110–150°C, light and brown dextrins are formed followed by caramel. Dark-brown products of roasting require about 150–200°C.

The taste of the baked bread depends on dough composition, the fermentation schedule, and baking process. Important dough ingredients that contribute to the taste include salt and fermentable sugars. The roles of those components are not limited, however, to contributing to taste and flavor; they were described earlier in this chapter. The overall taste of bread depends on flour darkness (extraction). Thus, bread baked from white flour is blander than bread baked from dark or whole wheat flour, or from wheat-rye meals. Volume, crumb texture, and taste are modified by adding such optional components as shortening, milk, germ, and soy flour (see also Chapter 13).

Bread flavor is due to the combined effects of many substances. Thus far over 150 compounds have been identified, even though only a few likely

Table 11-17. Concentration of Aromatic Substances in Various Breads

	Wheat bread		Rye mixed grain bread		Whole grain bread		Pumper-nickel
	Crumb (mg/kg)	Crust (mg/kg)	Crumb (mg/kg)	Crust (mg/kg)	Crumb (mg/kg)	Crust (mg/kg)	(mg/kg)
Ethanol	3,900	1,800	3,400	1,100	2,300	1,000	1,600
5-Hydroxymethylfurfural	9	40	12	300	20	400	70
Acetaldehyde	4.3	12.8	4.7	22.6	4.6	26.2	7.1
Isopentanal	1.2	4.7	2.7	15.2	1.9	19.0	4.6
Furfural	0.3	5.5	1.5	12.4	2.3	28.7	27.4
Methylglyoxal	0.7	0.8	1.5	8.9	1.9	13.5	4.3
Isobutanal	0.3	2.6	0.9	6.0	0.8	12.9	1.8
Acetone	0.7	4.5	1.4	5.6	2.0	6.5	1.9
Acetoin	0.9	1.0	0.2	1.1	0.3	0.7	5.0
Diacetyl	0.2	0.9	0.2	1.3	0.2	1.3	0.7

Source: Rothe, 1974.

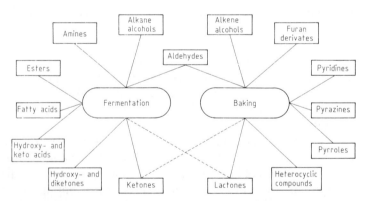

Figure 11-18 Relationship of fermentation and baking to the major classes of aromatic compound. (Rothe, 1980)

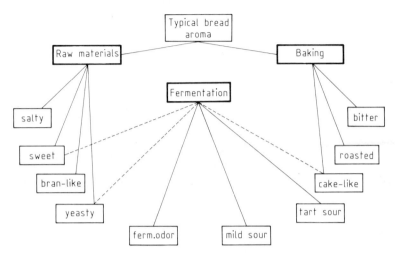

Figure 11-19 Effect of raw material, fermentation, and baking on the taste and aroma of bread. (Kohner, 1980.)

Table 11-18. Comparison of Nutrients in Wheat Flour and Wheat Bread in the United States

1 Pound (454 g)	Protein (g)	Fat (mg)	Calcium (mg)	Iron (mg)	Thiamine (mg)	Riboflavin (mg)	Niacin (mg)
Flour							
Unenriched (all purpose)	47.6	4.5	73	3.6	0.28	0.21	4.1
Enriched (all purpose)	47.6	4.5	73	13–16.5[a]	2.9	1.8	24.0
Whole wheat	60.3	9.1	186	15.0	2.49	0.54	19.7
Bread							
Unenriched (white) (3–4% nonfat dry milk)	39.5	14.5	381	3.2	0.31	0.39	5.0
Enriched (white) (3–4% nonfat dry milk)	39.5	14.5	381	8–12.5[b]	1.8	1.1	15.0
Whole wheat (2% nonfat dry milk)	47.6	13.6	449	10.4	1.17	0.56	12.9

[a]The new single level standard of 16.5 mg per pound became mandatory July 1983.
[b]The new single level standard of 12.5 mg per pound became mandatory July 1983.
Source: USDA, 1975.

Table 11-19. Composition of Baked Goods (100 g) in wt/%

Baked goods	Water	Proteins	Lipids	Carbohydrates	Added sugars	Added salt	Cholesterols
Breads							
Pumpernickel	34.0	8.7	2.1	51.1	0.7	1.8	—
Whole wheat	36.4	9.8	4.2	42.7	3.6	1.3	0.001
White pan:							
Old formula	32.3	8.4	3.5	47.4	5.8	1.4	0.002
New formula	35.8	8.3	3.2	46.7	3.5	1.3	0.001
Cakes							
Angel	31.5	6.2	0.3	14.3	45.1	0.5	—
White layer	24.2	4.3	10.8	22.2	35.6	0.9	0.002
Yellow layer	23.5	4.5	12.4	22.0	35.1	0.9	0.080–0.112
Pound	19.4	4.7	17.4	24.2	31.2	0.6	0.085–0.118
Old-style pound	17.2	5.7	26.0	21.6	27.6	0.4	0.220–0.281
Cookies (deposit)	2.4	4.7	28.0	37.7	23.8	0.5	0.043–0.052
Pie crust	14.9	4.0	34.8	41.7	0.5	0.8	0.022
Doughnuts							
Cake	23.7	6.5	16.3	33.3	17.0	0.6	0.083–0.108
Yeast-raised							
(without glaze)	28.3	5.4	19.6	42.5	0	1.1	0.037–0.050

Source: Chung and Pomeranz, 1984.

contribute in a significant manner to bread flavor (Table 11-17). The flavor components and precursors are the result of transformation of raw materials, the fermentation process, and the baking process (Figures 11-18, 11-19).

Microbial spoilage can result from survival of heat-resistant spores of *Bacillus subtilis* in bread low in acidity and high in sugar. The damage (so-called rope) is manifested by a bitter taste and fruity (objectionable) odor within 12–72 hours after baking. The contamination of baked foods with fungi takes place from outside through spores carried in the air or by contact with the surface of the equipment or shipping vans or utensils. Mold develops in contaminated bread within 3 days. The hazards of rope and mold are reduced by physical methods and/or use of propionates (0.15–0.30%, flour basis).

COMPOSITION OF WHEAT, FLOUR, AND BAKED GOODS

The composition of bread depends basically on the composition of the dough. Moisture is 35–45% of bread. The carbohydrate content ranges 45% to 58% and is higher in bread baked from white, low-protein flours

Table 11-20. Approximate Food Value of Common Wheat Flour Goods in the USA

Food	Approx. measure (g)	Water %	Food energy (kJ)	Protein (g)	Fat (g)	Saturated (g)	Unsaturated Oleic (g)	Unsaturated Linoleic (g)	Carbohydrate (g)	Calcium (mg)	Iron (mg)	Thiamine (mg)	Riboflavin (mg)	Niacin (mg)
Biscuits, baking powder from mix, 51 mm diam	(1 biscuit) 28	28	377	2	3	1	1	1	15	19	0.6	0.08	0.07	0.6
Rye bread, American, light (⅓ rye, ⅔ wheat)	(1 slice) 25	36	251	2	trace	—	—	—	13	19	0.4	0.05	0.02	0.4
White bread, enriched, soft-crumb type, 22 slices per 454-g loaf	(1 slice) 20	36	230	2	1	—	—	—	10	17	0.5	0.05	0.04	0.5
Whole-wheat bread, firm-crumb type, 18 slices per 454-loaf	(1 slice) 25	36	251	3	1	—	—	—	12	25	0.8	0.06	0.03	0.7
Muffins, with enriched white flour; muffin, 76 mm diam	(1 muffin) 40	38	502	3	4	1	2	1	17	42	0.6	0.07	0.09	0.6
Pancakes, wheat, enriched flour (home recipe) 102 mm diam	(1 cake) 27	50	251	2	2	trace	1	trace	9	27	0.4	0.05	0.06	0.4
Rolls, enriched, frankfurter or hamburger	(1 roll) 40	31	502	2	2	1	1	1	21	30	0.8	0.11	0.07	0.9
Wheat flour, whole-wheat, from hard wheats, stirred	(1 cup) 120	12	1675	16	2	trace	1	1	85	49	4.0	0.66	0.14	5.2
Wheat flour, all-purpose family flour, enriched, unsifted	(1 cup) 125	12	1905	13	1	—	—	—	95	20	3.6	0.55	0.33	4.4
Wheat flour, self-rising enriched	(1 cup) 125	12	1842	12	1	—	—	—	93	331	3.6	0.55	0.33	4.4
Wheat flour, cake or pastry sifted	(1 cup) 96	12	1465	7	1	—	—	—	76	16	0.5	0.03	0.03	0.7

Source: USDA, 1970.

Table 11-21. Average Nutritional Value of 100-g Baked Products

Milled product and type	Main nutrients (g)			Energy		Minerals/mg					Vitamins		
	Pro-teins	Fat	Carbo-hydrates	Calories (kcal)	Joules (kJ)	Na	K	Ca	P	Fe	B_1 (µg)	B_2 (µg)	Niacin (mg)
Rye, flour bread	5.2	0.9	47.8	220	935	520	230	35	130	1.9	160	120	0.5
Rye, whole grain bread	6.8	1.2	45.5	220	935	420	290	43	220	3.3	180	150	0.6
Wheat, white bread	8.9	1.8	51.6	258	1,104	385	130	60	90	0.9	90	60	1.0
Wheat, whole grain bread	7.2	1.2	44.9	219	931	370	210	95	265	2.0	250	150	3.3
Knackebread	11.4	1.7	76.2	366	1,554	460	440	55	320	4.7	200	180	1.1
Pumpernickel	5.2	1.2	44.9	211	916	370	340	55	150	3.3	50	50	0.5
Buns/rolls	8.0	1.0	56.0	265	1,126	490	115	24	109	0.7	70	35	0.9

Reproduced with permission (courtesy, AID, Ministry of Nutrition, Agriculture and Forestry, Bonn, 1982).

Table 11-22. Amino Acids of Wheat, Flour, and Bread (g/16 g N)

	Wheat	Flour	Bread
Alanine	3.25	2.78	2.93
Arginine	4.69	3.80	3.56
Aspartic acid	5.09	4.14	4.60
Cystine	1.97	2.11	1.88
Glutamic acid	28.50	34.50	31.70
Glycine	3.88	3.22	3.21
Histidine	1.92	1.88	1.89
Isoleucine	3.90	4.26	4.32
Leucine	6.48	6.98	7.11
Lysine	2.74	2.08	2.48
Methionine	1.76	1.73	1.90
Phenylalanine	4.42	4.92	4.80
Proline	9.85	11.70	11.10
Serine	5.06	5.44	5.45
Threonine	3.02	2.82	3.01
Tryptophan	1.09	1.02	0.97
Tyrosine	3.10	3.25	3.32
Valine	4.50	4.54	4.68

Source: Bradley, 1967.

Table 11-23. Vitamins of Wheat, Flour, and Bread (mg/100 g dry wt)

	Wheat	Flour	Bread
Thiamine	0.40	0.104	0.46
Riboflavin	0.16	0.035	0.29
Niacin	6.95	1.38	4.39
Biotin	0.016	0.0021	0.0029
Choline	216.0	208.0	202.0
Pantothenic acid	1.37	0.59	0.69
Folic acid	0.049	0.011	0.040
Inositol	370.0	47.0	53.0
p-Aminobenzoic acid	0.51	0.050	0.092

Source: Bradley, 1967.

than from dark, protein-rich flours. The protein contents can be significantly increased by adding protein-rich foods (milk solids, soy flour, germ) and is generally around 6% in white bread from a lean dough formula to 16% in special formulations. The amount of fat is 0.5–2% and is increased significantly by using shortening in the formula. Of 1.5–3.0% mineral components, salt is the major ingredient (1–1.5%). The amount of other minerals depends on the flour darkness (extraction) and optional ingredients.

Nutrients in wheat flour and wheat bread in the United States are compared in Table 11-18; composition of miscellaneous baked goods, in Table 11-19; and approximate food value of common wheat flour foods

Table 11-24. Mineral Composition of Wheat, Flour, and Bread

	Wheat	Flour	Bread
	(Percent dry basis)		
Potassium	0.454	0.105	0.191
Phosphorus	0.433	0.126	0.183
Magnesium	0.183	0.028	0.034
Calcium	0.045	0.018	0.127
	(Parts per million)		
Sodium	45.0	9.8	0.858%
Zinc	35.0	7.8	9.7
Iron	43.0	10.5	27.3
Manganese	46.0	6.5	5.9
Copper	5.3	1.7	2.3
Molybdenum	0.48	0.25	0.32
Cobalt	0.026	0.003	0.022

Source: Bradley, 1967.

(per indicated serving), in Table 11-20. Table 11-21 lists average nutritional value of 100-g baked products in West Germany. Amino acids, vitamins, and minerals of wheat, flour, and bread are given in Tables 11-22, 11-23, and 11-24, respectively.

BIBLIOGRAPHY

Allenson, A. *Bakers Digest* **1982**, *56*(5), 22–24.
Alwes, M.; Jolly, M.J. *Bakers Digest* **1974**, *48*(4), 28–31, 58.
Bond, E.E.. *Brot Gebaeck* **1967**, *21*, 173, 176–180.
Bradley, W.B. *Bakers Digest* **1967**, *41*(5), 66, 67, 70, 71.
Bruemmer, J.M.; Seibel, W. *Getreide Mehl Brot* **1979**, *33*, 135–137.
Chung, O.K.; Pomeranz, Y. In "Advances in Modern Nutrition," Vol. III. Kabara, J.J., Ed.; Chem-Orbital Publ. Co.; Park Forest South, IL; **1984**.
Cole, H.S. *Bakers Digest* **1973**, *47*(6), 21–23, 64.
Davis, E.W. *Bakers Digest* **1981**, *55*(3), 12, 13, 16.
Desrosier, N.W. "Elements of Food Technology." Avi Publ. Co.: Westport, CT; **1977**.
Dimler, R.J. *Bakers Digest* **1965**, *39*(5), 35–38, 40–42.
Dubois, D.K. *Cereal Foods World* **1981**, *26*, 617–619, 621, 622.
Elton, G.A.H.; Chamberlain, N.; Collins, T.H. Lecture R 807, Fourth Int. Bread Congress, Int. Assoc. Cereal Chem. (ICC), Vienna, Austria; **1966**.
Fowler, A.A.; Priestley, R.J. *Food Chem.* **1980**, *5*, 283–301.
Finney, K.F. In "Cereals '78: Better Nutrition for the World's Millions." Pomeranz, Y., Ed. Am. Assoc. Cereal Chem.; St. Paul, MN; **1978**.
French, F.D.; Kemp, D.R. *Cereal Foods World* **1985**, *30*, 344–346.
Kammann, P.W. *Bakers Digest* **1979**, *53*(1), 26–29.
Kulp, K. *Bakers Digest* **1983**, *57*(6), 20–23.
Matz, S.A. (Ed.). "Bakery Technology and Engineering." Avi Publ. Co.; Westport, CT; **1960**.
Moss, R.; Stenvert, N.L.; Pointing, G.; Worthington, G.; Bond, E.E. *Bakers Digest* **1979**, *53*(2), 10–12, 14, 16, 17.
Olsen, C.M. *Bakers Digest* **1974**, *48*(2), 24–27, 30.
Peppler, H.J. *Cereal Foods World* **1981**, *26*, 609–611.
Pomeranz, Y.; Meyer, D.; Seibel, W. *Cereal Chem.* **1984**, *61*, 53–59.

Ponte, J.G., Jr. In "Wheat Chemistry and Technology." Pomeranz, Y., Ed. Am. Assoc. Cereal Chem.: St. Paul, MN; **1971**.

Ponte, J.G., Jr. In "Variety Breads in the United States." Miller, B.S., Ed. Am. Assoc. Cereal Chem.; St. Paul, MN; **1981**.

Pyler, E.J. *Bakers Digest* **1969**, *43*(2), 46–52.

Pyler, E.J. *Bakers Digest* **1982**, *56*(4), 22–26.

Rothe, M. *Ber. 5. Welt-Getreide- Brotkongress* **1970**, *5*, 203–209.

Rothe, M. In "Handbuch der Aromaforschung—Aroma von Brot." Akademie-Verlag; Berlin; **1974**.

Rothe, M. *Nahrung* **1980**, *24*, 185–195.

Rothe, M.; Bethke, E.; Rehfeld, G. *Nahrung* **1972**, *16*, 517–524.

Schmidt, A.M. *Bakers Digest* **1973**, *47*(6) 29, 30, 57.

Schnake, L.D. In "Wheat Outlook and Situation." ERS-USDA, No. 11-14; Washington; **1981**.

Seibel, W.; Bruemmer, J.M.; Menger, A.; Ludewig, H.G. In "Brot und Feine Backwaren," Vol. 152. DLG-Verlag; Frankfurt; **1977**.

Seiling, S. *Bakers Digest* **1969**, *43*(5), 54–56, 58, 59.

Sheppard, R.; Newton, E. "The Story of Bread." Routledge and Paul; London; **1957**.

Smerak, L. *Bakers Digest* **1973**, *47*(4), 12–15, 18, 20.

Smith, W.M. *Bakers Digest* **1979**, *53*(4), 8–10.

Snyder, E. *Bakers Digest* **1963**, *37*(2), 50.

Spicher, G. In "Biotechnology," Vol. 5. Rehm, H.-J.; Reed, G., Eds. Verlag Chemie; Weinheim, W. Germany; **1983**, pp. 1–80.

Spicher G.; Stephan, H. In "Handbuch Sauerteig—Biologie, Biochemie, Technologie." BBV Wirtschaftsinformationen GmbH; Hamburg; **1982**.

Stein, E. vom. *Brot Gebaeck* **1952**, *6*, 165–168.

Stenvert, N.L.; Moss, R.; Pointing, G.; Worthington, G.; Bond, E.E. *Bakers Digest* **1979**, *53*(2), 22–27.

Sugihara, T.F. *Bakers Digest* **1977**, *51*(5), 76, 78, 80, 142.

Thompson, D.R. *Bakers Digest* **1980**, *54*(3), 28–30, 32, 34, 36, 37.

Thompson, D.R. *Bakers Digest* **1982**, *56*(3), 28–30, 32, 33.

Thompson, D.R. *Bakers Digest* **1983**, *57*(6), 11, 12, 14.

Tipples, K.H. *Bakers Digest* **1967**, *41*(3), 18–20, 24, 26–27, 75.

Tipples, K.H.; Kilborn, R.H. *Bakers Digest* **1974**, *48*(5), 34–39.

Tressler, D.K.; Sultan, W.J. "Food Products Formulary," Vol. II. Avi Publ. Co.; Westport, CT; **1975**.

Tsen, C.C. *Bakers Digest* **1973**, *47*(5), 44–47.

Ulrich, M.G. *Bakers Digest* **1975**, *49*(2), 43–45, 67.

USDA. "Nutritive Value of Foods," Home and Garden Bulletin No. 72. USDA: Washington; **1970**.

USDA. "Nutritive Value of American Foods," Handbook No. 456. USDA: Washington; **1975**.

Chapter 12

Bread Around The World

BREAD IN VARIOUS COUNTRIES

Around the world bread is the principal food and provides more nutrients than any other single food source. In over 50% of the countries bread supplies over half of the caloric intake; in almost 90% of the countries, over 30%. In most West European countries it is the source of half the carbohydrates, one-third of the proteins, over 50% of the B vitamins, and over 75% of vitamin E. The value of grain used for human consumption is over 2.5 times the value of world iron and steel production. Although only 15% of the grain in the world is handled through international channels, cereal grains make up for more than half of all goods in overseas trade.

Bread consumption varies widely. In Europe it is still highest in the southern and southeastern countries. It is low in Sweden and in The Netherlands, and intermediate in Central Europe and the Iberian peninsula. On the American continent bread provides about 50% of the nutritional requirements of Mexico, Chile, Peru, and Venezuela and is less important in the diets of the inhabitants of the United States, Canada, Paraguay, and Columbia.

Variations in bread consumption are particularly wide in Africa. It comprises over two-thirds of the total food intake in Egypt, in parts of North and West Africa, and in South Africa. In central Africa, starchy roots supply most carbohydrates and cereal grains are eaten to supply only about 25% of the total caloric needs. Great quantities of bread (up to 75% of total food intake) are eaten in several countries in the Near and Far East. Bread consumption is low in Australia and New Zealand (less than one-third of the total caloric intake).

The national average (per capita per year) bread consumption ranges from about 40 to 300 kg. Flour and its main baked product, bread, are the least expensive of our staple foods. Their consumption in recent years has been decreasing in most Western countries. The decline in bread consumption is related to rising standards of living, increased buying power, and availability of more expensive and more sophisticated foods.

Although meat, poultry, and dairy products are attractive, they are not efficiently produced. It requires 6–8 kg of cereal grains to produce 1 kg of beef. Thus, producing expensive foods is economically inefficient. Great strides in food production have taken us from a single nomad hunter who needed 8 sq mi to provide food for one person to having an equal area plowed and cultivated to support 6,000 people. In the nomadic stage of human civilization, there was enough food on earth to feed 30 million people; today, we can feed more than 100 times that many. We can do it only if cereals are the main source of nutrients in the diets of the people. With the dramatic increase in population, the role of cereals in feeding increasing numbers of hungry people in the world assumes increasing importance. Without entering into a polemic on the timetable for hunger to threaten most affluent nations, most scientists agree that in the not too distant future it will become increasingly difficult to provide foods from animal sources, so our dependence on cereals will increase.

Man mastered the art of bread making thousands of years ago. Excavations of the oldest bakers' oven in the world show that bread was known in Babylon 4000 BC. In the old kingdom of Egypt bread was baked in hot ashes or on heated stone slabs. At least as long ago as 2500 BC wedge-shaped bakers' ovens were known. Bread was baked on the inner surface of those ovens, and it still is in some parts of the Near East today.

Bread and grain played an important part in every phase of the life-span of the Egyptians—from the customs observed in connection with childbirth to the burial ceremonies. Consumption of bread in ancient Egypt was great. Grain and bread production dominated the economic life of the country and contributed significantly to its prosperity. The growth of a bread culture was possible as a result of several developments that took place in the Neolithic period. Farming, cultivation of plants, and plant breeding were introduced, and baking a cereal gruel into a fairly palatable flat bread was discovered. At the end of the Neolithic period (about 1800 BC) dough fermentation was accidentally discovered. Fermentation of a dough provided after baking a leavened, light, and porous product that was more digestible and that tasted better because baking improved its flavor.

From Egypt, bread making including fermentation, spread to other Mediterranean countries. Originally, the staple food in Greece was a flat cake. At first, it was placed on heated bricks, then on clay tubes and plates or on a cylindrical pan, and finally in hemispherical ovens. The flat cakes were round or oblong and made from ground barley. Leavened bread was introduced into Greece from Egypt in the eighth century BC. Greece had commercial bakeries as early as the fifth century BC.

The leaven used in baking was made in two ways. When large amounts were needed in commercial bakeries, the ferment was prepared from millet mixed with must. The mixture was stored for up to 1 year. The leaven was also made from a kneaded mass of wheat gruel and white must that was allowed to stand for 3 days; the fermented dough was dried. To make bread, a portion of the dried ferment was added to a dough. For small amounts of bread, the baker relied on airborne yeasts or bacteria to leaven his dough. Water suspensions or gruels of cereals, allowed to stand at room temperature, were an ideal substrate to develop microorganisms in spontaneous fermentation. The Greeks also made bread without leaven by using grape juice as a ferment. Leavened bread was a delicacy reserved for special feasts. It was made from wheat, barley, or millet. The art of bread making was highly developed, with numerous types of bread available. Bread was important in the culture and religion of the Romans. They developed large commercial bakeries in the second century BC to meet demands of increasing bread consumption. The masters of the Roman trade were Greeks, who established a flourishing milling and baking industry. The Romans carried bread making wherever they went in Europe.

During the Bronze Age (1800–800 BC), wheat was replaced by rye in most European countries. In the Iron Age (800 BC to AD 1), flatter, more sour bread was consumed until it became the main European food. During the early centuries of the Christian era, the advanced Roman baking influenced many colonies. Migration of people during the fourth and fifth centuries ended the development begun in the Roman provinces. From the middle of the sixth century rye bread was produced. White bread was rare; in the middle of the eighth century it was unknown in England and was eaten in France only at princely banquets. A note from the year AD 794 indicates that bread was made from a mixture of spelt, rye, and wheat or of groats, buckwheat, barley, and oats. Oat bread was regarded as the bread of the poor.

In the early Middle Ages, loaves were round or semicircular and flat and were marked with a cross—a Christian symbol that made it easier to break the loaf. White bread continued to be a delicacy. The best bread was made in monasteries. In the ninth and tenth centuries some monasteries produced several types, including unleavened bread, regular leavened bread made of wheat, spelt bread, bread baked in ashes, ring-shaped rolls, moon-shaped confectionery of fine wheat flour, roasted bread, wafers, and special bread.

In far northwestern Europe (Lappland, Norbotten), unfermented and unleavened bread, baked directly on embers or stones, or more often on an iron plate, were and still are produced. In the Scottish Highlands and in certain parts of Ireland unyeasted round loaves from oats and barley were baked. Loaves of the ancient bannock type were fried on iron plates. That kind of baking stems from primitive European houses, hearth houses having no built-in ovens. The fire was made directly on the earthen or stone floor or on an open low hearth in the middle of the living room. Food cooked in the oven and oven-baked fermented bread are typical of the diets of East European peasants.

During the Middle Ages and far into modern times, bread in northern

Europe was baked twice yearly. In the autumn, winter bread was made from grain ground on a water mill and baked to last all winter. In spring, summer bread was made for the entire summer from grain ground after the spring flood. In the latter part of the 19th century a major change occurred in the bread culture of peasants in the North. New fireplace and oven types were installed; the iron oven became widespread, and baking was frequently commonplace.

Bread from fermented wheat meal or flour doughs progressed from the Mediterranean to the North. North of the wheat area was the rye area, in which rye slowly replaced barley and oats. In northern and central Europe, bread was also made from mixed cereals. Mixed bread was characteristic of areas in which the soil and climate were less suitable for rye or wheat. Unfermented flat or thin bread made of barley and oats still is common in the Carpathian mountains, in some Balkan countries, in the Alps, and in Ireland, Scotland, the Nordic countries, and northern Russia.

In England several types of bread were developed in the 15th century. The very best was lords' bread made from white flour. A coarser, wheat-rye meal was used to make maslin, a brown bread for the middle classes. The poorer people made bread from rye, oats, and barley often mixed with legumes. Bran bread was the bread of the very poor.

The bread in France was mainly produced from wheat, but rye was used in those parts of the country where wheat did not grow well. Wheat bread developed into the excellent *pain blanc* and *pain mollet*. Its quality was improved about the time of Louis XIV by baking it in oblong loaves and brewing yeast as a leavening.

German bread was mainly made from a sour-fermented rye. *Herenbrot*, the best, was made from rye flour; *Hausbachenbrot* or *Gesindebrot* was from four parts of rye flour and one part of barley flour. Dark or black rye bread, fermented by a sour dough, is still baked as *Schwarzbrot* in some districts. Pumpernickel originated in Westphalia. The rye bread of Central and Eastern Europe is part of a diet that includes beer and butter. Oil and wine are part of a wheat bread diet characteristic of southern European countries. The Nordic countries were strongly influenced by the German rye bread, and only in the 18th century were French cooking and French baking (including white wheat bread) seen in Nordic courts and homes of the affluent. As cities increased in size, home baking was replaced by commercial bakeries. One of the major events in bread making in more recent times was the development of a method of bakery-yeast production by the Viennese engineer Mautner, which laid the foundation of the now worldwide famous Vienna bread making process. For many years bakers relied for leavening on spontaneous fermentation.

At a later stage, pieces from the previous batch of dough were used as inocula; this practice is still used to make sourdough bread. The underlying principles of fermentation remained a mystery until microscopic studies of Leeuwenhoek in the 17th century demonstrated the presence of yeast cells.

Pasteur studied biochemical activities of yeasts in the 19th century. A yeast industry was in existence around 1800. Brewers provided yeast to bakers from lager beer, a so-called "bottom fermenter." Because that yeast did not perform as well as "top-fermenter" yeasts, which the brewers had previously supplied, bakers turned to a more satisfactory supplier of yeast—the distiller. Over the years, distillers often produced compressed bakers' yeast from yeast they used.

The first industrial yeast plant was built in The Netherlands. From there, manufacture spread to England, France, Germany, and other parts of central Europe. Grain extracts from brewing and distilling furnished the substrates for the production of compressed yeast. Technology improved rapidly following Pasteur's discovery and subsequent work on breeding pure yeast lines by Hansen in Denmark. Introduction of aeration around the end of the 19th century increased yeast growth and laid the foundations of an economically viable industry. During the first decade of the 20th century, sugar-beet molasses slowly replaced grain extracts as a substrate for yeast. That was followed by numerous bioengineering developments and improvements. Industrial production of yeast in the United States began around 1870. By 1930 it spread to Canada, Central and South America, Australia, and New Zealand. Japan has an appreciable yeast industry, and yeast plants have been built in many countries in Asia, Africa, and South America.

Production of bread by a continuous process was "bioengineered" primarily in the United States. Modern commercial bread making by a batch or continuous process involves several basic operations or stages. Flour, bakers' yeast, salt, and other optional ingredients (such as sugar, shortening, and yeast food) are mixed with water to develop a dough with desirable viscoelastic properties. The dough is fermented in bulk under controlled conditions (at about 30°C). The fermented dough is divided into uniform pieces of suitable size, and the pieces are molded and shaped. The dough pieces are "proofed," generally in baking pans, in a final fermentation stage designed to provide a fresh supply of carbon dioxide and a properly developed dough. The proofed dough is baked at about 230°C. During early stages of baking, "oven spring" (rapid expansion of dough) takes place. It is followed by gelatinization of the starch and coagulation of the proteins, accompanied by formation of a rigid structure of a baked loaf. Chemical changes during baking produce the desirable flavor and crust color.

AFRICA

Africa is a continent of great extremes in bread culture and bread consumption. In some parts of the continent bread has been known for about 6,000 years; in others it has been introduced fairly recently.

Bread consumption is very high in Egypt, North Africa, West Africa, and South Africa. In Egypt, cereal grains provide 75% of the calories and 90% of the proteins in the diet. In Kenya, Uganda, and the Congo, cereal grains constitute a relatively small amount of the total food consumed. The main food of the Near East and North Africa is wheat; barley and corn are important supplementary cereals in some of those countries. People in parts of tropical Africa eat sorghum; in some of the coastal areas of West Africa and in Madagascar, rice predominates; and corn is the staple food in the east and south of the African continent.

Egypt

There are about 100 types of bread baked today in Egypt. European-type breads (mainly French bread and sliced and wrapped pan bread) are popular in cities and comprise up to one-third of the bread consumed. In villages, flat bread is the most common. Whereas in cities almost all the bread is made in commercial bakeries, in villages only 20% is made by professional bakers. The local flat *balady* bread with a diameter of 25 cm is the most widely accepted. It appears in two forms: the *maui* and *mayar* types. In addition, there is the Syrian bread that goes through a second baking stage at 200°C for 2–3 minutes. The Syrian bread is often made from an Arabian *balady* dough. Bread is the important food of the Copts, the Christian minority in Egypt. The bread is baked in bell-shaped ovens. It is called *batavah* in upper Egypt and *marahavah* in lower Egypt. The round loaves, up to 7 cm high, are made from sour-fermented doughs. Thinner loaves are baked from rather stiff doughs that are sheeted to a diameter of up to 75 cm.

Sudan

Cereals provide about half the calories and over half the proteins in the food of the Sudanese. Much of the wheat is imported. About 90% of the locally grown wheat is durum, and only 10% vulgare. Both in cities and on the farm, 80–90% of the bread is produced in commercial bakeries. The *balady*-type bread baked in Sudan is called *shamsi*. It is 35 cm in diameter and is baked at about 250°C for 10–15 minutes. In the city of Khartoum, capital of the Republic of Sudan, there are many dozens of small, relatively primitive, and in part hand-operated bakeries. Most of them prepare bread by a straight dough formula that includes flour, water, salt, and active dry yeast or sour. The dough is fermented for 90 minutes and scaled to produce loaves that weight 200 or 400 g. The dough pieces are rounded, placed on wooden trays with a coating of dusting flour, proofed for 1 hour, flattened, and reproofed for 30 minutes before baking. If the sponge process is used, about one-third of the total amount of flour in the form of a sponge is fermented for 4–5 hours. Then the remaining ingredients are added and mixed in the usual manner. In production of "sunny" bread in Sudan and in upper Egypt, the flattened piece is exposed to direct sun for final proof. The intense heat imparts a special flavor to the baked bread and is preferred by older people.

North Africa

One of the strong points in the development of ancient Carthage (around 815 BC) was the availability of a strong indigenous agriculture that included growing of cereals by the local Berbers. In the fifth century BC, the plantation system employed many thousands of slaves to increase production of cereal grains. Majs, one of the founders of the Carthaginian empire, presumably authored 28 books on various aspects of agriculture. With the destruction of Carthage by the Romans, there started an era of Roman influence in agriculture, milling, and bread making. The prominent role of North Africa in supplying the Roman Empire with wheat is well established. The milling and baking methods of today's peasants or nomads (Bedouins) are rather primitive and part of the housewife's obligations. The wheat is moistened and pounded in a large mortar, the bran is separated manually by sieving, and the dehulled kernel is air-dried.

The Libyans eat mainly European bread of the French and Italian type. During religious festivals, especially fasts, the Arabic *balady* bread is baked.

Inhabitants of Morocco eat four main types of bread: *kesra*, a relatively flat 400- to 500-g loaf; *pain boulot*, an 800-g loaf; the oblong *pain flute* that weighs 500 g; and the small, oblong *pain flute* of 250 g.

Other African Countries

Teff is a special bread eaten in Ethiopia. The bread is baked from ground seeds of *Eragrostis teff* in the form of dark brown or black doughnuts. In the highlands of Ethiopia, bread is made from the tiny grass seed *Poa abessinicum*.

Introduction of bread to West African countries is relatively new. "Sugar bread" and "butter bread" are quite popular. All the bread—usually sold from baskets by women vendors—is well baked and is consumed as a fresh product. The bread is baked in small bakeries. The bakers employ low yeast levels to slow down fermentation in the hot climate. The bread is baked in some small bakeries in ovens fired with wood on the hearth; in others, more modern imported ovens are used.

Flat bread is an important item in the diets of workers in coconut and copra plantations on the island of Zanzibar. Its preparation starts at night, and it is eaten fresh for breakfast. The bread is made from wheat flour, water, and salt; the leavening is chemical (a mixture of sodium carbonate and bicarbonate). Late in the afternoon a dough is mixed from wheat flour, water, salt, and a small piece of dough from the last bake. The dough is allowed to stand in a large pan until midnight. Then a small amount of chemical leavening is dissolved in water and kneaded into the dough. At the same time the oven is being prepared. The oven is heated with wood in the inner chamber, which also serves for baking. The inner chamber resembles a flower vase about 80 cm high, and it has a small hole for ventilation and removal of ash. Loading the oven with wooden sticks 30–40 cm long is done from above. To decrease the hazard of fire, the oven is covered by a plate about 60 cm above. After the fire subsides, the ashes are extinguished with a stick that has at the bottom a piece of cloth previously immersed in a salt solution. The residual salt after evaporation of the water improves the taste of the baked product. The dough is allowed to stand for about 2.5 hours and shaped as flat, round, 90-g pieces; the outside of the piece is thicker than the center. The pieces are placed on a bag covered with coconut fibers, plastered against the wall of the oven, and baked for about 4 minutes. The baking is completed as soon as the pieces peel off the wall, and the pieces are removed with a nailed stick. The heat of one round suffices for about 25 bakes. The outside, thicker circumference is somewhat browned; the rest is pale. The bread is tough and leathery. The taste is flat. Some bakers improve their products by covering the dough pieces with fat.

The bread in cities of South Africa resembles that consumed in Western countries. Consumption of cereal grains in South Africa is quite high, though the methods of their processing are often primitive. The government has encouraged baking of yeast- leavened bread from wheat flour mixed with corn, peanut flour, kaffir (local sorghum), and sweet potatoes.

NEAR AND MIDDLE EAST

The significance of cereals, in general, and wheat, in particular, as a source of calories and proteins in the diets of people in most of the countries in the Near and Middle East is illustrated by the data in Table 12-1. The typical flat bread of the Near and Middle East requires wheat of medium-low strength. Durum wheat, grown extensively in the Middle East, is incorporated into the grist. Most of the flour is of high extraction. However, in practically all countries short-extraction white flours are available. The flour is baked in small, primitive bakeries. In large cities, however, there has been a large increase in highly modernized and mechanical bakeries. In the small bakeries, the bread is unleavened, yeast-leavened, or sourdough. Seldom more than 1% of yeast is used.

Bread is baked from soft doughs mixed by hand or in relatively simple mixers. In some cities, however, production of all types of bread (including flat types) is fully mechanized; the facilities include mixers, specially designed rollers, and modern ovens. The most primitive baking is on heated stones. Most widely spread baking is on round, slightly convex,

Table 12-1. Cereals as Sources of Calories and Protein in Human Nutrition in the Near and Middle East

Country	Calories (% of total)	Protein (% of total)	Wheat (% of total)
Egypt	74	89	50
Afganistan	66	91	74
Syria	59	81	80
Iran	52	72	96
Lebanon	48	62	90
Jordan	45	58	86
Sudan	45	55	7
Iraq	43	61	85
Saudi Arabia	26	35	54

Source: FAO, Rome.

metal plates. A somewhat more sophisticated oven is the *tannouri*, which is in the form of a cupola and is made from clay. It is heated at the bottom; the shaped dough is introduced through an opening at the top, which serves also as water vapor exit, and is plastered on the side walls. The square clay or brick ovens, with vents for gases from burning fuel and with side heating, are used in small commercial bakeries. In primitive baking on farms, any organic material that can be burned is used. In cities, wood is used to a limited extent and fuel oil is used commonly.

The two main bread types are *tannouri* and *balady*. *Tannouri* bread is generally unleavened, paper-thin bread of 35 cm diameter; its water content is 15–18%. The bread is actually a dried dough rather than baked; it has several dark spots from short baking. Sometimes a sour is used. The popularity of this bread is on the decrease.

Balady is the most widespread type. It has a diameter of about 25 cm, is baked 1–1.5 minutes at 500°C. It is a two-layer bread that can be filled with vegetables, cheese, or meat. It is made from soft-slack doughs fermented for from three-quarters of an hour to 2 hours and baked as 120 to 170-g pieces. *Balady* with a central hole is called *kubs* in Lebanon and *kmag* in Jordan. *Armany* is baked in Jordan; it is 5 cm high (compared to 1–2 cm in *balady*), has a diameter of 15 cm, and has a hole in the center. *Armany* is generally better leavened than *balady*. The *balady* types of bread have a water content of 40%. They have a slightly caramelized outer surface with brown spots and no true crust.

Following is a partial list of bread types baked in some countries of the Near and Middle East.

Iran

Several types of bread are baked. They include the dry *tannouri*-type bread of 30 cm diameter and 1 cm height. *Lavash* is a square, *tannouri*-type bread, which includes rice flour and milk. The Persian oven (about the size of a barrel) is built on the ground, the sides being smooth masonry. The fire is built at the bottom and kept burning until the walls of the oven are thoroughly heated. *Sanguak* is a triangular, poorly leavened product baked in a square oven, one side of which is filled with stones, and which is heated with fuel oil. *Barbery* is baked in regular ovens in the form of oblong loaves 60 cm in length, 20 cm wide, and 2 cm high. The European bread types that are popular in Iran include French white oblong, Viennese small baked goods, and pan and hearth-baked bread.

Syria

About 70% is the *balady* type of 25 cm diameter, known as *kubs* or *kmag*; the production of *tannouri* is relatively small (about 20% of the total). In cities, European-type bread is available.

Iraq

The *tannouri*-type bread is called *gubbuz*, has a diameter of 35 cm, and is baked 2 minutes. *Samoan* resembles a European-type bread. It is baked in the form of 150-g, well-leavened loaves.

Jordan

The Jordanian *kmag*, a *balady* type, has a diameter of 25 cm and is baked 1.5 minutes at 500°C. It is slightly better leavened than the regular *balady* and comprises the bulk of the bread consumed. The *tannouri* has a diameter of 40–50 cm. *Armany* has a diameter of 15–25 cm and a height of 5 cm, has a hole in the center, is baked 15 minutes, and is better leavened than the *balady*.

Lebanon

A light *balady* with a diameter of 25 cm and baking time of 1 minute is widespread; a brown *balady* type is baked to a limited extent. Both types have a hole in the center; their combined production covers the bulk of the total bread. The large, underbaked *tannouri* is not very popular. The balance, especially in cities, is French long bread.

Israel

Only within the past century has the bakery tradition been established in Israel. In older times people in towns prepared bread at home and brought their molded loaves to a public oven for custom baking. These customs have not yet disappeared, and in many families of Oriental origin bread making is still the housewife's obligation.

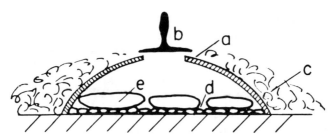

Figure 12-1. Cross section of a *tarboon* oven near Nazareth, Israel. Cupola wall (a); lid of cupola (b); ash and cinders (c); pebbles (d); and bread (e). (From Adler, 1958.)

Today, however, the Oriental type of bread no longer represents an essential part of bread production. European types of bread are common, not only in towns but also in small settlements. In Arab villages one may see small huts built of clay with low entrance doors. During the morning hours smoke rises through small holes and clefts in the walls of the hut. The hut is, actually, an oven called *tarboon* (Figure 12-1) which is heated and prepared for bread baking. This is the oven of the *fellahin*, or settled peasants. It consists of a cupola of burnt clay, about 3 ft in diameter, with a hole on the top that can be closed with a lid. The floor of the interior hearth is covered with a layer of pebbles for conservation of heat. The fire surrounds the cupola and is made of any available fuel. When the fire has burned down, ashes and cinders are heaped around the outside to keep the heat in, and the oven is ready for baking.

During the previous evening a soft dough has been prepared and left to rest overnight. Either a piece of old dough or compressed yeast is used as ferment. After the dough is fermented, it is divided and quickly shaped by hand into round, flat loaves which are immediately put into the oven without further proofing. The oven is loaded several times through the hole at the top, each batch consisting of 6–8 loaves. Bread of the first batch, exposed to flash heat, is sometimes burned on the outside while the inside is underbaked. The bread has a thin, soft crust, indented on the bottom from the pebbles on which it rests while baking.

Bedouin and Iraqui Bread Making

The above system of bread making is not suitable for the Bedouin, whose nomadic life dictates a more mobile baking tool—a curved iron plate called *sadj* which is placed on stones. While a fire is burning below, the soft dough is mixed and spread immediately over the plate. When one side is done, the bread is turned and baked on the other side. The *sadj* is used not only by the Bedouin but also by Kurdish Jews: it is assumed that the latter have brought it to Israel.

Ash tanour is the bread of immigrants from Iraq. It is similar in appearance to that baked on the *sadj*, although the baking process is quite different. The oven is similar to the *tarboon*—a cupola built of clay, about 3 ft high and the same in diameter. There is a fire hole near the bottom and a loading hole about 1 ft in diameter at the top. A small fire is maintained on the hearth. The worker spreads the soft dough over a pad made of flour sacks, to which a short handle is fixed. He puts this through the upper hole into the oven and sticks the dough against the inside of the arched cupola wall, where it bakes while adhering to the wall. The bread is very thin—only about a quarter of an inch thick—and is done in less than 1 minute. It is then torn from the wall and taken out with large, paddlelike implements (tongues). By the time the loading is finished, the first loaf is already done; it is removed and replaced by another sheet of dough.

Oriental Bread

The most common type of Oriental bread baked in Israli towns is the *kimaj*. It is a round disc about as thick as a finger, with an internal split into two layers. Sandwiches are often made with it, by cutting one side to expose the open space and filling it with meat or vegetables. To make this bread, a stiff dough is fermented by yeast. It is divided and rounded by hand, flattened with a roller pin, placed on a peel after a short proof, and baked for 1 minute. The oven is made of brick and is similar to a Dutch oven with a high arch. With the fire burning on one side of the hearth, bread is continuously put in on peels and removed from the other side. When put into the oven the dough immediately becomes inflated; after a short time it drops again and is almost as flat as before, but the inside is now split into two layers.

European Bread in Israel

The standard bread is a brown bread prepared from wheat flour of 86–87% extraction, fermented by sour, with some yeast added. It is baked on the hearth in 2-lb loaves, round or oblong. White bread baked from 75% extraction flour is available in various forms such as pan bread, Pullman loaves, French and Vienna bread, and in different, although fixed, weights. Generally it contains no other ingredients except yeast, salt, and oxidizing and rope-inhibiting ingredients. The traditional Sabbath bread and rolls called *challah* are baked with various additions of sugar, oil, and sometimes malt. Many bakers also produce specialty breads such as dark breads under the names of whole grain, caraway, pumpernickel, graham, and others. Most of the standard and regular white bread in Israel today is produced by the Chorleywood process.

Turkey

The principal cereal consumed in Turkey is wheat. Wheat is prepared and consumed in various forms. *Gendime* is made by separation of the wheat hulls and used for the preparation of various foods. *Kavrak* is made by roasting *gendime*; for bulgur, purified wheat grains are boiled and hulled.

Wheat is also used to prepare sweet dishes. The wheat is germinated for several days, dried, ground, and mixed with concentrated grape juice and consumed as *uhut*. *Tarhana* is made by lactic acid fermentation of a dough prepared from flour, tomato juice, paprika, yogurt, and various spices. It is dried and used in the preparation of soups.

In Turkish villages bread is generally baked from wheat; other cereals are used to a limited extent. The wheat is ground on stone mills, and the coarse bran is separated by sieving. The extraction of the flour used in bread making is about 95%. Both yeast-leavened and unleavened bread types are produced. The Turkish farm bread often contains one or more of the following: milk, cheese, minced meat, meat with onions, walnuts, sesame, eggs, poppy seeds, vegetables, fats, and sugars. The bread is baked in an earth oven on thin iron plates on the floor or in a tray above a wood-burning peel oven. Thin layers of unleavened bread are baked on a round iron or steel plate, 60–100 cm in diameter and 2–2.5 mm thick. Some bread is baked between two thin iron plates. A thin layer of dough is prepared from leavened dough, placed on a heated iron plate, and covered with another hot iron plate with hot ashes on top of it. Bread baked by this process is called *kapama* or *kastro*. *Gomme* bread is prepared from a stiff dough of flour and milk. The dough is 35–40 cm in diameter and up to 5 cm thick. The dough is placed on a hot stone and covered with a thin iron plate with ashes around and on top of it.

The earth oven is used in the north and high plateau of Turkey; it is called *tandir*. The oven is made from bricks directly in the soil of the living room or the kitchen. A thin air pipe extends outside for better draft. Thin layers of dough are pasted on the hot inside walls and baked. Bread is baked on pebbles that have been removed after heating in an air oven.

The most common bread is *bazlama*. Flour is mixed in a trough with saltwater and a piece of old dough. After 1.5–3.0 hours of fermentation, the dough is cut into 200- to 250-g pieces with a *cysiran* and rounded into a ball by hand. The ball is placed on a cake board, *yaslagac*, rolled into a dough 4–5 mm thick, and baked on a hot, thin, iron or clay plate. After baking, bread is removed with a tongue (*pislegec*) and placed on a plate between two fires for final baking accompanied by puffing.

Yufka bread is made from unleavened dough. The dough is kneaded in a trough with water and salt and allowed to rest for about 30 minutes. The bulk is cut into 150- to 200-g pieces that are made by hand into a ball. The ball is made into a thin sheet (40–50 cm in diameter, 1 mm thick) with a long, thin stick (*oklova*) on a cake board (*tabla*). *Yufka* bread is baked on a hot iron plate. *Pisili* is a *yufka*-type bread baked in a peel oven. About 7–10 dough pieces are placed on top of each other. The bread is often baked from a dough containing ground walnuts, eggs, or mint. The Iranian *lavash* is baked in Turkey from either leavened or unleavened dough. The dough is cut into pieces 30–40 cm in diameter, 1–1.5 cm thick, weighing 250 g. They are often brushed with eggs before baking. *Hamursuz* (or *corek*), *peksimet*, and *halka* are three types of bread from fermented doughs of the *bazlama* type baked in a peel oven. In cities, hearth bread is baked in European-style bakeries from bread wheat alone or bread wheat mixed with durum, rye, and other cereals. Yeast leaven and sour propagation are both used. Most bread is hearth-baked in the form of 550- to 750-g loaves of various European types.

FAR EAST

Burma's version of flat bread is rice cakes made of crisp thin sheets called *mon-le-bway*, meaning whirlwind cake. Made of rice flour batter, the sheets fry feather light and are filled with air bubbles.

In the village of Sargaz, 10,000 feet high in the Wakhan Valley, the inhabitants are cave dwellers. Women bake on a fire *nan* an unleavened bread from a coarse wheat meal that serves as an Afghanistan mainstay. The round, flat cakes of unleavened bread are more like pancakes, and are called *chapatties*. The cook shapes them between his hands and bakes them on a griddle or on coals.

The Pakistani *chapatti* forms the basis of every meal in West Pakistan. In poor households the *chapatti* is virtually the whole meal, and any other food is spread thinly over the bread like butter.

India

Cereals and legumes provide about 80% of the calories in the diet, though there are significant differences among the 15 states of India. Rice is the main cereal in West Bengal, Bihar, Orissa, Mysore, Madras, and Kerala. In the northern states (Madhya Pradesh, Kashmir, and Uttar Pradesh), consumption of wheat, legumes, and sorghum is high. In view of the high cereal consumption, the legumes in the diet are very important.

In the past 20 years, India has quadrupled the number of modern mills. Many of these mills process newly developed, high-yielding Indian wheat varieties. Much of the locally grown wheat is still ground on primitive stone mills to a wheat meal for *chapatti*. In the south, about 78% of the flour is removed and only 10% of the grain is *atta* for *chapatti*; in the north, 45% *atta* is made after 40% flour (*maida*; fines) has been removed. In addition, a 95% *atta* is made after coarse bran has been removed.

Flat baked products are the most important food, next to rice, in many parts of India. The most important is *chapatti*. It is baked from *atta* prepared by the farmer's wife daily on a primitive stone mill. The stiff dough from meal, water, and salt is made into 100-g pieces that are flattened by punching or by rolling. The baking, or actually roasting, is on a flat (slightly concave or convex) iron plate that is heated directly or from below. The *chapatti* is baked on both sides for several minutes, depending on the thickness of the dough. It is eaten warm with curry rice or cooked legumes. The *chapatti* has a somewhat raw taste and a rather stiff and coarse texture. It is produced in various types and forms, such as *paramitha, poori,* etc.

Tandoori Roti

These are thick *chapatties*, baked in a special clay oven that has an opening on the top for introduction of dough pieces that are pasted on the surface heated by burning wood. The oven has side holes for ventilation; the baking time is 5 minutes, and the product is baked more thoroughly than the regular *chapatti*.

Maka Ka Roti

This is baked in a flat, covered pan; the stiff flat dough pieces are made from corn, water, and salt. In addition to wheat *atta*, milled corn, sorghum, and legumes are added in varying amounts depending on availability. In some locations, up to 50% (normally 10–25%) of the other cereals and legumes are added. In other locations, up to 20% soybeans or tapioca flour, or 5% sweet potato flour is added.

Some *chapatties* are made from corn *atta* only; some are made from ground *jowar* millet. In addition, fermented rice products are eaten in some parts of India. They are made from rice and legume flours (3:1) that are made into a thick slurry and allowed to ferment at 35°C for 8–12 hours. They are then steam-cooked or baked-toasted. Some baked products contain legumes as the main ingredient. In addition to lentils, red gram, pigeon peas, black gram, and green grams are used. Some are sweetened, others are strongly spiced, and most are baked with fat.

People's Republic of China

The Chinese introduced the leavening of wheat flour at an early date (by about 1200 BC). A traditional Chinese bread called *man-t'ou* is made from a leavened dough steamed in a porous container above steam or boiling water. *Man-t'ou* is fermented for about 3 hours on a damp cloth at room temperature, scaled, molded into various sizes (25–100 g, generally), and steamed for 20 minutes. The final product has no crust. There are many modifications of the homemade *man-t'ou* dough formulation, size, shape, and preparation. *Hua-chuan*, steamed rolls that resemble *man-t'ou* are multilayered with oil between the layers; generally they are not sweetened. Baked bread is available in most cities. It is generally produced in batches of 250 kg flour mixed with 112.5 kg water in a slow mixer. Other ingredients (per 250 kg flour) include 0.5–3.5 kg salt, 12.5–37.5 kg sugar, 1.25–3.75 kg compressed yeast, and 3.5–10 kg peanut oil. The use of milk products or malt is small; some lard is added in small amounts. Bread is produced by the sponge method (1/3 + 2/3); fermented in each stage for 4 hours, followed by a 40-minute proof; then baked for 7–30 minutes at 180–220°C in a three-stage tunnel oven. In addition to the mixer and the oven, commercial bakeries have large fermentation vats, molders, and cutters. Completely modern, automated bakeries are being installed in the large cities.

Main bread types marketed are 100-, 150-, or 250-g panned plain (more recently also larger sizes) and 150-g *tien-yuan mien* pac (sweet round) or fruit bread (with nuts and plums). Much of the bread is freshly consumed, but more recently some is sliced and wrapped and the formulation is designed for improved shelf life.

Japan

The Japanese first came in contact with Portuguese bread about 400 years ago, but only for the last 100 years has there been any sizable production of bread. The Japanese word for bread is *pan*—from the Latin. Bread was introduced into Japan by Portuguese, Spanish, and French seamen. Most of the first bread sold in the vicinity of Nagasaki was of the French type. This type spread widely and was for a while considered the prototype of European bread.

The English domination of the seas and the frequent visits of Englishmen to Japanese harbors brought, at the turn of this century, some changes in Japanese eating habits. The English influence was so strong that most of the bread baked prior to World War I was of the English type. Bread baked in tall, narrow pans, a typical English bread, is still known by many. The Germans have taught the Japanese to bake rye bread, though its production is small. American-type bread was introduced to Japan at the turn of the century. After World War II, the baking of panned, 1-lb loaf and Pullman bread became widespread. Today, the types of bread made in Japan include American-type *kin pan* (1 lb), *kin shokupan* (Pullman type), *roru pan* (soft rolls); English-type *gin shokupan* (old English tin); French-type *bagetto* (French *baguettes*), *koppe* (French *michette coupée*), *genkotsu* or *oketsu pan* (French michette fendre type); Viennese-type *vin pan* (old Viennese); and German-type *doitsu pan* (hard rolls, *Broetchen, Semmel*). Figure 12-2 shows a variety of Japanese breads and rolls.

AUSTRALIA

For many years the types of baking that developed in Australia were patterned after those that prevailed in Europe, particularly in Great Britain. By 1803 wheat was produced or imported in sufficient quantity to warrant the construction of both wind- and water-powered mills in population centers. The grain was ground in public mills, and the flour was baked into bread in public bakeries. The bread was supplied three times each week, the day after the baking was completed. The full ration was 9.5 lb of bread per man per week.

Figure 12-2. Japanese bread types. (Courtesy U.S. Wheat Associates, Tokyo.)

Only one quality of wheat flour was allowed to be made during the early days of the colonies. It was composed of a meal made by removing 24 lb of bran from 100 lb of wheat. Regulations required that no bread be sold until it was 24 hours old. The baker was forbidden to charge more than the amount fixed by the Crown.

Toward the middle of the 19th century wheat became plentiful, numerous towns grew, and the baking industry became established along traditional lines. In 1901 the first federal regulation of bread manufacturing was established. Soon after 1900 the baking industry began to move slowly from hand operations to mechanization, and yeast became available to replace brewers' yeasts and sourdough methods. In 1908 a large bakery incorporating the machinery available at that time was opened in Melbourne, followed in 1910 by similar operations in Sydney. In 1914 the Brisbane Automatic Bakery was established which is considered to be the first fully automated bakery in Australia.

Bread remains a staple food in Australia despite increasing availability of new prepared foods, but consumption has decreased. Although the J. C. Baker process developed by the Wallace and Tiernan Processing Company in the United States was tried out early in Australia, the bread type produced proved to be unpopular in Australia. However, the concept of producing bread quickly without the need of bulk fermentation became an objective which stimulated a great deal of research. In 1962 the Bread Research Institute of Australia obtained a patent for a process known as Brimec, by which a range of bread types from the conventional crumb to a fine, close, and fragile crumb could be produced by varying operating conditions.

NEW ZEALAND

The history of bread making in New Zealand goes back to the beginning of the 19th century. The first wheat grown as a crop was from seed brought in 1814 by Samuel Marsden from New South Wales, Australia. Early in 1815, William Hall, one of the missionaries who came with Marsden, built a flour mill. By 1840, with the start of organized immigration, wheat acreage increased and many watermills and windmills were built. The Partington mill in Auckland, which operated for over 100 years, goes back to that period.

Women baked bread in regular camp ovens and in large camp ovens ("colonial ovens"). In addition to bread, unleavened potato scones were popular. Another cereal food, "damper" was made from a dough baked in hot ashes.

Commercial bread making started in New Zealand around 1840 with the foundation of new settlements. One of the 57 French immigrants to Akaroa was a baker, and it is likely that he started to bake as soon as flour from the first wheat crop became available. A trough and an oven were the main equipment of an early New Zealand bakehouse, and there was also a cask for the leaven. The top of the trough served as a table and a workbench. All work, including dough mixing, was by hand. The output of a bakery by this method until 75 years ago was a batch of about 1–2 sacks of flour. The earliest settlers made bread leaven or "spon yeast" from boiled hops and scalded malt which were fermented, and then mashed potatoes were added. Later bakers and housewives used brewers' yeast. Up until the 1980s, bakers near towns used to boil and mash about 3 gal of potatoes and add brewers' yeast to produce 6 or 7 gal of ferment which was allowed to stand for half an hour. Most bakers used an overnight sponge with a dough time of 1–5 hours.

Country bakers made a "patent yeast" or "bakers' yeast" from hops, malt, and potatoes and often added bran. Doughs made with the yeast ferment required 6–8 hours of fermentation. Sponge bread was also made, mostly from an overnight half or three-quarter sponge which in the morning was doughed up 1 hour before scaling and baked 2–2.5 hours after scaling. Since 1914, the introduction of commercial compressed bakers' yeast has largely removed the other types of yeast from commercial use. Some 70 years ago, most of the bread made in New Zealand was of the hearth type. The main loaf was called a "French" or "turnover." In addition, there was a "barracouta," which was oval in shape and had a split center, and the round-shaped "scone." In recent years there was a gradual shift from hearth to sliced and wrapped pan bread. The consumption of bread in New Zealand continues to be relatively high. Bread manufacture has been largely mechanized. Bakeries are rapidly aggregating into large plants, distributing in a radius of up to 100 miles. Continuous breadmaking and "no-time" dough processes are rapidly replacing conventional fermentation.

EASTERN EUROPE

The Soviet Union

Bread in the Soviet Union has a central role in the diet; it is part of practically every meal. Bread and salt are symbols of hospitality.

The Soviet Union is basically a grain-producing country, and wheat is the most important crop (almost half of the total grain acreage). Most of the wheat and rye produced in the Soviet Union are used in bread making. Bread is the basic food of the Soviet citizen, who eats about six times as much of it as the U.S. citizen (500 vs 80 lb per capita per year). Of all bread consumed in the Soviet Union, 80–85% is from wheat flour, the rest from rye. Nearly all bread making in large cities is done in large, well-mechanized bakeries (Figure 12-3). Most bread is sold unsliced and unwrapped.

The baking industry in the cities of the Soviet Union is a highly mechanized branch of the food industry. There are over 2,500 bakeries which in some cities produce an average of 500 tons of bread per day. Some bakeries use specially prepared flours and liquid yeast produced in the bakery.

Standardization of bread manufacture in the Soviet Union encompasses flour production, dough composition and baking, and bread manufacture. Wheat flours include fancy white, straight white, dark (85% extraction), and whole wheat (96% extraction). For manufacture of rye and mixed-rye bread, rye flours of 63, 87, and 95% extraction are produced. White bread and white rolls are baked with the following additives (parts, flour basis): 0.8–1.5 yeast, 1.25–1.50 salt, and 0–6 sugar. Specialty baked goods also contain butter, margarine, shortening, milk, or eggs. Rye and mixed wheat-rye breads contain about 1.5% salt and 0.003–0.5% of a pure yeast culture. Various special rye breads contain 3–7% dark malt extract, 5% light malt extract, 1–5% syrup, 0.1–0.5% spice extracts, or seeds. Most wheat bread is made by the sponge dough method which saves yeast and also yields a product that is preferred by the consumer. The sponge rests for 3–4 hours at 28–32°C; the final dough rests at 28°C for 1–1.5 hours.

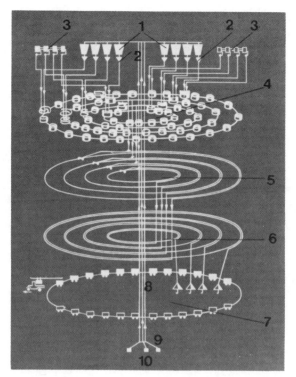

Figure 12-3. Schematic diagram of a Russian circular bakery. Ingredients are blended on the top floor (1-4), dough makeup (5) and baking (6) take place on lower floors, and delivery of bread and supplies (7-10) on bottom floor. (Courtesy Periscoop, Inc.)

Straight doughs are fermented 2–4 hours (depending on baking strength of the flour) at 28–30°C.

Part of the rye bread is made by a four-stage sour fermentation process, though a shortened one-stage fermentation is more and more popular. For both wheat and rye breads, various pure yeast cultures are produced in special departments of large bakeries. There are many types of rye bread; most are hearth oven, a few are panned.

The following are some of the types: whole rye and whole grain (55–65% rye, 35–45% whole wheat); light rye and special breads such as *Borodino* bread (80 parts whole rye, 15 parts dark wheat flour, 5 parts dark malt extract, 4 parts syrup, 6 parts sugar, 1.5 parts salt, and 0.5 parts spices); Riga bread (85 parts light rye flour, 10 parts light wheat flour, 5 parts light malt extract, 5 parts syrup, 1.5 parts salt, and 0.4 parts caraway seeds); and Minsk bread (similar to Riga bread). Wheat bread is made from white, dark, or whole grain flours.

Buns and rolls are made from wheat flour and contain (on a 100-kg flour basis) 0.7–1.5 kg yeast, 1.25–1.5 kg salt, 0–6 kg sugar, 0–3 kg syrup, 0–3.5 kg margarine, 0–0.8 kg eggs, and spices, raisins etc.

Typically Russian are the Moscow *kolatsch* (a pocket-shaped bun) and pretzel-type *suschki*, *baranki*, and *bubliki* which retain their excellent taste for months after baking.

Caucasian home bread is made from butter-yeast dough in the form of round, flat pieces (diameter 170–175 mm, thickness 30–35 mm). The dough is prepared manually or in a mixer. Yeast is dissolved in milk at 30–32°C, and butter, sugar, vodka, salt, alcoholic tincture of safran, eggs, and flour are added. The dough is fermented at 27°C for 90–105 minutes. The fermented dough is cut into 300-g pieces, rounded, and pressed into round pieces (diameter 140–150 mm, thickness 15–18 mm). The surface is covered with egg. The pieces are given a final proof of 15–20 minutes on large, fat-covered pans. Baking time is 8–10 minutes at 190–200°C.

Bulgaria

The inhabitants of Bulgaria consume large amounts of bread—about 700 g per day. Until the end of World War II, practically all farmers and many city dwellers baked their own bread. With recent industrialization and development of cooperative settlements, the picture has changed and most bread production is done in state bakeries. The bread is produced commercially in 32 types that are officially approved. In addition, 97 types of breakfast baked products are on the market. The bread types *Sofia*, *Zagora*, and *Dobruja* are the most popular. Sofia bread is baked from dark wheat flour; the loaf is large and has a coarse texture. The bread is made by the sponge method; a sour starter in the first stage is used to make the main dough. The sour is slack (100 kg flour and 80 L of water) and contains all the yeast. The dough is fermented for 6–8 hours at a starting temperature of 27–29°C, mixed with additional 5% flour, and fermented for another 40 minutes. The final dough is prepared by mixing the starter sour with the rest of the flour and 1.4–1.7% salt (on flour basis) and about 10–15% of dough from a previous bake to increase acidity. The dough is fermented for 30–40 minutes at 26–29°C, remixed, and fermented for another 20–30 minutes. The dough is scaled, rounded, allowed to rest 5 minutes, shaped into oblong loaves, placed on long wooden pallets, and given a final proof of 30–35 minutes at 30–32°C at a relative humidity of about 85%. The bread is baked for 40 minutes at 220–230°C.

The bread types *Stara Zagora* and *Dobruja* are baked from relatively white flour (0.7% and 0.5% ash in dry matter, respectively) by the sponge method. The manufacture of the two white breads has increased substantially at the expense of the dark *Sofia* bread. The breakfast baked products vary widely in size, form, filling, and coating. There are 25 types of buns that have several optional ingredients—jams, jellies, cheese—or may be sprinkled with poppy, sesame, salt, sugar and various spicy seeds. The

Figure 12-4. Bulgarian spiral cross-bird *kolatsch* bread. (Courtesy Brot und Gebaeck.)

typical Bulgarian products include *banitza, milinki,* and sesame seed–covered *gevretzi* and *mekizi.*

Banitza is made from thin layers of folded dough filled with dairy products, meat, fruits, or vegetables. The dough is basically unfermented. It is prepared from white flour, water, salt, fat, sunflower seed oil, and eggs. The thin dough containing all ingredients and half the amount of eggs is allowed to stand at 19–21°C for 15–20 minutes. The rested dough is shaped by hand or machine on a fat-coated table into small round pieces and allowed to rest for 30–40 minutes. The rested pieces are drawn into 2- to 3-mm thin layers of 25–30 cm in diameter. Up to four layers may be placed one on top of the other. The dough is worked and flattened into almost transparent (0.2-mm) layers. The filling is placed in the center, and the dough is folded and then baked at 250–270°C on flat pans.

Mekizi is baked from thin layers of yeast-leavened dough coated with vegetable fat. The slack dough (1 kg flour, 700–730 ml water, yeast, and salt) is fermented for 3.5–4 hours at 26–27°C. After scaling and punching, the dough is proofed for 30–40 minutes, and the 36- to 38-g pieces are baked in hot sunflower oil for several minutes at 220–230°C. Excess oil is allowed to drip. The center of the round or eliptical baked product is thin; the outside thick. The baked product is generally sprinkled with fine sugar and served hot.

The kolatsch (kolac) is a discuslike bread cake that is found on the table of any festive group. The bread has various ornaments, the motifs for which are taken from agriculture or religion. An example of a spiral-cross bird *kolatsch* bread is shown in Figure 12-4.

Yugoslavia

Most of the bread produced in Yugoslavia is from wheat flour. In cities most bread is produced in plants with a capacity of 10–60 tons per 24 hours. About 40 types of bread are made by the straight dough procedure. Most of the bread is baked on the hearth; relatively little is panned.

A short sponge is baked in villages on the Adriatic coast, and an overnight sponge is baked in the center of the country. In some villages in Macedonia, Serbia, and Croatia, an air-dried sourdough is used as a starter to produce an excellent bread. The oblong *vekna* is made from wheat flours of various extractions. The round *pogaca* 2-kg loaves are baked in two stages (at about 225°C for 30 minutes and at 190°C for 40 minutes). Mixed bread is made from wheat-rye mixtures or from wheat-corn flour mixtures.

In some areas an extract from *Carum carvi* is added to improve taste and flavor. Dough formulations include mainly wheat flour, water, salt, yeast, and occasionally a little fat; ascorbic acid is used as a bread improver. In many villages of Monte Negro and Herzegovina, bread is baked by a sponge process from corn or barley flours. *Mlinci* is a thin, *chappatti*-like bread baked in Croatia. Average bread consumption in Yugoslavia is high.

Hungary

Most bread consumed in Hungary is made from wheat; special rye breads are produced on a small scale only. The farmer's wife still bakes the true Hungarian bread with bran and hops. In September she prepares a supply of bran and hops for a year. One liter of dried hops, one-half liter of clean barley, and several small onions are cooked with water for 1 hour. A liter of wheat flour is added to the brew, and the somewhat crumbly dough is allowed to cool. A piece of old fermented dough is dispersed in lukewarm water, and juice is expressed from several grapes. The mixture is fermented in a warm cabinet. The fermented product is called *koficz*. Additional water is added to the hops, and the mixture is boiled vigorously after adding clean wheat bran. After cooling, the *koficz* is stirred in and the mixture is fermented for several hours. The dough is divided into small pieces, air-dried, and stored in the form of small to large, nut-sized chunks. A small number of the chunks are added before baking.

The chunks are moistened with lukewarm water, crumbled, and allowed to ferment. An extract is screened and mixed with a sourdough and fermented for 4 hours; a salt solution is mixed in, and the flour is kneaded with the dough. After fermentation for 1 hour, the dough is punched, placed into baskets, allowed to proof for about 45 minutes, and baked in a hot oven to a delicious bread.

The types of bread baked by the Hungarian baker of today have developed over a period of five centuries. About 50 types of bread are on the market. In recent years, bread consumption has decreased and the percentage of white bread has increased. The white baked products include a large selection of buns. The small bakeries have been replaced by mechanized plants.

Some of the bread is of the rye-wheat type, with small amounts of wheat and butter bread and small amounts of rye bread. The rye-wheat formula includes 80 parts dark wheat and 20 parts medium-dark rye flour, 2 parts salt, and 0.4 parts yeast. In making oblong rye bread, 90 parts of rye flour and 10 parts of wheat flour are used together with 2 parts of salt, 1 part of yeast, 1.5 parts butter, 0.2 parts caraway seed, and 25 parts milk. Many dough formulations in Hungary are rich and differ substantially from those of other East European countries. Coffee bread, coffee buns, and an assortment of about 45 types of rolls in various sizes, shapes, and formulations are available. The dough formulations are rich, and milk, butter, sugar, and malt are present in relatively large amounts.

Czechoslovakia

The consumption of bread in Czechoslovakia is on the decrease, even though it is quite high compared with many West European countries. About 60% of the bread is consumed as mixed rye, 15% mixed wheat, 15% fine rye, and the rest white of various types.

Mixed rye is made from a mixture of 80% dark rye and 20% dark wheat flours; the bread is sour-leavened; the baking temperature is 240–280°C. Mixed wheat is made from 60% wheat and 40% rye flours. To each ton of flour, 2.5 kg yeast and 16 kg salt are added. This type of bread is made mainly in Slovakia. Both mixed rye and mixed wheat breads are baked on the farm with the addition of caraway seeds. These and any of the other bread types are made in the form of round or oblong (hearth or panned) loaves. An old-time round loaf type is known as *pogaca*.

For rye bread, the baker uses a darker rye flour than for the mixed rye bread. *Vita* bread contains 10% germ oil and seeds (caraway or others) added to rye flour and is prepared from sourdough. Graham whole wheat bread is made in the form of small loaves (200 or 500 g) from 90% ground wheat and 10% wheat flour and is leavened with yeast. A specialty, sour-type bread, is made from 50% white rye flour, 24% whole ground rye, and 18% wheat flour of medium extraction.

Poland

Rye bread in Poland is produced from four types of rye flour of 70–98% extraction. The light rye is used mainly in the production of mixed bread. There has been over the years a consistent decrease in rye and an increase in mixed bread consumption. The decrease in rye bread consumption resulted mainly from decreased availability of low-yielding soils for rye cultivation and increased use for feed and alcohol production. In addition, the milled grist includes increasing amounts of imported grain that is mainly wheat.

Rye Bread

There are several types of rye bread that vary in processing and composition.

Pytlowy

This is made with a multistage sour, from rye flour of the 1074 type (intermediate extraction and color). The bread is made as 1- or 2-kg loaves (generally in baskets) and has a moisture content of 48% or lower and an acidity below 8.

Whole Meal

This is made from whole grain ground finely or coarsely. Generally, the finely ground rye is baked with 5% wheat flour and the coarsely ground rye with 20% wheat flour of intermediate extraction. The coarsely ground type is baked on the hearth in the form of round 0.5- to 1.0-kg loaves. The finely ground type contains maximum 51% moisture and has an acidity below 11.

Honey Whole Meal

This bread is gaining increasing acceptance. The following is a recommended formula:

Parts rye flour (type 2000)	90
Wheat flour (type 800)	10
Sugar	6
Natural and artificial honey (1:1)	2
Potato syrup	1
Salt	1.5–1.7

Generally the loaves weigh 1 kg, are prepared by sour fermentation, and have a maximum moisture of 49% and acidity of 11. The bread has a mild taste, pleasant flavor, and longer shelf life than the other bread types.

The 82% Rye Type

This is made from flour of the 1400 type and is produced by a multistage sour fermentation. It is baked on the hearth or in forms as 1- or 2-kg loaves. It is often baked from flour mixed with up to 2 parts (per 100 parts of flour) of dry, ground bread. The final baked bread has a maximum moisture of 50% and acidity below 10.

Starogradzki

This is made from flour of the type 1850 and peeled rye. Loaves of 1 kg are baked in forms, from 95 parts of rye flour and 5 parts wheat flour of type 950. The baked bread has a maximum of 51% moisture and acidity of 10.

Other Rye Breads

According to another classification, the following types of rye bread are produced in Poland: *razowy* (coarse rye), *sitkowy* (fine rye), *miodowy* (honey), *litewski* (Lithuanian), *mleczny* (milk), *wytrawny* (nutritious), and *finski* (Finnish). The milk rye is made from 100 parts light rye flour, 1.5 parts salt, 30 parts fluid skim milk, 1 part yeast, and 0.5 parts potato flour. Only a small portion of the rye bread is baked in forms; most is baked on the hearth.

Mixed Bread

There are six types of mixed bread. The most popular ones are *sandomierski* and *praski*.

Sandomierski

This is a sour-type rye-wheat bread made by a sour fermentation procedure. *Sandomierski* resembles typical rye bread. It is baked from 30 parts wheat flour (type 800) and 70 parts rye flour (type 1074). It is almost always hearth baked in the form of 1- or 2-kg loaves. The moisture content is up to 47%, and the acidity up to 8. The fermentation is multistage sour.

Praski

This is the second type of mixed bread. The flour includes 60 parts wheat flour (type 800) and 40 parts rye flour (type 1074). The bread is baked from a soft rye sour supplemented with bakers' yeast. The mild, sour taste of well-leavened loaves has contributed significantly to the popularity of the bread. *Praski* is baked in 1.0- or 1.5-kg oblong loaves or bread in baskets. The bread has a moisture below 46.5% and an acidity below 6.5.

Mazowiecki

This type of mixed bread is closer to wheat bread than rye, as it is baked from 80 parts wheat flour (type 800) and 20 parts rye flour (type 1074). The formula also includes 1.5–1.7 parts salt and 1.7–2.0 parts yeast. *Mazowiecki* is valued because of the variety of forms in which it is produced. The best *mazowiecki* is produced from a yeast-leavened wheat flour sponge and a rye sour. This provides an appetizing, mild-flavored bread with improved shelf life. The following is a popular formula: The sponge is mixed from the wheat flour (40 kg), water (33 kg), and yeast (1.5–2.0 kg). The homogenous, well-developed dough should have a temperature of 27°C, and it should ferment for 180–210 minutes. The full rye sour (31 kg from 18 kg rye flour and 13 kg water) is mixed with 15 kg water (29–30°C) in which the salt is dissolved, and the two are added to the sponge. Then 40 kg of wheat flour is added, and the whole is mixed thoroughly and fermented another 30 minutes. After scaling and forming, the rounded pieces are proofed for 35–45 minutes, during which time the loaves are lightly moistened twice—once immediately after forming and the second time before putting into the oven. The loaves are cut slightly before the second moistening and dusted with poppy seeds. The bread is baked in ovens saturated with steam. The loaves are moved 15 minutes after placing in the oven. The initial baking temperature should not exceed 220–230°C. Total baking time is 38–43 minutes for a dough piece weighing about 935 g and baked down to about 830 g. The *mazowiecki* resembles a wheat bread, has a moisture of less than 46% and acidity below 5, and is baked as oblong loaves or in baskets.

Naleczkowski, Leczycki, or Zakopane

The first two are of limited production and in certain regions only. These are yeast-leavened breads made by the sponge method. They have the character of wheat breads. Naleczkowski is made from 50 parts of wheat flour (type 800) and 50 parts of rye flour (type 580). The composition of flour for *leczycki* is similar, except that the wheat flour is lighter in color.

Zakopane is a traditional product of the mountains (Tatry), and the bread is named after the well-known skiing resort Zakopane. The bread is made from 50 parts rye flour (type 800) and 50 parts wheat flour (type 580). It is prepared from a rye sponge and is made with fat-free sour milk (about 40 parts per 100 parts of flour). The 0.5-kg loaves are fermented on wooden boards. The bread has a thick, hard crust and a specific taste.

CENTRAL AND WESTERN EUROPE

Fresh bread is consumed in all countries of Central and Western Europe. In France, bread is eaten before it is more than 4 hours old. Many bakers thus bake at least four times a day, and the consumer buys bread twice daily. In Belgium, the consumer eats bread up to 2 days old, though generally within 1 day after baking. The Dutch bread has a relatively soft texture and is made from a dough that includes fat and emulsifiers to lengthen acceptability up to about 48 hours. The Germans demand fresh buns and rolls; dark rye bread, on the other hand, is eaten up to 14 days after baking.

Germany

Germany is a land of widely diversified flour and bread types. The 200 types are divided into four classes: wheat breads containing a minimum of 90% wheat components, mixed-wheat breads containing 51–89% wheat components, mixed-rye breads containing 51–89% rye components, and rye breads containing a minimum of 90% rye components. Specialty breads are a separate category. The classification of the bread types is reproduced in a modified form in Figure 12-5. Many types of small baked goods (below 250 g) are governed by the same general rules and requirements as breads, except for weight. Breads and small baked goods contain a minimum of 90% grain products and maximum 10% sugars and/or fats (including those from milk or milk products). Fresh, nonsliced bread must weigh at least 500 g and may be sold in various sizes that are multiples of 250 g; the most common is the 1,000-g (1-kg) bread. Family loaves over 2 kg are increasingly less popular. Sliced and wrapped bread is available in 125-, 250-, or 500-g packages.

Bakeshops and Bread Factories

Bread and rolls are produced today in West Germany (population about 62 million) in about 30,000 bakeshops (*Handwerksbetriebe*) of various sizes and in 200 large baking factories. It is estimated that the bakeshops produce (based on flour weight basis) about two-thirds of the total breads and rolls sold to consumers. For years, the shop adjacent to, and part of, the bakery was the source of rolls and whole bread loaves. Today the grocery store sells sliced, wrapped bread and increasing amounts of whole bread as well as rolls. The bakery shop sells also sandwiches ("covered" rolls), making it possible for the consumer to select a combination of a few slices from several bread types. The price differential between the bakery shop and factory products is often more than 2:1. The high price of the bakery shop products is predicated on freshness, quality, service, variety, and reputation.

The number of very small bakeshops is decreasing, but the bakeshops are increasing in size and are holding their own in total production. The large factories produce increasing amounts of fine breads, with associated problems of distribution costs, costs of display in groceries and supermarkets, and shelf life (staling and microbial damage).

Consumption, Preferences, Expectations

In the past 100 years, grain yields in West Germany increased and flour consumption decreased as follows:

	Yields (tons/ha)			
	1880–85	1935–38	1955–57	1980
Wheat	1.29	2.59	3.05	4.97
Rye	0.99	1.93	2.51	3.84

	Flour consumption (kg/year/capita)			
	1890	1935–38	1954–55	1979–80
Wheat	63.4	61.0	63.2	50.3
Rye	112.6	47.0	29.2	14.1

The large decrease in rye flour consumption is of particular interest. Total flour consumption decreased from about 94 kg/year/capita in 1955 to about 57 in 1971. It has since increased slowly. In 1979–80, the consumption was wheat flour 47.5 kg, rye flour 14.1 kg, and durum flour–semolina 2.8 kg. The changes in prices of milled and baked wheat and rye products (deutsche mark per kg) in the past 30 years were as follows:

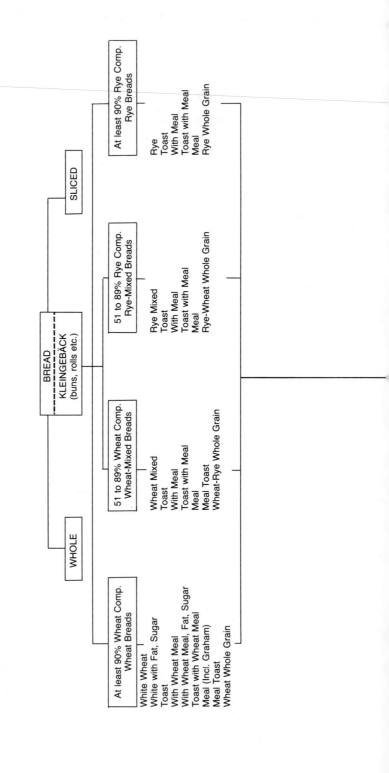

SPECIALTY BREADS

With Special (other than wheat/rye) Grains
i.e. Three-grain bread (each 5% min)
Four-Grain Bread

With Special Milled Products
i.e. Steinmetz (Rye or wheat 93–94% extn.)
Schlüter (Whole grain rye, steam, heat treated)

With Special Dough Handling
i.e. Simons (Rye and/or wheat whole grain, special swelling and wet milling)
Loos (Rye whole grain, slow swelling)

With Special Baking Process
i.e. Wooden oven
Stone oven
Steam chamber (no crust)
High-heat final bake (Gerster)
Dry flat (Knäcke)
Pumpernickel (whole rye, baked at least 16 hrs; max. 10% wheat and rye flours, no sugars or colours added)

With additives of Animal Sources
Milk—at least 50 l whole fat milk (or equiv.)/100 kg grain products
Milk Protein—at least 2 kg milk protein (or 6 kg NFDMS) per 100 kg grain products
Sour Milk—at least 15 l per 100 kg grain products (i.e. sour milk, yoghurt, buttermilk, kefir)
Soft Cheese—at least 10 kg/100 kg grain products
Butter—at least 4.1 kg butterfat/100 kg grain products
Whey—at least 15 l/100 kg grain products

With Additives of Plant Origin
Wheat Germ—at least 8 kg with a min. of 8% fat d.m. per 100 kg grain products
malt—at least 8 kg per 100 kg grain products

With whole fat oilseeds
(i.e. sesame, linseed) at least 8 kg/100 kg grain products
Raisins—15 to 40 kg air-dry raisins, sultanins, or corinths/100 kg grain products
Spices—in quantities to impart distinct taste-flavor odor
bran—at least 10 kg with max. 15% starch d.m./100 kg grain products

With Modified Nutritional Value
protein enriched—22% protein $(N \times 6.25$ d.m.) min.
carbohydrate reduced—30% less
calorie reduced—20% less

Dietetic
protein low—starch rich
gluten free—gliadin free
diabetics — max. 200 Kcal/100 g bread
Na-low—max. 120 mg Na/100 g bread
Na v. low—max. 40 mg/100 g bread
vitamin enriched—through addition of germ, yeast, or vitamin supplements to restore to whole grain levels

Figure 12-5. Classification of German bread types. (Courtesy DLG Verlag, Frankfurt.)

	1952	1960	1970	1980
Wheat products				
Local wheat (at the mill)	0.45	0.45	0.40	0.50
Wheat flour (white, type 550, wholesale)	0.60	0.60	0.60	0.75
White bread (consumer price)	0.85	1.09	1.70	3.50
Buns and rolls (consumer price)	1.20	1.65	2.15	4.20
Rye products				
Local rye (at the mill)	0.42	0.43	0.35	0.51
Rye flour (type 997, wholesale)	0.51	0.54	0.50	0.75
Rye bread (consumer price)	0.62	0.77	1.38	2.52
Rye-mixed bread (light, consumer price)	0.68	0.85	1.35	2.59

The figures speak for themselves. The daily bread plus roll consumption is 201 g (73.3 kg/year) of which 154 g is eaten in the home. The consumer in the restaurant, café, hospital, school, etc eats more mixed-wheat bread and rolls than the consumer in the home. At home, the main baked goods (% of total) are rye or mixed-rye bread, 25.4%; wheat-rye mixed (about equal proportions) bread, 23.0%; grain-meal breads (wheat, rye, or mixed), 16.1%; wheat rolls, 11%; and mixed wheat bread, 10.6%. White wheat bread comprises only 5.5% of the total. Consumption of breads and rolls is highest among the 30- to 39-year-old consumers and decreases as household size increases from one to three or more.

Whereas only 36.9% of the consumers shop for food once a week, 48.2% purchase bread and rolls at least once every week. The percentage of those who purchase bread every day ranges from 2% in Berlin to 30% in Hamburg. Rolls are purchased daily by 56% of Hamburg consumers and only 5% of Berlin consumers.

On a scale of 1 (most important) to 8 (least important), buyers of bread for restaurants, cafés, etc ranked the important factors in bread in the following order: quality, 1.2; price, 2.4; reliability of supply, 2.7; delivery service, 4.5; multitude of varieties, 6.1; packaging size, 6.7; and payment conditions, 7.4. In the eyes of 39% of the consumers, ideal bread should be well baked and have a good crust; to 23% the most important bread attributes are good quality and a full rounded-out taste; freshness was of greatest significance to 17% of consumers. The above characteristics were found by 27% of consumers in mixed, by 23% in whole grain, by 22% in dark, and by 21% in black bread.

It is believed that shelf life of bread depends on the rye:wheat ratio, type and extraction of flour or meal, and on the type and levels of bread additives. Hard rolls should be consumed freshly baked. Shelf life of wheat flour bread can be increased through addition of fat from 1–2 days, 3 days, and 1 week in bread baked without fat, with 1–2% fat and 5% fat (toast bread), respectively.

As the proportion of rye increases, shelf life increases from 3 days in mixed wheat to 5 days in mixed rye breads and to 1 week in rye breads. The shelf life can be prolonged by 1–2 days as the meal:flour ratio increases. Pumpernickel has the longest shelf life among the high-moisture and *Knackebrot* among the low-moisture breads.

Bread types in East Germany are similar to those in West Germany.

The Netherlands

Much of the bread in The Netherlands is still produced in small bakeshops. These bakeries produce 50 kinds of bread out of a list of over 300 that were baked some time or other in Holland. More than 80% of the bread is panned. Generally, bread is sold in multiples of 400 g. Almost 90% of the bread contains water as the main or only fluid. Except for small baked goods, milk bread that was popular before World War II has almost disappeared from the market. More and more bread is sliced and wrapped.

There are large differences in the types of bread consumed in various parts of the country. Rye bread is popular in Groningen and Limburg, especially in rural areas, and to a lesser extent in Friesland. White, lean bread is preferred in Friesland, and brown bread in the rural areas of South Holland and Zealand. The following are some types of bread baked.

1. Plain white bread (water white bread) made from flour of 70% extraction; baked in the form of 800–g panned loaves that have an incision on the top; the major product of large bakeries. Small amounts of this bread are shaped round. The 70% flour is used also to bake "casino" or "sandwich" bread; oblong, hearth-baked bread; "strand" bread, which has a thin strand of dough added on top of the bulk dough; and "tiger" bread, that has a mottled appearance from brushing the top of the dough with an aqueous suspension containing rice flour, salt, yeast, and fat.
2. Light and dark gray bread made from flour of 80% extraction, alone or mixed with 25–75% whole wheat meal. Special breads of this class include whole wheat and "Tarvo" or "Allison" breads made from whole wheat meal and 5–10% wheat malt.
3. Rye bread from whole or high-extraction rye products, varying in particle size and milling method. The bread is made mainly in the northern parts of the country. Most rye bread is made from yeast-leavened dough, as the Dutch prefer mild- to sour-tasting bread. Characteristic of this bread is the spontaneously acidified *Friesen* bread. It has a mild taste, the texture is open but not gummy, the color is quite dark, and the crust is relatively soft.
4. Luxury breads include the large- and small-sized goods. The bread is made from flour, water, yeast, and salt supplemented with milk powder, fat, eggs, lecithin, malt extract, sugar, and cream. Some rich formulations call for dry fruits, marizpan, and sweets. The small-size

luxury goods may be baked with (*Koffiebroodjes*) or without (*Kadetjes* or *puntbroodjes*) fruits and spices.

Most of the bread consumed in Holland is of the plain white type. Only 5–10% are dark breads, less than 5% are rye breads, and about 10% are luxury baked goods. The regular white bread is produced by a straight dough fermentation from 100 parts of flour, 54–56 parts water, 2 parts salt, 2 parts yeast, and 0.5 part cream. Milk bread is made with milk as the only fluid or a suspension of whole milk; the amount of yeast is increased to 3 parts and of cream to 1–1.5 parts (on flour basis). Casino or sandwich bread is similar in composition to milk bread, but the doughs for the latter are mixed longer and baked in steam. For making whole wheat or dark wheat bread, 100 parts flour, 54–56 parts water, 1.6 parts yeast, 2.7 parts salt, 0.5 part fat, and 0.2 part malt extract are used. The dough is mixed slowly for almost 45 minutes at 25°C. The fermented, scaled, punched, and proofed dough is baked in steam at 240°C for 30–35 minutes.

To increase bread consumption, the bakers in Holland have developed a rich breakfast bread and special glazed buns, *Tielerbol*, made from 100 parts flour, 10 parts butter, 2 parts milk solids, 2 parts salt, 2 parts sugar, and 3.5 parts yeast.

Belgium

In 1959, Belgium had over 12,000 small bakeries and about 100 larger bakeries. However, the number of small bakeries is decreasing, and the amount of bread produced by automated large bakeries increases constantly. At the same time, per capita bread consumption is on the decrease. About 50 kinds of bread are produced, which can be divided into five classes.

1. Home bread (*pain de menage*), plain bread in the form of round 500-, 1,000-, 1,500-, or 2,000-g loaves. The bread is hearth baked, and each piece is placed so close to the other in the oven that each loaf has 4–6 crust-free surface patches. It is made from medium dark wheat flour or from a mixture of wheat flour and *Kneipp* meal of 85% extraction or with whole rye or wheat meal. The latter additions are quite popular, and the bread is sold as farm bread, mixed bread, Ardennes bread, etc.
2. "Fantasy" bread is from 450- to 900-g round, oblong, or square loaves hearth baked or panned. It is sold under the name *pain de galette* or *pain boulot*, depending on whether oblong or round. Generally the top crust has diagonal or crossed incisions.
3. Special "fantasy" breads are made from white wheat flours milled from a mixed grist of wheats. They contain at least 2% fat (on flour basis). They are generally 700- or 800-g panned loaves, sliced and wrapped, of nice appearance and good taste.

4. Diet breads, such as for diabetics, salt-free bread, etc.
5. Small baked products, below 300 g in weight.

The plain bread is made from 100 parts flour, 50–52 parts water, 1.5 parts yeast, and 2 parts salt. The dough is mixed, fermented at 26–27°C, scaled, shaped, and baked (after proof) at 230–235°C for 50–60 minutes. The regular "fantasy" bread is similar in composition to plain bread, but baking schedules are more optimized.

Before World War II, about half of the bread was whole grain or mixed wheat rye. A more recent distribution is the following.

Type	Percent of total
Fantasy (450–900 g)	40
Plain (menage) (500–2,000 g)	30
Special fantasy (Ameliore)	
(600, 700, 800, 900 g)	25
Diet (Regime) (600–800 g)	1
Buns	4

Belgian bread is relatively voluminous with a good crumb grain and texture. About 85% is white, 10% medium brown, and 5% whole grain. A large proportion (up to 65% in towns) is sliced and wrapped.

France

Bread to the Frenchman is what rice is to the Chinese and what potatoes are to the German. An old, seasoned French sailor was asked what was the most surprising thing he saw in his voyages around the world. He answered that nowhere had he found bread of the quality baked in his small village. The French are extremely fussy about their bread. It must be freshly baked, and the crust must be thick and make razor-sharp pieces when broken. Bread is eaten at all three daily meals: in the morning with hot, milky coffee or chocolate; at noon throughout the main meal; and late in the evening with soup.

Some 20 years ago there were over 50,000 bakeries in France. Today there are more and more large, fully mechanized bakeries that produce bread for institutions, but the small consumer still buys bread from his favorite family bakery. Figure 12-6 illustrates the steps in the manufacture of French bread.

Consumption of bread has decreased, but the French still eat more bread than most Europeans. Whereas before World War II, in most districts people ate 50–85% of plain large bread, today the consumer buys mainly special bread. Only 5% of the bread is panned.

Two-pound, round, oblong, or long bread (such as *boulot, marchand de vin*) comprises 40% of the consumption, and the smaller (300–700 g)

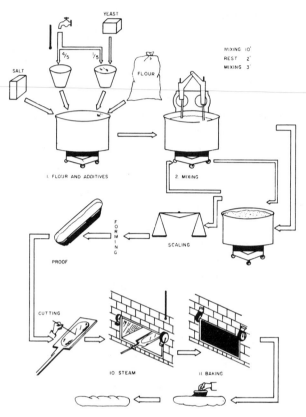

Figure 12-6. Schematic diagram of the manufacture of French bread. (From Bure, 1961.)

specialty breads (flutes, baguettes) are slowly reaching the 50% mark. Small buns (*croissants*) are more and more popular. There is very little demand for dietetic breads.

There are about 30 kinds of French bread of four main types:

1. Household bread, *gros pain*, is sold by weight, generally in 1.0-, 1.5-, or 2.0-kg loaves. In various parts of the country it is long (*marchand de vin*), oblong (*boulot*), or round (*miche*). Loaves smaller than 700 g are considered specialty breads.
2. Specialty breads are known all over the world as French bread. They include *baguette* (700 g, 70 cm long), *batard* (700 g, 35 cm long and 300 g, 20 cm long), *flute* (300 g, 40–50 cm long), and *ficelle* (100 g, 20–30 cm long).
3. Dietetic breads—for diabetics, those on salt-free diets, etc—are generally made from white flour and contain 1–3% malt extract and 2–4%

milk powder. In addition, toast bread that has 3–5% fat, whole wheat, mixed (wheat-rye), and small- to medium-size bread (*luxus-baguette, joccos, empereurs*) are available.

4. Small baked goods, also known as Vienna type, are well accepted.

The typical, long French bread is made from 100 parts soft French white flour, 60 parts water, 1 part yeast, and 1.75 parts salt. The bread is made from a dough of medium consistency. The direct fermentation, without sourdough or sponge, is finding increasing acceptance. The ingredients are mixed in two stages for 10–15 minutes to obtain a homogenous mass—5 minutes of rest and 5–10 minutes of intensive mixing to incorporate air. To obtain optimum consistency, water or flour can be added before the second mixing. The dough at 22°C is allowed to stand 3 hours, punched, and fermented for 0.75–1.0 hours. The long fermentation is essential in producing acceptable bread. The dough is scaled, formed, and proofed. Shortly before baking, parallel incisions are made into the dough. The doughs are slack; the bread is baked in a steam-rich oven at 270–280°C. The temperature of 100°C in the center is reached within a short time. Baking time is 12–13 minutes for a 300-g loaf or 25–30 minutes for a 700-g loaf.

The unique qualities of French bread are derived to a large extent from the methods employed in its production. At present, most bakers use commercial bakers' yeast in the straight dough process, in which yeast is incorporated into the dough in the mixing process. The other techniques of fermentation are the *sur poolish* and *sur levain*. The *sur poolish* involves a prefermentation or sponge stage. The prefermentation medium includes a semiliquid mixture of 1:1 flour and water, with about one-third of the total flour, and 0.5–1.0% yeast (flour basis), and a 5-hour fermentation at 22°C. In the *sur levain*, the long prefermentation depends entirely on the natural microflora of the flour (wild yeasts and bacteria—mainly lactic acid—that produce short-chain organic acids). The *levains* or preferments are doughs that undergo natural fermentation and are "renewed" daily by adding water and flour. The *poolish* imparts a strong and pleasant flavor; the *levain*, a distinct acidic taste. It is generally accepted that the superior bread characteristics are the result of prolonged fermentation which enhances intensive secondary fermentation and production of a variety of volatile organic compounds.

During the past 35 years, procedures for bread making in France have undergone considerable changes. Intensive mixing is being more and more widely adopted. The new techniques subject the dough to intensive mechanical development, and the mixing action is aided by adding small amounts of fava bean flour and/or ascorbic acid. This reduces the mixing time, increases loaf volume, whitens crumb grain, and produces a uniform crumb grain and texture. It also, however, impairs (or at least alters) bread flavor. Consequently, there is interest in bread with real or alleged qualities made by traditional methods in the production of *pain de campagne* (country bread) or *pain au levain* (leavened or sourdough bread). The latter two require long fermentation times or rely on spontaneous dough fermentation, respectively.

The popular *brioche* is made from 100 parts flour, 2 parts salt, 4 parts yeast, 2.5 parts sugar, 70–75 parts fat, 1.5–2 parts malt extract, 60–70 parts eggs (or 50 parts eggs and 10–20 parts milk). The malt extract is dissolved in lukewarm water or milk, the yeast is suspended, and sugar, salt, four-fifths of the egg mass, and the flour (sometimes with the rest of the milk) are added. The dough is mixed intensively with the rest of the egg mass added without interrupting the mixing, and mixing is continued until a smooth, homogenous mass is obtained. The fat is dispersed in a small part of the dough and incorporated with the rest in a mixer. The dough ferments for 8–10 hours during which it is punched three times, the last time before scaling and forming. The formed pieces are proofed 30 minutes, during which the pieces are brushed twice with a suspension of egg yolk. Large *brioches* receive an incision with moistened scissors at the periphery. The *brioches* are baked at 230°C without steam. There are various types of *brioche*. The most popular is *brioche à tête* in which a small dough sphere (about one-fifth of the total) is placed on a flat piece. *Brioches* range in weight from 40 to 350 g.

Luxembourg

The inhabitants of Luxembourg buy their bread from hundreds of small bakeries that produce 20 kinds of bread of three main types. The annual per capita consumption is about 90 kg.

The regular, standard bread is made from a mixture of 88% wheat flour (70% extraction) and 12% rye flour (50% extraction) fermented by a sour fermentation or a sponge method. Dry sour starters are also used. The bread is generally hearth baked. It is round or oblong (*galette* or *boulot*) and seldom panned. Weight units are 500 g.

Loaves smaller than 1,000 g can be sold under various "fantasy" names. They are made by sour, seldom by direct yeast fermentation, from 70% extraction flour. The bread is hearth baked, oblong or round, with incisions on the top.

Rye bread is made from flours of 50% extraction. About four-fifths of the bread is of the regular, standard type; the amount of rye bread is very small and is sold mainly in towns.

Switzerland

Switzerland is a country of many peoples and of many breads. In Aargau, the daily house bread is an oblong loaf with a deep incision. For holidays, a band-decorated ring dating back to the 1712 Second Villmerger War is baked. Bread in Appenzell is in the form of round loaves. The specialty bread, *Filebrot*, is baked at the beginning of a new year. The bread has a unique shape that may suggest the coming of spring or the source of which may be Greek symbolism. The town of Basel is proud of the pretzel-like *Fastenwahen* and of the *Bolweggen* bread that dates back to the 14th century. In farms around Basel, oblong loaves (with or without cuts) are the most popular. The opposite is true for Bern. The regular bread is round, the old time *Vèques de Noël* is round and flat, resembling a thick pizza. The regular rye bread (*Rua*) and especially the religious (*Wasteln*) bread of Fribourg are among the most colorful in Switzerland. The *pain rond* (*campagne*) of Geneva is a somewhat shapeless loaf; the *couronnes de rois* are thick, doughnut-shaped products into which are baked small figures of the Three Kings. The canton of Glarus has two types of bread: an oblong loaf and a rather flat, round product. The barley bread of Graubunden has lost much of its popularity; however, the large, doughnut-shaped *Brascidela* that can be hung on the wall, dried, and eaten a long time after baking is still made to a limited extent. The large, braided *gotti* bread of Lucerne is baked by the housewife for the New Year; during the year, the *Mandatbrot*, a two-part oblong loaf, is seen. The typical bread of Neuenberg is a large, hearth-baked *pain Neuchatelois*; the decorative (braided and shaped in various forms) *tauillaules* are reserved for more festive occasions. Both the plain and the more festive, ornamental, rich raisin bread of Obwalden are oblong. In Nidwalen, the daily bread is a round loaf with a deep diametrical incision; the raisin ring is a well-known delicacy. Schaffhausen bread is a rather irregular loaf baked from 2–3 attached spheres; the *gige* are long, oval-shaped pretzels with a prominent knot at the end. The *Kopfbrot* of Schwyz has the shape of a panhandle; the *Einsiedler-Schafbock* are round, flat buns with ornamental religious decorations. The Solothurn bread is a slightly oblong, round bread; the *Grittibenz* are baked figures of soldiers, saints, etc. The *Sargaserland* flat bread and various excellent white breads originate from St. Gallen. The Thurgau *Schiltbrot* is a starlike, large bun. The *mica mendrisiotto* and *sciampa locarnese* apparently originated in Tessin; out of reverence for bread, the latter is made into loaves composed of segments that can be easily broken without a knife. The regular breads of Uri are round, two-piece *Halberli* or the *Doppelmutschli*; on holidays a large *Unernpastete* is baked. The round, three-segmented *pain en tête and four-segmented pain à 'la croix* are popular in Waadt. A famous flat rye bread with various inscriptions and impressions can still be found in Wallis: the decorative, oblong, braided *Mitscha* is baked for Holy Communion. *Rosenbrottli* is an attractive bread that resembles an open rose; an elaborate bird-shaped (*Vogel*) egg bread is reserved for special occasions in Zug. The oblong bread of Zurich, with 3–4 short incisions, is simple in shape but excellent in taste. *Fastnachtsweggen* bread is a two-piece, rounded loaf that has been handed out in the past to children before Holy Communion.

Austria

The Austrian bakery is well known throughout the world, mainly because of its white breads. The main contribution of the old Viennese school of baking was the use of compressed bakers' yeast instead of brewers' yeast. This permitted shortening fermentation time and eliminated undesirable off flavor and off taste, poor crust texture, and crumbliness. The use of bakers' yeast was followed by the introduction of diastatic malt that further improved the quality of baked products. Finally, recognition of the importance of flour granulation, the control of fermentation and baking conditions, and the use of indirect heating methods at the baking stage led to the production of new and excellent breads, buns, and rolls of the Vienna type. In addition to the common Vienna type, white bread baked from a relatively rich formula and numerous special types of breads are baked.

The important types of bread in Austria are round (*Leib*) or oblong (*Wecken*) rye or wheat-rye breads. They are either smooth or have cuts on the crust. Whole grain or coarse meal breads are often baked in pans. *Zeppelinwecken* is the term often used to denote an oblong wheat flour bread with deep incisions. In Salzburg and surroundings, oblong bread with many small holes or notches in the crust is baked. The round *Vintschgau-Laibchen* is baked mainly in Tyrol: it is a yeast-leavened, light rye flour bread and contains anise seeds. The round *Schweizer-Laible* is from rye sour and wheat flour and is baked mainly in Vorarlberg.

Farm bread is made from dark rye flour by a sour fermentation. In small bakeries or household bakes, a small amount of fermented dough is refreshed once, and bread is baked from the whole sour. Whole grain bread is baked from sour and rye meal. In recent years, use of rye meal is not practiced as the addition of some rye flour improves texture, and the mixing of a meal dough gives the best results with small doughs only. In some cases, small amounts of wheat flour or wheat meal are added. For a whole grain bread, a dough of 60 kg contains three-fourths medium rye meal and one-fourth medium dark rye flour. To make this, 10 kg of a ripe sour from normal breadmaking is refreshed with 12 L water at 35°C, 4 kg rye flour, and 5 kg rye meal; the dough is fermented for 2 hours. For the final dough, 9 L water at 28°C and 20 kg meal are added. The dough contains 3% salt, calculated on the liquid basis. The dough should be on the slack side to avoid a coarse crumb grain. The scaled doughs are rounded, made somewhat oblong, fermented for 45–50 minutes, and placed in an oven at 220°C maximum. An excessively hot oven causes excessive oven spring and damages the crust. The bread is baked in an oven with little steam, and the temperature falls somewhat during the baking. Some bakers make bread from a dough that includes whole meal (about 25–30% of the total) that has been scalded with hot water to gelatinize the starch. This increases the amount of fermentable sugars and improves taste and freshness retention.

Steinmetz bread is baked from debranned grain that contains the germ and aleurone layer and is finely ground. It is made essentially a whole meal bread. However, the baking is at relatively lower temperatures and for longer times.

The American physician, Graham, introduced the use of wheat meal in baking. Graham bread is baked with 15% or more wheat flour from a slowly fermenting dough with little salt. The bread is well baked. *Kletzenbrot* is prepared for Christmas from dark rye dough that includes available fruit, mainly dried pears (*Kletzen*). It is popular in the Alps, but it is also baked in Vienna with other fruits and spices. Hopped bread in the south is a result of Hungarian influence. It is prepared 3 days in advance. About 1 kg hops is boiled in 36 L water until a concentrated extract is obtained. The concentrate is filtered through a cloth and cooled, and the so-called *Gomlo* is used to make a stiff dough that is allowed to ripen for 36 hours. The dough is made slack by adding half the amount of water or milk and fermented for 12 hours. The rest of the fluid is added with the flour, and fermentation is continued for 4–5 hours. The flour used is dark, often whole meal. The bread has a sweet, aromatic taste and remains fresh and moist for a long time.

The crisp, round, specially shaped hard Kaiser or Vienna roll is an Austrian specialty. It is made from medium white wheat flour by a straight or sponge fermentation. *Laberl* or *Schusterlaberl* is made from soft dough prepared from a medium dark flour. The baked product is less crusty than the Kaiser. Simple, yet delectable, baked products are the *Wachauer Laberl* or *Bierweckerl* baked from a mixture of 75% wheat flour and 25% light rye flour. The dough fluid contains, per liter, 30 g yeast, 30 g salt, and 10 g malt; the dough is fermented 90 minutes. The strongly baked and crisp *Bierweckerl* often contains caraway seeds. The small baked products for which Austria is famous include caraway seed, salt, and poppy seed products. The *Jourgebaeck* is half the regular size of the corresponding baked product; it is hand-made, crisp, and served fresh in exquisite restaurants.

Italy

Bread consumption in Italy is high (about 100 kg per capita per year). There are approximately 75,000 bakeries. Small baked goods are very popular; they are baked from white wheat flour of 60% extraction. Most products contain milk, fat, butter, and malt extract. They are most popular in the northern industrial part of the country and in cities; the most well known are *grissini* and *cracker*. The mainstay of farmers is white bread of relatively high (up to 85%) extraction. The loaves are 300–500 g, and are consumed by inhabitants of southern Italy and Sicily; in Tuscany, the farm bread is baked in 2-kg loaves. Bread made from whiter flour and in the form of smaller loaves is popular in central Italy and in Rome. Specialty breads, generally small in size, from white flour are popular in the north. They include the round buns of the Belgian type (*pistolets*) and the French bread types up to 50 cm long. The consumer likes a bread that is rather dry and has a thick crust.

Only a small portion of the bread is panned. Many Italians like a fine-textured bread; to accomplish that, some bakeries use folding machines that work the dough to produce a silky grain. With increased mechanization of large bakeries, only farmers produce the old type of bread. Turin was famous for years as the city that produced the excellent *grissini*—the white, long stick bread. Today this type is produced by machines and is of inferior quality. Interestingly, *grissini* is recently popular in Roumania. It is produced from a dough made from 100 parts flour, 3.5 parts yeast, 1.2 parts salt, 3.4 parts malt extract, 1.9 parts malt powder, and 4.0 parts margarine. The yeast is dispersed in a salt solution, the margarine is melted, and all ingredients are mixed with water to a stiff dough. The dough is allowed to ripen for 30 minutes. The fermented dough is rolled out several times, cut into strips 200 mm long, proofed for 60 minutes, baked, and packaged.

Panettone is an Italian cupola-shaped festive bread that has a fine, silky texture and a yellow-gold crumb; it keeps well for months and has an excellent, relatively sweet taste and a specific fruity aroma. The bread is yeast-leavened, but it has no yeasty taste or flavor. Its production spread from Italy to many other European countries (Switzerland, Austria, France, England) and to the United States. Best *panettone* requires low yeast levels and long fermentation times, which result in sour fermentation and distinct flavor development. The yellow color results from egg-rich formulations. This also facilitates strong mechanical development during mixing.

Greece

Primitive baking continues in some villages of Greece. The flat types resemble those produced in old Greece. Most of the bread today is made from milled wheat products. The special festive breads are rich in raisins. *Lambropsomo* is a Greek Easter bread that has hard-cooked, colored eggs set in the dough.

The Iberian Peninsula

Spain

The Vienna bakery is well known and recognized throughout the whole world. The development of bakers' compressed yeast and its use by the Austrian baking industry have contributed more than any other factor to the exclusive position of light, bold, and appetizing yeast-leavened bread in Austria, Germany, Switzerland, the Scandinavian countries, and some eastern and southern European countries. In some countries such as Spain, sour fermentation is still widespread.

The Spaniards like and eat white bread. Some parts of Spain, mainly Castile, produce large amounts of acceptable wheat; rye is used as a feed. The poor may replace part of the relatively expensive wheat bread by corn (*vorona*). In addition to white bread, less expensive dark bread (the so-called *panchon*) from high-extraction flour is produced. Both types of bread (in sizes from about 1 lb to 6 lb) are produced by a sour fermentation. In addition, buns (*bollos*) are produced from sourdoughs. Yeast-leavened white bread is produced mainly in large towns. It is generally made from a lean formula that includes flour, water, salt, and yeast.

The farmers' sour bread is made in a fermentation process that involves four stages: basic sour, refresher, sponge, and final dough. The basic sour, a piece from the last bake, is used for making bread on the following day. The doughs in all stages are cool (below 23°C) and firm. The final dough is tough and is premixed in a *masadora* (a mixer with two blades rotating in opposite directions) that disperses the water, salt, and flour. The content of the *masadora* is then passed several times through a *bregadora* equipped with rolls to give a fairly homogenous dough. That dough is scaled and molded on a smooth table from hardwood that has been covered with a little oil to enhance dough handling. The shape of the dough pieces varies in different parts of the country (*pan de Grijon*, *pan de Oviedo*, etc).

The dark bread *panchon* is made in three stages of sour fermentation. The dough is generally of a softer consistency than the one made for production of white bread. The dough is scaled and placed as round pieces into dough containers; it has generally a higher acidity than the white bread. The texture of *panchon* is coarse, and it has rather large holes. The white bread, on the other hand, which is made from dough that has passed through the rolls several times, has a rather uniform, fine, porous texture. Manufacture of the sour-fermented bread is quite an art. The baker uses a small amount of sponge in bread making and must make sure that the sponge is of the right activity to leaven the bread dough. Overripe sour is undesirable, especially in the hot climate, and leads to a sour, unacceptable bread. In many parts of the country, semolina from durum wheat is added to the flour.

The bakers in Spanish villages are conservative and slow to adopt yeast fermentation. The farmer believes that sourdough bread has more staying power, is better for overall health, and prevents dental caries. In recent years, the government has provided the bakeries with uniform and high-quality liquid ferments.

In towns, in addition to the typical Spanish white bread, baked products patterned after Viennese and Parisian styles are available. In addition, various types of buns are widespread.

In recent years Spain has seen considerable mechanization of bread manufacture. The development has been most significant in the installation of modern ovens. In medium-size bakeries, the trend is to introduce efficient mixers.

Portugal

The coastal strip of Portugal between Minho in the north to Algarve in the south offers a rich selection of decorated religious and festive baked products. Most common are small statues of saints carved from wheat or rye doughs, or formed by pressing doughs into wooden or metallic forms. In addition, themes from folklore, music, nature, etc are used in preparing various forms of bread.

THE UNITED KINGDOM AND IRELAND

Today, most of the bread in the British Isles is baked from milled wheat products, about 85% being white and the rest a mixture of dark (including whole meal) and special breads. Originally, emmer wheat was used for bread making in England. It was gradually replaced by common wheat, but for many years barley and rye were widely used and flours made from peas, beans, vetches, and acorns frequently found their way into the dough. For many years, barley was the main crop in some areas, especially in the southwestern counties, in the north, and in Scotland. In the 17th century, wheat was 38% of the grain grown for human food; percentages of rye, barley, and oats were 27, 19, and 16, respectively.

Although commercial bakeries appeared in cities from the 11th century onward, most of the bread eaten as recently as the beginning of the 19th century was home-baked or baked in communal ovens. In the Middle Ages and later, white bread was a sign of superior social status, wealth, and substance. For the majority of the population, bread was made from the other grains. The common loaf was made from maslin, a mixture usually compounded of wheat and rye, and sometimes barley. By the beginning of the 18th century, wheat was established in England as the main bread grain. Yet barley grain continued to be the main food of the poor of Cornwall, and in the north and in Scotland barley and oats were used more than wheat for bread.

In early English and Norman times, the best-quality white bread was known as *pain de maigne* or *pain main*, and loaves were often stamped with a representation of the figure of Christ. By the 15th century, this bread was generally known as *manchet*. But for centuries the bulk of bread eaten in England was brown or gray. The latter was made from flour sifted on coarse bolting cloth. Brown bread, known as *tourte*, was from whole wheat meal or flour. For at least three centuries (until 1645), there were in England two branches of the bakers' guild—for white and for brown bread. By the end of the 18th century white bread spread from the south to most of the country. During periods of wheat shortage, the government authorized the production and sale of standard bread stamped "S" and containing a higher proportion of bran than white wheat bread. It was cheaper but never popular. Before World War II most of the bread in England was made from flour of 70–72% extraction. In March of 1942, the extraction rate was raised to 85%. This increase was dictated by the difficulties in importing adequate amounts of wheat, but also for nutritional reasons. The Medical Research Council recommended that a higher extraction rate was desirable for maintaining the nutritional level of the restricted wartime diet. Only in 1953 was the control of milling extraction largely revoked.

Today, most of the bread in cities is baked in large, mechanized, automated bakeries in the form of panned loaves from white wheat flour and a yeast-leavened, relatively rich dough formula. Yet in many farms and coastal areas, special types of bread are still produced.

For years bakers used a ferment, a stage of bread making preliminary to a sponge or dough. The ferment was a concoction of flour, potatoes, sugars, and yeast foods. Most ferments were allowed to stand 5–6 hours; some (Scotch barms), up to 13 hours. A ferment usually contained a small proportion of the total liquid and all of the yeast, no salt or fat, all yeast food (potatoes, sugars, scalded flour), and a small proportion of untreated flour. In Scotland, where the system of ferments (or barms) was developed, there were special soft flours for that purpose. The sponge was another preliminary stage to the finished dough. It played an important part in the Scottish system. In England, especially in London, it used to be universal to use three stages: ferment, sponge, and dough. They were later replaced by sponge and dough. In Scotland and Ireland the sponges were divided principally into quarter and half sponges, being respectively one-quarter and one-half the total liquor. The quarter sponge, which lay for 12–14 hours (up to 16 hours occasionally) was usually stirred in tubs and was known in the south as a ferment tub. It was moderately stiff and stiffer than the short sponge into which it was made before being made into a dough. The use of two sponges for one batch of final dough of the north differed from the slack ferment and tight sponge of the south, even though the two served the same purpose.

There was also a Scotch ferment, sponge, and dough system adapted to Scotland. Various amounts of salt were added at each stage. The strong flour was added to the quarter sponge, and the sponge and soft flour were added to the dough. The quarter-sponge system was popular in the west and in Glasgow; the half-sponge system on the east or Edinburgh side.

Sponges in England were usually made in troughs and not in tubs, and they were usually stirred and broken up for dough making by hand and not by the vertical sponge stirrers used in Scotland. The breaking was considered essential if the bread was to be free of large holes. Virgin and Parisian barm were the leading types of Scotch barm. The difference between the two was that while virgin barm was set without yeast and allowed to ferment on its own, the Parisian barm was stored or started with

Figure 12-7. English bread types. (a) Bloomer; (b) farmhouse; (c) milk loaf; (d) cholla; (e) cob; (f) sandwich; (g) long baton; (h) French stick. (From Sheppard and Newton, 1957.)

barm from a previous brew. For the virgin barm, a mash of malt was infused with a hop extract and the whole was strained. The liquid was mixed with flour and scalded with boiling water, and the tub or barm was left uncovered for 20 hours. The barm was transferred to a second tub, and salt, sugar, and little flour were added. It was best after 4–5 days of fermentation and was stirred every 12–24 hours. Some bakers discarded malt and hops in their barms and used flour and sugar only. About two-thirds of the flour was scalded, and one-third was raw.

The "spontaneous" virgin barm was quite unpredictable at times and rapidly lost popularity. For Parisian barm, instead of adding salt and sugar, a 2-day brew was added. Active fermentation was over within 1 day.

A description of English bread invariably includes the terms "crumbly" and "crusty." A crusty loaf is one completely bounded by a crust and is usually baked on the sole of the oven. A crumbly or crumby bread is baked from dough pieces that touch each other at the sides at which there is no crust (Figures 12-7 and 12-8 show types of English bread). There are three main Coburg breads: plain, crusty, and split. They are all round. In the crusty, two cuts are made across the surface of the molded round dough piece to form a cross about half an inch deep. In parts of northern England, deeper cuts are made in the surface of the dough piece and the

Figure 12-8. English bread types. (a) twist; (b) pork pie; (c) skip; (d) lemon; (e) Cornish brick; (f) buster; (g) long loaf; (h) fancy brick; (i) crumby brick; (j) Cornish turnover; (k) notched cottage; (l) farmhouse. (From Sheppard and Newton, 1957.)

bread is somewhat flat. In the split Coburg, a rolling pin is pressed across the dough to divide it into two equal pieces. There are many variations of the baton type. They vary in length, number, shape of surface cuts, and type of end (square, tapered, or pointed). The English crusty baton is a squat, square ended-loaf, about 9 in. long and 5–6 in. wide, with eight diagonal cuts along the top surface. It is baked in steam.

The Belgian baton is similar to the English crusty baton except that it is boat-shaped. The ends are rounded, the top surface is cut diagonally five times, and the bread is baked in steam and finished off in dry heat.

The bloomer is a popular baton in cities. It is 12–14 in. long, about 4 in. across the middle, and 3.5 in. deep. It is generally made from a rich formula that includes milk, sugar, and fat. The loaf is molded in a long shape and proofed, and the top is cut at 1-in. intervals at an angle of 45° to the top surface. This imparts an S-shaped scroll along the surface of the baked bread. A herringbone design is obtained by means of a longitudinal cut from one end to the other, followed by smaller cuts. The dough pieces are placed in steam-loaded baking chambers, and baking is completed in dry heat. The leopard (or tiger or "papped") baton is similar to the bloomer, except that before setting in the oven it is brushed heavily with a special paste that includes starchy material, sugar, and oil.

The Danish baton is an oblong (8–9 in. long), 2-lb. loaf. The proofed dough is heavily dusted with flour. One deep cut is made along the length of the top surface, and the bread is baked at a relatively low temperature to develop a special flavor.

The lemon loaf is first molded as a round Coburg and then lengthened with a sharp knife, brushed with a 1:1 mixture of eggs and milk, and baked in dry heat. Buster baton is somewhat shorter than the lemon baton and has a deeper central cut and additional small side cuts.

The French long is made from batons 18–36 in. long. The dough used in its preparation is rather soft, and the pieces are preferably proofed in cloth-lined baskets, though long boards and perforated wire trays are also used. The surface is brushed with diluted milk before cutting and baking. The large crusty surface has contributed much to its popularity. The Italian long is shorter than the French long and has more and smaller surface cuts. The Underseller long from the East End of London is 14–16 in. long and has two sets of 9–10 diagonal cuts to give it a diamond pattern appearance.

Great Britain has many panned breads (called tin-shaped). The most widespread is the commercial pan loaf baked in a 9 × 4½ × 4½ in. pan. It is made from a soft dough enriched by fat, sugar, milk solids, and diastatic supplements. The bread is baked 45–50 minutes, though large baking plants produce thin-crust breads after 40 minutes of baking. A split pan bread measures 12 × 3½ in. and is cut from one end to the other before baking. Square pan loaves are baked in 7 × 5 × 4½ in. pans; if cut longitudinally, they are called tribly. Sandwich loaves are baked either in individual pans fitted with a lid or in blocks of 4–6 pans covered with a close-fitting large lid. Commercial sandwich pans are made to hold 2, 4, or 8 lb of dough. The last is popular in Scotland, where one loaf can be made in a 2 ft 8 in. × 5 in. pan or four 2-lb pieces of dough can be baked in one container. The ends of the loaves are brushed with melted fat before they are placed in a pan. The smaller types of this bread are called Scotch pan. A pan Coburg is made from 1- or 2-lb dough pieces that are molded round and placed in shallow, slope-sided circular pans measuring 8 in. top diameter, 6¼ in. bottom diameter, and 3–3½ in. deep. The tops can have various types of cuts. Some breads have a steam glaze; others, a dry crisp crust.

Liverpool bread is baked in a special pan fitted with a hinged lid which is attached firmly by a pin during baking. The musket, or toast loaf, is baked in round pans about 10 in. long and 4–4.5 in diameter. To prevent the pans from rolling during baking, they are made in pairs hinged by a metal piece. The crinkled musket loaf (or the lodger's loaf) is made from a slightly stiff dough baked in pans with a crinkled, fluted, or concertina-like appearance. For the long plait, the dough is divided into two pieces. The 1½ -lb piece is molded to fit the long pan; the half-pound piece is divided into threee sections, plaited, and laid on top of the larger piece.

The *quartern sans* is made in a 4-lb sandwich pan. The dough is molded long, the top is cut with a sharp knife to give a diamond pattern, and the dough placed in the pan. If the cuts are even and not too deep, they extend across the surface and halfway down the sides of the baked loaf. The molded twin (or composite) loaf is popular with many bakers. Two or four 1-lb pieces of dough are molded round and placed side by side into a pan. The pieces knit together during proofing and baking but can be readily separated, if desired, after baking.

In addition to the panned varieties, there are many two-piece loaves. For the plain cottage loaf, an attractive and impressive bread, the dough is divided into 1¼-lb and 3/4-lb pieces. Both are molded round and slightly flattened, and the smaller piece is placed on top of the larger. Midway in proofing, the center of the top piece is pressed carefully into the lower one. After 10–15 minutes of proof, the dough has recovered and can be baked on the sole of the oven with or without steam.

In the notched cottage loaf, sharp cuts are made at about 2-in. intervals before baking. Plain brick is basically an oblong cottage loaf that is sometimes notched on the sides to give a fancy brick. The Jewish *challa* is a plaited loaf, with or without seeds, made from a rich formula and is generally egg-washed before baking. The English *challa* is prepared from a 1-lb, 10-oz piece that is first molded round and then elongated into a boat-shaped piece about 10-in. long with tapering ends. A smaller, 6-oz piece is divided into three sections and made into a plait that is long enough to cover the top surface of the loaf. The loaf is brushed lightly with water to make the pieces stick together. The two pieces can also be fermented (proofed) separately, a deep lengthwise cut made in the larger piece and the plaited piece placed on the incision. The loaves are baked 10 minutes in a steam-loaded oven and then in dry heat. The Dutch *challa* is similar to the English except that a thin dough rope (1–2 oz) instead of the braid is placed on top of the large piece. There are quite distinct variations in types of bread produced in various parts of the British Isles.

The main loaf from the Midlands to the south is the crusty cottage; other crusty sorts and tinned loaves are less numerous. The crumby loaves are baked little. Farther north, particularly in Lancashire and on the coast of Yorkshire, most are panned. In Scotland there are many bread types. Although there are "French" loaves and pans and cottage bread, the full-faced, crumby, all-round bread is typical. The Scotch cottage breads are "squatty" with a big head and relatively smooth and pale crust, and they are more salty and less sweet than cottage bread from the south.

Ireland, especially in the north, is more similar in style to Scotland than to England and has big, highly fermented bread and a good deal of crumby bread. As a rule, its bread is from tighter dough than in Scotland and is less fermented and less well finished. The crumby, plain, or batch loaf in the north is not of the oblong Scotch type, but often with four square sides. The crumby loaf in other parts is long and narrow; in Dublin the crumby bread is made from tight doughs and baked into hexagonal loaves. In the south of Ireland the crumby loaf is usually of the turnover type, molded in one piece after the style of the Scotch French, more particularly that of Guernsey. The Irish panned bread resembles the Scotch. As in Scotland, the crumby is the rule and the cottage an exception, but crusty types made from shorter fermentation systems are quite popular. The Welsh cottage bread resembles that of the English Midlands. It is whiter and has a firmer crumb texture than cottage loaves from the south.

There is a large variation in the crust of the Scotch breads; that of the west coast is darker than that of the east. In the Scotch plain, there are only top and bottom crusts; the sides are completely crumby. These crumby sides—that is, all the outsides (except the one next to the other half) of each pair of half loaves or half-quartens (a loaf is a quartern)—are greased with a brush with melted lard or vegetable oil to smooth the sides. This is also done in some parts of Ireland, but seldom in England. The plain loaves that give the good pile and volume (they are usually smaller on the east than on the west coast) are from slacker doughs than the Scotch crusty such as Coburg and cottage. The Scotch cottage breads are bigger, lighter, riper, more salted, and with smoother and paler crust (indicative of longer fermentation and use of lard) than the English cottage.

Brief mention should be made of the special types that were and are baked in parts of the British Isles only. The typical Scottish bread originated as a bannock—a round flat cake baked on a "girdle," or griddle. Later, the bannocks and scones were "finished" or baked entirely in the oven. Bannock usually means a large, round, dinner-plate-size scone or cake. When the bannock is cut into "farls" or wedges and these triangles are baked individually, they are referred to as scones, as are the smaller, round, cut or dropped varieties. Drop scones are thick pancakes that are chemically leavened, made into a batter that includes buttermilk, and baked on a griddle. Other types of scones include white oven, griddle, treacle (with cinnamon, ginger, and molasses), cream, and potato. Buttermilk bread is made from a recipe similar to that for Irish soda bread, but apparently when made by a Scot it is different. Baps are yeast-leavened dough rolls, a Scottish specialty. Oat bread is a yeast-leavened, oven-baked loaf made from a recipe that includes 5 parts of wheat flour and 2 parts of rolled oats; the other ingredients are milk, molasses, salt, and yeast.

Yeast-leavened Scotch bread baked from wheat flour is well piled (bold) and has a silky texture; it is relatively acidic and salted. Batch bread, the most popular in Scotland, is made from oblong pieces, 7 × 3½ in. Two such pieces are laid side by side on the table and the ends are brushed with melted fat. The pairs are proofed in a wooden box. The dough is somewhat underproofed, and final proof takes place in the oven. The bread is baked slowly and has excellent texture, flavor, and shelf life. The Scottish French bread is baked in a special pan that holds five loaves and is 34 in. long, 5½ in wide, and 5 in. high in the back and 2½ in. high in the front. The ends of a 12 × 6-in. piece from a relatively stiff dough are folded inward and meet in the center. A rolling pin is pressed over the joint, and then half is folded on top of the other. The ends are buttered, and a slight brushing of fat between the surfaces of the last fold makes it possible to produce an attractive, upright loaf, sometimes called a "piano."

The Irish grinder is similar to the Scottish French. The Irish brack is a panned fruit loaf baked in a rather cool oven. Irish soda is aerated by using a combination of buttermilk and soda. The Irish baking powder bread is made in Northern Ireland. Leavening is accomplished by action of acid on sodium bicarbonate. Generally, a soft dough containing about 0.6% sodium bicarbonate (on flour basis) is mixed with sour buttermilk. The acid frees the carbon dioxide from the carbonate; the free carbon dioxide is essential in leavening. Baking takes place immediately after dough mixing. Thus, long and expensive fermentation is eliminated. The process is particularly suited to baking bread from soft and weak Irish flours. Actually, if the flour used is too strong and too rich in protein, the bread will be poor. Generally, the bread is made from strongly bleached flours. If the flour was milled from sprouted wheat, the bread is soggy and has water lines. An 1⅛ kg loaf is baked at 220°C for 75 minutes.

A few of the Welsh-type breads are still on the market. For making the flat *planc* bread, the baking iron *(planc)* is placed over a fire and the circular proofed dough piece about 1½ in. thick is baked for 20 minutes on each side. *Bara brith* is made from a rich formula that includes lard, sugar, eggs, fruits, and spices; it is proofed a long time and baked at a relatively low temperature (180°C). *Bara cymareg* is made from a soft dough piece that is flattened slightly before proofing. A small hole is made in the center of the loaf which is dusted with flour and baked.

In addition to white bread, various types, shapes, and sizes of whole wheat bread are available in Great Britain. Some proprietary products include bran, others germ, and still others generous amounts of malt flour or malt syrup. Some of the special types include starch-reduced or practically starch-free breads (for diabetics), gluten-enriched breads, rye breads (including pumpernickel), chemically aerated (by incorporation of carbon dioxide under pressure or by chemical leavening), fruit breads, cheese bread, pulled bread, salt sticks, Queen's bread, *brioche* (both sweetened and unsweetened), babas, *savarins*, and tomato bread (a rich formula that includes 25% tomato pulp and is shaped like a flat Coburg or crinkled musket). In all parts of the country one finds small, flat-type breads. They include the Scotch bap, Devon flat, barm cakes, Welsh muffin tea cakes, turnovers, bakestones, and many others.

SCANDINAVIAN COUNTRIES

In northern and western Scandinavia, barley and oats rather than wheat and rye were the most important bread cereals for the peasants from prehistoric times till the 19th Century. Around the year AD 900, two kinds of bread were produced. The poor man's bread was doughy, heavy, and filled with bran; the noblemen ate "thin loaves white with wheat." The bread was baked on bread irons. The round irons had a diameter of about 10–15 in. Unfermented, relatively thin wheat loaves fried on bread irons are still eaten today by the Lapps. The wheat required for making bread for the Norwegian nobility was imported from England and Denmark. The poor man's bread was generally unfermented and contained little if any wheat; the bread staled and deteriorated rapidly and was consumed within a few days.

In the beginning of the Middle Ages, at the time of the introduction of Christianity, bread-making practices changed. This resulted from improved milling techniques (from the introduction of water mills) and the availability of larger amounts of wheat and flour. Thin bread was produced in larger quantities, and the dry, crisp product was stored for longer periods. This required large bread irons to produce *flatbrod* which is still widespread in western Scandinavia. The seagoing Norwegians introduced at that time the twice-baked ship bread.

In the northern and western parts, barley was the dominant cereal and was slowly replaced by oats. In the south and east, rye production was encouraged by the church. The rye bread was fermented and baked. A baking oven with a rectangular floor was developed in the east. An oven with a round or oval floor originated in the south.

The agrarian policy of the church and the prohibition of meat during numerous fast days increased the consumption of bread. Oven-baked bread, thin bread fried on irons, dried cakes with holes in the center, thick and hard bread made from rye, and twice-baked dark rye bread were baked on a large scale for people from various economic classes as well as for the army and navy. The oven-baked products, especially rye bread, spread farther to the west and to the north. In the interior of Norland the rye bread became known only in the 18th century; the unfermented barley bread is still widespread. In Halland and Bohnslan, rye bread replaced barley and oatmeal bread only in the 19th century. During the first half of the 18th century, the many wars, epidemics, crop failures, and increased population dictated production of famine bread. This was the time when potatoes were introduced—originally for the production of famine bread, later as an independent food competing with bread.

In the 18th century, the government regulated the supply of ship biscuits and a rye bread, called *ankarstock*, for the army. The price of white bread, *franskbrod*, has been controlled since 1768. Brewers' yeast was originally used for baking white bread, buns, and cakes, and later also for ordinary rye bread. During the 19th century use of brewers' yeast and later of compressed bakers' yeast became widespread.

Sweden

"Pease bread" originated in the chalky regions where it is baked in the form of thin loaves that contain a considerable proportion of peas. Bean bread is characteristic of part of western Sweden. Potato bread became more generally known in northern and central Sweden around 1880.

In large parts of Sweden, fermented, dried, and hard rye bread is widespread. The bread is round and has a hole in the center for hanging. In certain bread types that are eaten undried and soft, the hole has no functional role and has been retained as a decoration. Fermented loaves that are eaten undried are rolled out into flat bread in some districts. Generally, however, it is customary to bake thick, round loaves that are thickest in the center, so-called *limpa*. The round loaf almost completely dominates the Swedish bread culture. Only a few bread types with oval or rectangular shapes are seen. The oval shape was developed under the influence of medieval continental (European) bread and ship bread. Some oval loaves, *kavving*, are made by putting two halves together, one on top of the other, with an intermediate layer of fat; the bread is baked twice.

Today, consumption of wheat-based baked goods in Sweden is relatively low (about 55 kg per capita per year, calculated as flour). Home baking is decreasing, and production by large, commercial bakeries is increasing. In 1963 there were about 6,000 bakeries, but large bakeries and commercial baking plants are taking over more and more bread production. A continuous process was developed by Nemo Ivarsson and has gained limited popularity. The most common type of plain bread in Sweden is the so-called VR (mixed wheat-rye), a yeast-leavened and relatively sweet loaf. The French bread baked from wheat flour, water or milk, shortening, and often some sugar is increasing in popularity. Whole rye bread is still made in the south. The old Swedish bread products include a characteristic crisp bread that is consumed at a rate of about 6 kg per capita per year. In addition, many millions of killograms of bread are exported from Sweden all over the world. The rye-crisp bread is made from coarsely milled rye, water, salt, and yeast. In some types of crisp bread, such as *Ryvita*, no yeast is used. Instead, the rye flour is mixed with ice, water, and salt to form a cold, porous dough which, like the normal rye-crisp dough, is sheeted, notched, and scaled. It is baked without fermentation or proof. Another type of "hard bread" is a biscuitlike wheat crisp bread with wheat flour, milk, sugar, salt, and shortening.

Norway

The road of bread development in Norway, as in the other Scandinavian countries, was long and tortuous since it was first discovered that grain could be roasted over an open fire, until today's assorted choices of fresh bread, packaged biscuits, crisp bread, and flatbread became available in bakeries and supermarkets. At first, flour from grains and from all available flour-yielding plants was used either boiled or fried as porridge or bread. Unleavened bread can be made from almost any kind of flour, but only wheat and rye flour can be used for yeast bread. However, because of climate, wheat and rye were late arrivals in Norway.

It is not known when the first bread was baked in Norway, but it was most likely a compact lump of dough with a burnt crust and an underbaked, doughy center. When a dough of flour and water is formed by hand, a round ball is the natural result. This is the form the dough will retain even after baking in the embers or ashes of fire. A step forward was made when it was discovered that the dough piece could be pressed against a flat stone and the stone heated over the fire. Such flat stones have been found often in Norway. They have a diameter of 15–30 cm. Baking progress was made by learning to press the dough into thinner layers against the stone on the underside and thus produce a softer product. *Lefse Lompe*–type breads (which resemble the Mexican tortillas) were the common products resulting from this kind of baking.

When the use of iron was discovered, small plates replaced the flat stones. Handles were added to the iron pans in the period of the Vikings, and this probably became the forerunner of the flatbread griddle. There is not much written about the development of flatbread in Norway, but apparently it was not earlier than in the Middle Ages. The first time flatbread is mentioned by name in Norwegian sources is in the account book of Bergenhus in 1519. Flatbread was described by the Swedish Archbishop Olans Magnus in his book *Historia Degentibus Septentrionalibus* in about 1400. From 1700 on, references to flatbread occur regularly in the literature, and the reputation of this bread for taste and keeping quality became well established.

A survey conducted in recent years in Norway showed flatbread to be quite popular over the whole country. The home baking of flatbread, however, is too time-consuming. Norwegian flatbread factories of today retain the old national tradition, excellence of product, and continue to bring this type of bread into the Norwegian home as well as into the world market as an export product.

Finland

Suomi (Finland), the land of a thousand islands and the midnight sun, is not only picturesque but also a land of unusual bread varieties. Bread is part of practically every meal, especially the warm noon and evening meals. The main meal includes meat, potatoes, vegetables, a cereal, a glass of milk, and several slices of bread. Despite the rather unsatisfactory climate for wheat growing, most kinds of Finnish bread contain a substantial amount of wheat flour. In addition to mixed rye-wheat and barley-wheat, a white wheat (French) bread is also produced. The bread is hearth baked (round or long) or panned and made round and flat, or round and flat with a hole in the center. The latter is typical farm bread, and the flatbread is dried on poles until it attains the taste of rye-crisp. The amount of buns is small. Until 1700, barley was the chief bread cereal in Finland. Relatively little rye was grown and used, and practically no wheat. Some buckwheat was grown in the southeastern region. In about 1750, rye replaced barley as the main bread cereal, and only during the past 20 years has wheat become a major crop.

During years of great famine, cereal flours were partly replaced by other available foods. In 1868, bread was baked from a mixture that included reindeer moss, which grows wild mainly in Lappland. Some pine bark has been used during the "hunger" years. A certain kind of turnip has also been used as a flour substitute in the midwestern part of Finland. Turnip bread was baked in the autumn and stored through the winter at below-freezing temperatures.

The per capita consumption of flour and other milled products has decreased from about 220 kg per year at the turn of the century to about 100 kg. The decrease was mainly in rye, which earlier comprised almost 80% of the bread grist but today is only about one-fourth of the bread cereals. Total bread consumption on farms is significantly higher than in towns. In addition, cereal products are consumed by farmers and farm workers in the form of various breakfast cereal foods and by children in rural areas in the form of a dish served at lunchtime.

Although wheat bread is becoming more popular and some barley and oatmeal breads are still made, the typical Finnish bread is made from rye. In eastern Finland, bread is baked from whole ground rye in the form of round loaves; in the western part of the country, bread in the form of rings is baked. The bread has a fairly acid taste, and its pH is 4.0–4.5. This has resulted from the fact that much of the rye is sprouted, and the only means of making bread with an acceptable texture is by the use of an acid fermentation. The bread stores well. On farms bread is baked in large quantities, pierced with a rod, and hung near the warm roof of the living room.

There is a great variety of different types of rye-crisp breads (*Knackebrot*), varying as to the grain used (mostly 100% whole rye, some made of mixtures of rye and wheat, and a special type that includes graham flour), additives (spices, mainly caraway seeds), shape and surface structure, and manufacturing technique (Figure 12-9). Most rye-crisp is yeast-leavened, but some so-called ice bread has a whitish surface color and is made without yeast by trapping air during vigorous mixing of dough with ice or in a refrigerated mixer. The modern rye-crisp manufacturing plants employ a totally continuous mixing and baking system. Normally, the characteristic taste of 100% rye bread is preferred, and the Finnish connoisseur believes that even small amounts of wheat impair the taste. A water formula is generally used, but in some types milk or milk powder may be used. Similarly, fat is included in some formulas. The dough is mixed, allowed to

Figure 12-9. Various types of Finnish bread. (Courtesy Brot und Gebaeck.)

rise, pressed in thin layers, given a desired surface structure, and baked quickly in large ovens (for 10 minutes or less) to a product with 5–7% moisture. The large cakes are then cut to desired size. The yearly per capita consumption is about 3 kg. Rye crisp stores well and can be used in army rations. The amount of crisp rye produced in Finland during the years of World War II was three times the amount baked in years following the war.

Bread from barley only is practically extinct, but mixed bread is still baked. The Finnish mixed bread is made from dark wheat flour and white barley flour. Sometimes barley flakes or oat flakes are added. Among white breads, the most popular is the "French" bread which in Finland has a lengthwise cut of the crust to form a "comb" upon baking. The crust of the baked product is relatively crisp, although modern wrapping practices affect the desirable crispness adversely. The other, equally popular type made of a wheat flour dough is the "Polish" type (in Finnish *polakka*), which has several diagonal short cuts in the crust. A slightly different proof schedule and baking make the crumb more open and the crust less crisp.

Several specialty products are baked by the Finnish farmer. *Mammi* is an Easter dish baked from a soft rye flour dough that contains malt flour to improve taste and browning. *Talkkuna* is made from a cooked mixture of rye, barley, oats, and beans that is dried and ground to a flour. The flour is eaten with cultured milk. *Kalakukko* is rye bread that contains small fish and lard. The fish are baked with the bread, and the product is a "preserve" in which the container is also consumed.

Denmark

Bread contributes about 30% of the calories and protein in the Danish diet.

Smorrebrod is a typical Danish dish; the word means buttered bread. It is commonly made from one of the several types of rye bread: dark rye (*fuldkorns rugbrod*), light rye, and crisp rye. Others include the French (*franskbrod*), the Danish sour bread, and a white bread made from rye flour (*sigtebrod*). Any of these can be used in making *smorrebrod*, spread with butter, and served with at least two ingredients (fish, fowl, egg, vegetable, or meat; sometimes even with some fruit).

MEXICO, SOUTH AMERICA, AND CENTRAL AMERICA

Corn is an important item in the diet of Mexico, the northern Andes, and northern and central Brazil. Rice predominates in southern Brazil and the Caribbean islands. People in Argentina, Uruguay, and Chile eat wheat products.

In most cities in Mexico, South and Central America, and the Caribbean islands, large, well-mechanized plants produce a substantial proportion of the bread. Generally, however, the degree of mechanization does not approach the extent found in English-speaking countries or some parts of Europe. Where labor is available readily and pay scales are low, shops are mainly hand-operated except for the dough mixer. Dough mixers are often the slow-speed type.

New, modern flour mills have been constructed throughout Latin America, and these have generally supplied a better grade of flour to the baking industry than when flour was imported. Clears or stuffed-straight grade flours are often used for bread production. Exceedingly high-protein, strong wheat flours are often required for some baking operations in the Caribbean area because of the punishment given the dough and the long fermentation time. Most flours used for hearth bread of the type with "wild" crust breaks are heavily bromated. In Columbia, bread is produced from wheat flour of relatively low protein content. Most bakers use straight doughs with a relatively high percentage of yeast (about 4% on flour basis). Most baked goods are rolls in various shapes, types, and sizes. Typical and most looked-for is the *media luna*, or half-moon. French bread is produced in large quantities, and the larger bakeries also market Pullman bread.

Bread in villages of many countries of Central America is baked in the form of tortillas. Tortillas are prepared from Indian corn which is first parboiled to make it clean and soft. The meal is then crushed into a paste with a stone rolling pin on a small stone table, after which it is baked on a plate of iron or earthenware but not enough to brown the tortilla, which is best when served hot.

In the larger population centers of Mexico, modern, mechanized bakeries produce an American-type pan bread, soft buns, and rolls. In the majority of small bakeries, the most common breads are *bollilos* (hard rolls) and *pan dulce* (sweet goods). The bread is baked from relatively lean doughs containing low yeast levels. *Cassawa* is a native plant of tropical America, but it has been extensively introduced into Africa and many tropical countries. It grows in bush form, usually 6–8 ft high, and its roots, which grow in clusters, vary in size from a few inches to 3 ft long and sometimes weigh as much as 35 lb. *Cassawa* roots form the principal food of the common people of tropical America. It is generally handled commercially in the form of a meal sometimes resembling oatmeal, but it is made into thin, round cakes by the natives, into cassava bread. The meal is exported from some parts of the West Indies to Europe, where it is used for starch manufacture.

In the absence of wheat, the early Portuguese colonizer in Brazil had to change his food habits, a change that was to last for over 200 years. Bread was introduced into Brazil only in the 19th century. The early settlers adopted the use of manioc and corn from the Indians. The results are still evident in present-day Brazilian food habits, as there are many regions in Brazil where corn and manioc are preferred to wheat. Wheat was never raised in Brazil to the same extent as corn. Today Brazil ranks third in the world production of corn and is the largest importer of wheat in Latin America. When the first colonists arrived, the use of manioc was highly developed by the Indian woman, who prepared a cake (*carima*) for children or a paste or porridge (*mingan*) for the whole family. Northern and western Brazilians still prefer corn and manioc to wheat. Southern Brazil, on the other hand, has long enjoyed the use of wheat, and so have the large Brazilian cities, even though it has to be imported. In the four wheat-producing states of southern Brazil, which contain one-third % of the total population, wheat products have been a staple item in the diet of both urban and rural populations. The most common bread resembles the French type and is called *pao Frances*. It is baked in several sizes. The slang word *bisnaga*, or tube, designates the smallest size; it is often sliced, oven-dried or toasted, stored in tins, and served as Melba toast. The baker also sells sandwich bread that is sliced lengthwise for rolled sandwiches or crosswise for toast or regular sandwiches. The twist is baked little and is called *tranca* (braid). Rolls (*paezinhos*) are either yeast- or chemically leavened. Corn bread (*broa de micho*) is widely baked. Special types include cassava bread, an oven-baked, yeast-leavened bread from a dough composed of a mixture of wheat and cassava flour. Tchikao women of the primitive tribes of Brazil pulverize maize in a crude trough. They bake the crushed grain into bread or steep it to make a soup. Scraping, crushing, squeezing—hours of labor by village wives—go into the preparation of manioc, the bread of the Tchikao. They dig up the cultivated roots when they reach the size of large sweet potatoes. With shells and pieces of metal the women scrape off the skin and pound the roots into a pulp. To rid the manioc of its deadly prussic acid, the women strain the pulp through a sieve. A bark trough catches the juice. In another step to remove the poison, the women put pulp in a palm fiber tube to be squeezed as dry as possible. Baking the white mass on a ceramic griddle drives off more acid, leaving only harmless traces, and yields the final product, a staple of tribal life. The Tchikao, to whom pottery making is unknown, probably took the griddle from the nearby Waura, the ceramists of the Xinger tribes.

Baking plants in the large population centers of Brazil use both sponge and straight doughs. One can find Pullman bread, French bread, Vienna-type rolls, and practically any familiar bread variety. A typical bakery in a small country town in Brazil uses a lean dough for making bread of the French type. The same dough is used to make the somewhat tough, hard roll.

Originally, corn was used in baking bread in Chile. It was replaced, however, almost entirely by wheat. Use of rye has been on the decrease. Limited amounts of wheat are ground by milling stones and sieved for production of *harina tostada* (roasted flour) that is mixed with water or milk and made into appetizing bread substitutes. Flours produced by the primitive mills constitute only 1% or less of the total production today.

The daily bread of the city dwellers of Chile differs little from the bread consumed in cities of most Western countries. It is produced mainly from wheat in modern bakeries and is consumed within a day or so after baking. The Chilean family makes several types of bread. They include the *pan amasado* (kneaded bread), *pan de grasa* (fat bread), and *pan de mujer* (housewife bread that is made and baked by women only). The baked loaf has a diameter of about 4 in. and is 2–3 in. thick with a conelike top. The top is notched from puncturing the dough with a nail before baking. Dough composition is highly variable. Generally, 5 kg white flour is mixed with 0.5 kg lard, 20 g yeast dispersed in a cup of lukewarm water, and 3 cups of a salt and sugar solution. Lard is often replaced by tallow, butter, margarine, or shortening. The farm laborer (*inquelino*) is entitled to a daily ration of two 0.5-kg freshly baked loaves from dark wheat flour baked by a lean formula. The farmer himself prepares a basic bread from 1 kg white flour, 3 eggs, about 5 g lard or butter, 5 g yeast, sugar, and salt. This dough is supplemented for holidays and special occasions with raisins, milk, and confectionery.

In cities and farms, the popular *hallula* bread resembles, in composition, the typical farm bread. It is rich in fat and is made from a dough passed through narrowly spaced rolls as many as 20 times. The rolled-out sheets are dusted with flour and are made into 90- to 100-g square loaves that are notched by a pin roll and brushed with water (to provide an attractive shine) before baking.

The *empanada* is the Chilean Sunday bread in the form of a half-moon 12–14 cm long, a baked product that was introduced by the Spaniards. The

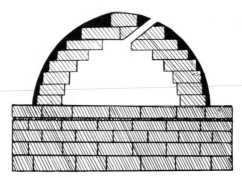

Figure 12-10. Early type of oven used in Chile. (Courtesy Brot und Gebaeck.)

dough, resembling that used in making *hallula*, is cut into round slices. Half of the slice is sprinkled with finely cut onions; small pieces of meat and spices are added, and the whole is heated for about 15 minutes. After adding hard-boiled eggs and olives, the dough is rolled into bread, baked, and served hot.

Early inhabitants of Chile mixed millstone-ground corn with water and salt and baked the flattened dough on a hot stove or in ashes. The oven was gradually modified; an improved type is shown in Figure 12-10. The beehive-basket-type structure is made of clay, bricks, or natural stones; the inside is rough. The arc is unlike the Roman type; it is made from units shifted from the center as in structures of the Aztecs and Mayas in Mexico and Guatemala and the Incas in Peru. The oven is often covered with a simple roof for protection from rain. The oven is generally heated with wood, seldom with coal; the ashes are removed before the dough is placed with a wooden paddle on the clean hot surface. The oven is closed, and the bread is baked for about 15 minutes. The clean, uniformly baked loaves are stored in white sheets. On many Chilean farms use of this oven has been discontinued in recent years with the availability of relatively inexpensive natural gas.

It is difficult to characterize the baking industry of Argentina by a simple statement because of the diversity of baking practices. However, the major population of the country is urban rather than rural, and since wheat has been a major crop, the production of wheat bread in commercial bakeries has a long history. The population of Argentina is predominantly of Spanish and Italian extraction, although there are strong English, French, and German influences. Therefore, as expected, baking methods and bread types are of European origin. There are many ultramodern bakeries with the latest types of machinery and ovens imported from all over the world or produced locally. Argentina is in a favorable position from the standpoint of availibility of the raw material required by the baking industry, because wheat flour, sugar, salt, shortening, and yeast are all produced within the country.

The simplest type of baking is done in *campo* (ranch) for the preparation of *galleta criolla*. This is a hard, flat loaf made by mixing a very stiff dough composed of sourdough, flour, salt, and water. After a variable fermentation period, the dough is baked in a crude, small, mud or brick oven called *hornos*. The ovens are heated with wood or charcoal.

All kinds of bread are made in the cities. The most popular bread types are the hard French rolls, although there is a demand for sandwich bread baked in covered pans. Per capita consumption of bread and pastries is high in Argentina.

As in the capital cities of all Latin American countries, there are modern bakeries in Lima, Peru, but in the mountainous villages baking is often done under a number of handicaps such as high altitudes and variability of ingredients. The majority of bread in Peru is the hard roll type made by the sponge baking procedure. Modern equipment in bakeries is often limited to a slow-speed mixer and bun cutter. Labor often resists mechanical improvements.

The appearance of bread in Peru is often not good because of lack of satisfactory volume caused by overproofing. Usually, lean formulas are used consisting of only flour, water, salt, and yeast. The long fermentation times used result in bread of excellent taste and flavor.

NORTH AMERICA

Baking in Early America

Indians were the sole inhabitants of the North and South American continents for many centuries before immigration from other countries occurred. Therefore, any consideration of baking in the Americas must consider the food habits of the Indians before the arrival of settlers from Europe. Wheat was not indigenous to either the North or South American continents, but corn and potatoes were, and these are foods the Indians gave to white men. Since corn was extensively grown in the Americas, most Indians were accustomed to its food use and showed the white men how to make such products as corn grits and how to prepare hominy, popcorn succotash, and tapioca from corn grits. Some remnants of Indian types of bread can still be seen in the United States today.

When drought struck in the 13th century, the Indians from Mesa Verde left their Colorado cliff homes and moved south into certain regions of New Mexico and Arizona. Today, a visit to Santa Ana Pueblo in New Mexico shows that the tools and the products are similar to those discovered in Wetherill Mesa.

In grinding bins, women make corn meal by rubbing dry corn between hand stones and slanted slabs or metates. Differing textures of stone produce various degrees of fineness. The new bins are identical to those of the 13th century except that wood has replaced stone in the partitions. The ground corn is used to make paper-thin bread. To the uninitiated, "paper bread" is of rather poor taste, but the Pueblos consider the bread a treat reserved for festive occasions.

To make the Indian delicacy, a thin batter of corn meal is spread on a smoking-hot griddle, allowed to bake a few seconds, and lifted off. Much practice is needed to smear the batter without burning the fingers. The stone griddle, one of a Pueblo woman's most treasured possessions, is handed down from generation to generation.

The early settlers, as they moved across the northern part of North America, used corn meal as a main food. It was used in some way for almost every meal in the form of a mush, johnnycake, hoecake, or corn bread. All of these corn products were made from meal, salt, and water, and it was the way in which they were cooked that gave them different names. The usual mode of baking was to use the hot ashes from the outdoor fireplace or fire. Hoecake was cooked on a hot griddle, and when baked on a smooth stone before a fire it was called johnnycake. Dutch ovens or square openings built into the chimney adjacent to the fireplace became common in pioneer days. A fire was built in the oven, the coals were raked out, and the corn bread was then baked in the oven.

Tortillas, a flat pancake-shaped product baked on an ungreased griddle, became common in the southern part of North America and parts of Central America. For tortillas, the corn was generally softened by treatment in hot lime solution, then boiled and ground into a meal. The meal was then made into tortillas.

Introduction of Wheat to the Americas

Wheat is thought to have been introduced into Mexico in 1530, when several grains of wheat were accidentally imported mixed with rice. The first wheat to be planted in what is now the continental United States occurred in 1585, when wheat was brought from Mexico and planted on one of the Elizabeth Islands off the coast of North Carolina. Later, immigrants to the new continent brought wheat seed with them, and thus many different types of wheat were introduced. Wheat production became common in the New England area and along the northern Atlantic coast, even though neither the climate nor the soil was suited to this crop. The first windmill for grinding grain was erected in Jamestown in 1621. New York City, called New Amsterdam at first, became the central point for the wheat and wheat processing market. Wheat was produced in the upper Hudson area, the Connecticut valley, and in Virginia. Gradually, production shifted westward as settlers moved to the interior of the country.

As early as 1640 commercial bakeries were established in Plymouth and New York, but because of the small and scattered population there was little need for bakeries. Even at the time of the American Revolution there were only about 2.5 million persons in all the colonies combined, with about 5% of the people living in cities. Baking did not become established during this period as an important industry but remained a practice restricted to small shops and homes. Both shops and homes baked the type of bread with which they had previous experience. There was really no typical type of American bread, since in colonial days immigrants from many countries were arriving in America, bringing with them their own ideas and experiences of bread production. As wheat flour became available, bread of many types and shapes was produced, but baking differed little from what it had been in Europe. Even bread laws regulating size, price, and quality were passed as early as 1640 by the Massachusetts Bay Authority after laws prevailing in England.

Colonial Baking

Colonial baking remained a continuation of conditions as they existed in Europe. Baking remained a century-old handicraft operation. The baker worked from sunrise to sunset, and the quality of his product was the test of his skill. The baker sold his wares in his shop, and in many instances baking was done only on order by a system known as custom baking.

From the beginning, baking has been an art in Pennsylvania. Pioneer women made their own yeast, used whatever kind of flour was available, and baked excellent bread in primitive ovens. Bread was the staff of life, and bread and soup were often the whole meal. The homemade yeast was kept in the "sotz crock" on the kitchen shelf and stirred into flour at night. Then the dough was set to rise. For this purpose the early settlers wove baskets with small holes in the center so that the basket could be emptied easily after it was turned over. Later the dough trough came into use which could be placed next to the oven for control of temperature. The dough was kneaded early in the morning, and the loaves were shaped and carried to the baking oven.

Pioneer bakehouses were of plastered mason with a tile-roofed shed across the front to protect the baker and the bread from bad weather. Shelves ranging down the side of the shed were for the loaves as they came from the oven. The oven, elevated for convenience, was floored and arched with brick and had an iron door. Early in the morning a cordwood fire was built inside and allowed to burn until it was reduced to ashes. The ashes were raked out with the "kitch," and the oven floor was swabbed out with the "huddel lumpa." The loaves were transferred from the paddle-shaped, long-handled "Schliesel" to the bottom of the oven for baking. Many of these old bakehouses still stand on farms. The breads varied and generally included sugar and butter, and some types included milk solids, mashed potatoes, and eggs. In addition to wheat breads, raised cornmeal bread, mush (corn) bread, and rye bread were baked.

In other parts of the country the housewife was essentially a craftsman, just as the bakeshop baker was. Procedures and equipment were the same for each. Ovens used for baking in the very early days were often made by simply lining pits dug into the ground with hot stones. By this means a crude type of oven was provided for cooking and making baked products, including corn bread. Later, when homes became substantially constructed, the great kitchen fireplace was constructed with a separate oven chamber with intake into the chimney and an ash pit at the bottom. Ovens in the bakeshop, as in the home, were fired with wood until the brick walls were thoroughly heated. Then the coals and ashes from the fire were removed from the oven, and the bread dough was placed in the oven to begin baking.

There were several types of bread produced, but the traditional one was the hearth loaf produced from white flour. However, brown bread was popular, and breads baked from a mixture of corn and wheat flours were common. Hard breads, known as ship's bread or hardtack, were commonly made in the colonial period. This type of bread was made from unleavened dough and was called ship's bread because it would keep for long periods without becoming stale. With the development of early sea commerce in the colonies and the long journeys that were being taken, this type of bread was in popular demand. In fact, soon after the Revolutionary War there began the considerable development of wheat production and milling capacity that permitted the establishment of trade with the West Indies and southern Europe. Among the exports were hard breads (ship's bread) made only from flour, salt, and water, and this development became so extensive that for a time the Colonies were referred to as "bread colonies." The basis for the modern biscuit-and-cracker industry that has flourished so successfully was laid by the reputation created by the hard bread manufactured in the 1700s.

Baking in the 19th Century

Bread was an important consideration during the Revolutionary War. Bread was issued to soldiers on the basis of 1 lb per day. Sometimes flour was issued in place of bread when the army was on the march. Often the flour was pooled by the soldiers and traded at homes for bread. On the retreat of the soldiers from New York and afterward, the bread available was a hardtack or ship biscuit. Soft white bread made from refined flour was a luxury. There were, however, portable baking units constructed during the war that could be put on wagons and moved to supply bread for the army when on the march. In 1778, Congress ordered the formation of a company of bakers because of the dissatisfaction in the army with the poor quality of bread. Changes occurred rapidly in the 1800s, when what had been villages began to become towns and new villages were springing up as the frontiers were extended. Philadelphia had 91 bakeries in 1857, nine of which were cracker plants; Cincinnati, which was served by only one bakery in 1780, had 40 bakeries in 1820. According to the Census of Manufacturers published in 1850, there were at that time 2,027 bakeries in the United States producing over $13 million worth of baked goods. The majority of bakeries were one-man shops and probably accounted for less than 10% of the total bread consumed, because most bread was made at home. Bakeries were located in cities, but the major part of the population was rural.

Perhaps the most challenging and difficult problem confronting the baker during this period was the preparation of the yeast mixture or brew. Compressed yeast was as yet unknown. Therefore, the baker cultivated the growth of yeast cells in the fermented brew made from flour, sugar, potatoes, water, and old stock yeast or portions of dough. Brewers' yeast, when available, was the liquid, foamy stock from the top of ale fermentation vats. There were many types and kinds of brews and many so-called secret recipes, and, of course many of these gave bread a distinctive aroma and taste. A variety of bread products could be made from the same sponge, and this was common practice.

Before 1850, the practice of distributing baked products was unknown. Bakers generally sold only at their shops or delivered to a consumer's home. There was no advertising or use of any means to build up sales. Automation in baking operations started in the late 1800s with the introduction of a bread machine consisting of a set of rollers with automatic dusters that could discharge 80–90 loaves a minute. The testing of flour and other ingredients was begun during this period, and the first electric oven was put in operation. In 1913 the first traveling oven was installed in Montreal.

Automation in the Baking Industry

At the turn of the century the first dough divider was installed in a Detroit bakery. It had a capacity of 40 loaves per minute. Bread was sometimes wrapped in wax paper, bread coolers came into use, as did the overhead proofer. The baking industry was becoming science-oriented, and the Ward-Corby Company of Pittsburgh in 1903 employed the first chemist in the baking industry as Director of the Division of Chemistry and Technology. This company made an offer to the National Association of Master Bakers to help establish a school for training bakers, but no action was taken.

The passage of the National Pure Food Law made bakers more conscious of quality standards and better control of shop conditions. Most states passed bread weight laws. In 1900, about 25% of the total bread consumed was made by commercial bakers; therefore, regulatory laws were becoming more essential.

In the early 1900s high-speed mixing, automatic bread-making machines, bread slicers, and twist bread were new inventions, as were vitamin additions and refrigerated doughs. The promotion of health diet breads by the industry caused the Food and Drug Administration to investigate whether or not false claims were being made. On May 27, 1940, fortified (enriched) bread became available throughout the United States. Since the public generally refused to eat 100% whole wheat bread, the enrichment program recommended by the National Research Council was that bread made from refined flour be enriched with thiamine, niacin, riboflavin, and iron to an extent that would replace the vitamin losses caused by processing wheat. Optional enrichment ingredients that could be added were vitamin D and calcium. Thus, bread in the United States has been a leading food in the experimental development of improved human nutrition.

Recent Developments

Bread, crackers, cakes, and sweet goods are produced in a vast assortment of tastes, sizes, and shapes and are available in grocery stores, supermarkets, and retail bakeshops. These developments represent gradual change.

The bread most commonly eaten in the United States is a white, enriched, sliced, and wrapped loaf. But a variety of other breads are also made. They include whole wheat, cracked wheat, rye, pumpernickel, French, Italian, and Scandinavian crisp rye. Corn bread, or corn pone, is popular in the south of the United States. Partly baked rolls and bread and refrigerated doughs are widely sold. These are first heated and browned or baked in the oven before serving.

Sourdough breads were popular in the far west of the United States with old-time mountain men, shepherds, prospectors, and miners. They have lost their popularity. Sourdough French bread, however, is one of the more popular types that have gained significance in recent years. The main changes that have occurred in recent years have been in the processing of baked goods and in the modified or new ingredients now available. Flour quality standards are closely maintained, shortening types have undergone profound changes and have been replaced to a large extent by blends of oils and surface active agents-emulsifiers, yeast has become a carefully controlled product in various shape and forms (including fast-acting, dry, etc), and additives such as proteases and amylases, mold-growth inhibitors, and oxidizing and reducing agents are now common adjuncts in the baking industry.

In colonial times about 95% of baking was done in the home. Today 95% of all baked goods consumed is produced by the baking industry. In the history of baking in the United States there has been a complete reversal of the importance of commercial baking. It is now one of the major industries in value of products produced.

U.S. BREAD PRODUCTION—RECENT DEVELOPMENTS

Bread in the United States is produced in about 1,900 large, wholesale, automated bread plants and in about 31,500 small, retail bakeries and in-store supermarket bakeries. About 90% of the white bread is produced by the large wholesale plants which operate a distribution system over a radius of 400 km from the bakeries. The bread is sold in supermarkets and bakery thrift shops. The American wholesale baker generally produces a soft loaf with a good flavor and uses either a sponge dough or brew system of bread production. The rapid and no-time dough processes have met with little acceptance.

Wholesale bread plants produce from 1,500 to 3,000 kg of white pan bread per hour and are equipped with dough mixers that handle up to 750 kg of dough per batch. The retail bakeries include about 18,000 small bakeries, employing 2–10 persons. A typical batch is about 40 kg of dough. In addition, there are about 14,000 in-store bakeries located in a section of a large supermarket. The latter have increased from 1,000 in 1970 at a rate of about 1,000 per year. The wholesale plants account for 75% of the total bakery sales, the retail bakeries for about 10%, and the in-store bakeries for about 15%. White bread comprises over 50% of the total bread purchased; the only other bread types purchased in quantity are wheat (about 24%) and rye (about 10%). Small amounts of other bread types are French, whole grain, raisin, bran, pumpernickel, stone ground, fiber, and a whole plethora of miscellaneous variety breads. The popularity of variety breads in the United States is on the increase. Still, white bread continues to be the staple bakery food in the American diet.

Many wholesale bakeries use liquid-brew systems in the conventional batch system to produce pan white and variety breads and rolls. Brews containing 40–50% flour are used with minor modifications in tanks, piping, meter, or nozzle sizes. At the 60–70% flour level, the brews are so viscous that special heavy-duty pumps are required along with heavy duty-squirrel cage mixers and enlarged tanks, piping, nozzles, etc. Most water brews are in a concentrated form and contain all the yeast but only about 2% sugar and only 10–20% water and are buffered with calcium carbonate. Massive horizontal mixers for dough capacities of about 1 ton are used. Mixing schedules are often 2–3 minutes at low speed to incorporate the ingredients followed by 9–12 minutes at high speed to develop the dough. With straight dough procedures that do not involve prior fermentation, 16–18 minutes at high speed may be required for optimum dough development. Reducing agents (such as L-cysteine hydrochloride) or high levels of proteolytic enzymes are used for flours that require a long mixing time.

In continuous-mixing schedules, closed developer heads holding about

200-kg batches of dough for an average of 2½ minutes retention time are used. Massive, 50–100-HP motors are used; extreme shear takes place, along with temperature rises of 5–8°C. Some Tweedy-type mixers (developed for the Chorleywood process in England) are used to a limited extent in the United States. In addition, some ultra-high-speed mixers are used on a small scale. The small batches are mixed for 2–3 minutes and the doughs are uniformly developed.

Variety Breads

Standards of identity of the Food and Drug Adminstration describe and define foods used in baked goods. They include standards for enriched wheat bread, milk products, egg products, milled wheat products, fats and oils, and several additives.

The standards of identity established conditions for standardized baked goods. Standardized baked goods must contain certain specified ingredients. Other defined ingredients are optional. There are weight specifications for standardized bakery foods. Standardized variety breads require only flour, a moisturing agent, and a leavener. Bread products must weigh half a pound or more per unit, and rolls less than half a pound per unit.

Standards of identity have been established for the following: bread, rolls, and buns; whole wheat bread, rolls, and buns; raisin bread, rolls, and buns; milk bread, rolls, and buns; and enriched bread, rolls, and buns. Additional standardized products are French, Italian, and other hearth breads. English muffins, wheat bread that contains some white flour, multigrain breads, rye breads, and fiber breads are not included in the standardized classification. Whole wheat bread must contain 100% whole wheat flour and no white flour; the weight of raisins in raisin bread must equal at least 50% of the weight of the flour. Ingredients must be listed in decreasing order of weight in the total formula.

The White Pan Bread Industry

A survey of the white pan bread industry showed that practically no nonfat dry milk is used in white pan bread production. Many bakeries have eliminated all dairy ingredients, and most use milk replacers composed of soy and sweet dairy whey blends. The amount of fat has been reduced from the 3% to an average 2.3% within the past 10–15 years. Solid shortening and lard have been replaced in about 75% of the bakeries by fluid oils in combination with dough conditioners. During the past 15 years, high-fructose corn syrup has replaced sucrose or corn syrup in U.S. white bread formulations. The salt level has been reduced from 2.1% (flour basis) to 1.7%. Lowering the level to 1.5% affected taste and flavor perception according to most consumers.

Ingredients

Bread is produced from hard red winter, hard red spring, or hard white flours or blends. White pan bread is made from flour containing at least 10.75% protein; flour for soft rolls (ie, hamburger buns) contains 11–12% protein and for hard rolls and hearth breads 12–14% protein. Most flours are treated with azodicarbonamide for maturing and benzoyl peroxide for whitening.

The most commonly used yeast is of the compressed cake type; active dry yeast is used little. Instant active dry yeast is used in some small wholesale and retail shops. The most popular oxidant continues to be potassium bromate. Ascorbic acid is used little, mainly in brew systems. To shorten the mixing time, L-cysteine is used, generally in combination with potassium bromate, to counteract the action of cysteine in the late fermentation and early baking stages.

Barley malt is added at the mill as a source of alpha-amylase of intermediate thermostability. Fungal proteases are used in the sponge or in the flour brew systems to reduce mixing requirements and mellow the gluten. Combinations of dough softeners (mainly mono- and diglycerides) and strengtheners (lactylates, polysorbate 60, ethoxylated monoglycerides, succinylated monoglycerides, and diacetyl tartaric acid esters) are used extensively as powder concentrates, fat-based products, and hydrates in the production of white bread. Over the years most of the water conditioning agents have been removed from yeast foods. Most of today's formulations contain ammonium salt(s), potassium bromate, and (some) monocalcium phosphate as an acidifying agent. Two types of special products are available for brew systems. One contains a buffering agent, calcium carbonate, and ammonium salts and is added at the brew stage. The second contains potassium bromate and monocalcium phosphate and is added at the dough-mixing stage. In a commercially available new product, urea is used in concentrated tablet form as nitrogen source for yeast.

The most commonly used retarder of mold growth is calcium or sodium propionate. Sorbic acid is used to a limited extent as preservative. Because of its large effect on yeast, it is encapsulated by fat that melts at high temperatures in the oven.

New Technologies

During the past two decades, microwave or radiofrequency power has been investigated as the heat source in baking bread. When microwave power is used to transmit heat to bread dough, the temperature of the entire dough rises, which may considerably shorten the required baking time. However, without the higher outer temperatures there is no crust formation. The quality of the bread is affected as well. The most promising approach to baking with microwave is a combination of microwave and conventional heat sources. One of the most likely processes uses a conventional heating system during the first stage of baking followed by an intermediate stage of microwave heating to speed up the internal temperature rise and then finishes with a final stage of conventional heating to obtain the desirable browning effect. Another development being evaluated is the ionization process, where a conventional oven is retrofitted with a grid network that can be charged with high levels of voltage to ionize the atmosphere inside the oven. In this device the boundary layer around the dough piece is reduced to accelerate the heat transfer mechanism. The claim of proponents of this equipment is that a better product can be produced with less energy (see Lange, 1984).

Frozen and Refrigerated Baked Goods

Frozen dough has been gaining increasing acceptance in the United States. It was reported that conventional formulations for regular and continuous bread production required only minor changes in the production of frozen doughs. Winter and spring wheat flours of good quality may be used to produce high-quality frozen doughs. The level of yeast food in the formulation requires no change, but an increase in the level of potassium bromate (above the normal) in the yeast may improve volume, grain, and texture of breads made from frozen doughs. The level of yeast should be increased to 4–5%. When yeast levels of 3% (conventional) were used, proof times were higher (about 20% above the 4% yeast level), the resulting loaf volume was lowered, the crumb grain was more open, and the texture harsh. At 6–8% yeast (flour basis), the aroma and flavor of the bread were impaired.

BIBLIOGRAPHY

General

Adrian, W. "And Thus Bread Was Made from Stock and Blood." Ceres-Verlag; Bielefeld, Germany; 1959.

Akademiya, N. Biochemistry of Grain and Breadmaking. U.S. Dept. Agric.: Washington, D.C.; 1965.

Anon. "Bread Cook Book." Better Homes and Gardens, Meredith Press; Des Moines, IA; 1965.

Anon. "The Sunset Cook Book of Breads." Lane Magazine and Book Co.; Menlo Park, CA; 1966.

Anon. "Betty Crocker's Breads." Golden Press; New York; 1974.

Anon. "Brot und Feine Backwaren." DLG Verlag; Frankfurt, Germany; 1977.

Anon. "Ireks—ABC der Baekerei." Ireks-Arkady Inst.; Kulmbach, West Germany; 1978.

Barrows, A.B. "Bakery Specialties." Elsevier Applied Science Publ.; London; 1984.

Bennion, E.B. "Breadmaking: Its Principles and Practice," 4th ed. Oxford Univ. Press; London; 1967.

Bluemel, F.; Boog, W. "5000 Jahre Backofen." Deutsches Brotmuseum, e. V. Ulm; Donau, West Germany; 1977.

Campbell, A. "The Swedish Bread." A.B. Bennel & Co.; Boktryckeri, Stockholm; 1950.

Cassella, D. "A World of Breads." David White Co.; New York; 1966.

Claiborne, C. "The New York Times Cook Book." Harper & Row Publishers; New York; 1961.

Clayton, B. "The Complete Book of Breads." Simon and Schuster, New York; 1973.

Dunlap, F.L. "White Versus Brown Flour." Wallace and Tiernan Co.; Newark, NJ; 1945.

Hanson, H.; Borlaug, N.E.; Anderson, R.G. "Wheat in the Third World." Westview Press; Boulder, CO; 1982.

Horder, T.J.; Dodds, C.; Moran, T. "Bread: The Chemistry and Nutrition of Flour and Bread." Constable; London; 1954.

Inglett, G.; Munck, L. (Eds.). "Cereals for Food and Beverages. Recent Progress in Cereal Chemistry and Technology." Academic Press; New York; 1980.

Jacob, H.E. "Six Thousand Years of Bread, Its Holy and Unholy History." Doubleday Co.; New York; 1944.

Kent, N.L. "Technology of Cereals with Special Reference to Wheat." Pergamon Press; Oxford, England; 1966.

Kent-Jones, D.W.; Amos, A.J. "Modern Cereal Chemistry," 6th ed. Food Trade Press; London; 1967.

Kosmina, N. "Biochemie der Brotherstellung." VEB Fachbuchverlag; Leipzig, Germany; 1977.

Martens, V. (Ed.). "Grains and Oilseeds: Handling, Marketing, Processing." Can. Int. Grains Inst.; Winnipeg, Canada; 1975.

Matz, S.A. "The Chemistry and Technology of Cereals as Food and Feed." Avi Publishing Co.; Westport, CT; 1959.

Matz, S.A. "Bakery Technology and Engineering." Avi Publishing Co.; Westport, CT; 1960.

McCance, R.A.; Widdowson, E.M. "Breads, White and Brown; Their Place in Thought and Social History." J.B. Lippincott Co.; Philadelphia; 1956.

Neumann, M.P.; Pelshenke, P.F. "Bread Grain and Bread." Verlag, Paul Parey; Berlin; 1954.

Pomeranz, Y. *Adv. Cereal Sci. Technol.* **1978**, *2*, 387–413.

Pomeranz, Y. "Functional Properties of Food Components." Academic Press; Orlando, FL; 1984.

Pomeranz, Y.; Shellenberger, J.A. "Bread Science and Technology." Avi Publ. Co.: Westport, CT; 1971.

Pyler, E.J. "Baking Science and Technology." Siebel Publishing Co.; Chicago; 1952.

Pyler, E.J. "Our Daily Bread." Siebel Publishing Co.; Chicago; 1958.

Schaeffer, W. "Brot Backen." Otto Maier Verlag; Ravensburg, West Germany; 1980.

Schneeweiss, R.; Klosse, O. (Eds.). "Technologie der Industriellen Backwaren Produktion." VEB Fachbuchverlag; Leipzig, East Germany; 1981.

Schuster, G.; Adams, W.F. *Adv. Cereal Sci. Technol.* **1984**, *6*, 139–287.

Seibel, W.; Bruemmer, J.M.; Stephan, H. *Adv. Cereal Sci. Technol.* **1978**, *2*, 415-456.

Seibel, W.; Steller, W.; Wuerziger, J. "BBV Backwaren," Handbuch 4. Behrs Verlag; Hamburg, West Germany; 1982.

Sherman, H.E.; Pearson,C.S. "Modern Bread from the Viewpoint of Nutrition." Macmillan Co.; New York; 1942.

Spicer, A. (Ed.). "Bread." Applied Science Publ.: London; 1975.

Spicher, G.; Stephan, H. "Handbuch Sauerteig; Biologie, Biochemie, Technologie." Behrs Verlag; Hamburg, West Germany; 1982.

Sultan, W.J. "Practical Baking, 2d ed. Avi Publishing Co., Westport, CT; 1969.

Sultan, W.J. "Elementary Baking." McGraw-Hill Book Co.; New York; 1969.

Wahren, M. "Bread Through the Ages." Publishers to the Swiss Association of Master Bakers and Confectioners, Berne (no date).

Williams, A. (Ed.). "Breadmaking—The Modern Revolution." Hutchinson, Benham; London; 1975.

Africa, Near East and Middle East

Adler, L. *Cereal Sci.Today* **1958**, *3*, 28, 30, 32.

Ballschmieter, H.N.B. *Brot Gebaeck* **1968**, *22*, 66–71.

Ballschmieter, H.N.B.; Vliestra, H. *Brot Gebaeck* **1961**, *15*, 153–158.

Ballschmieter, H.N.B.; Vliestra, H. *Brot Gebaeck* **1969**, *23*, 28–32.

Bauer, G. *IGV Mitt.* **1966**, *2*, 56.
Dalby, G. *Bakers Digest* **1966**, *40*, 64–66.
Englebert, V. *Natl. Geographic Mag.* **1968**, *133*, 850–875.
Faridi, H.A.; Finney, P. L. *Bakers Digest* **1980**, *54*(5), 14, 19, 20, 22.
Forsyth, G.H. *Natl. Geographic Mag.* **1964**, *125*, 82–108.
Freund, O. *Cereal Sci. Today* **1961**, *6*, 320–324, 329.
Heinrich, G. *IGV Mitt.* **1965**, *1*, 154–155.
Hoover, W.J. *Cereal Sci. Today* **1974**, *19*, 153–156.
Meckel, R.B.; Holmes, W.H. *Cereal Sci. Today* **1965**, *10*, 220–223, 233.
Meinardus, F.A. *Brot Gebaeck* **1964**, *18*, 210–211.
Pelshenke, P.F. *Cereal Sci. Today* **1961**, *6*, 325–327, 329.
Pelshenke, P.F. *Cereal Sci.Today* **1966**, *11*, 291–299.
Pelshenke, P.F. *Brot Gebaeck* **1967**, *21*, 185–189.
Pelshenke, P.F. *Brot Gebaeck* **1967**, *21*, 205–208.
Refai, F.Y. *IGV Mitt.* **1966**, *2*, 150–151, 169–172.
Tekeli, S.T. Proc. 4th Int. Bread Congress, Vienna, 1966.
Tekeli, S.T. Private communication. Ankara Univ., Turkey, 1967.
Tekeli, S.T.; Seckin, R. *Brot Gebaeck* **1964**, *18*, 38.
Wahren, M. "Bread Through the Ages." Publishers to Swiss Association of Master Bakers and Confectioners, Verlag des Schweizerischen Baker-und-Konditorenmeister Verbandes; Bern, Switzerland (no date).
Wahren, M. "Our Daily Bread in History and Folk Use." Verlag des Schweizerischen Baker-und-Konditorenmeister Verbandes; Bern, Switzerland (no date).
Wahren, M. *Brot Gebaeck* **1959**, *13*, 21–29.
Wahren, M. *Brot Gebaeck* **1961**, *15*, 1–12, 28–32.
Wahren, M. *Brot Gebaeck* **1962**, *16*, 12–20, 30–38.

Far East

Abercombie, T.J. *Natl. Geographic Mag.* **1968**, *134*, 297–344.
Adams, R.J. *Br. Baker* **1964**, *155*(11), 22–26.
Akutsu, S. *Brot Gebaeck* **1961**, *15*, 101–103.
Akutsu, S. *Arkady Rev.* **1968**, *45*, 1–5.
Akutsu, S.I. *Getreide Mehl Brot* **1977**, *31*, 113–118.
Arnott, M.L. *Brot Gebaeck* **1971**, *25*, 227–230.
Dubois, D.E. *Bakers Digest* **1982**, *56*(4), 14–20.
Keating, B. *Natl. Geographic Mag.* **1967**, *131*, 1–47.
Moore, W.R. *Natl. Geographic Mag.* **1963**, *123*, 153–199.
Nagao, S. *Cereal Foods World* **1979**, *24*, 593–595.
Newman, J.M. *Cereal Foods World* **1981**, *26*, 395–398.
Nicholson, J. *Cereal Sci. Today* **1970**, *15*, 279–281, 326.
Schneeweiss, R. *IVG Mitt.* **1966**, *2*, 68–72.
Uchida, M. *Cereal Foods World* **1979**, *24*, 596–597.
Zimmermann, R. *IGV Mitt.* **1966**, *2*, 74–78.

Australia

Anon. *Aust. Bakers Millers J.* **1968**, *71*(12), 12.
Bond, E. Aust. Bakers Millers J. **1953**, *Nov.*, 39–40, 43, 45–46, 49, 51.
Bond, E. Lecture R 8.06, Proc. 4th Int. Bread Congress, Vienna, 1966.
Bond, E. *Arkady Rev.* **1968**, *45*(2), 26–33.
Marston, P.E. *Aust. Chem. Process Eng.* **1967**, *Febr.*, 1–4.

New Zealand

Anon. "Breadmaking in New Zealand." New Zealand Wheat Res. Inst.; Christchurch, New Zealand; 1949.
Anon. "Report of the Committee on the Wheat, Flour, and Bread Industries." DSIRO, Christchurch, New Zealand; 1963.

Anon. "Associated Bread Manufacturers of Australia and New Zealand—The First Sixty Years, 1904–1964." Sydney and Melbourne Publishing Co.; Sidney, Australia; 1967.
Bycroft Limited Flourmillers. "Bread Making in New Zealand." Whitcombe and Tombs; Auckland, New Zealand; 1934.
Cawley, R.W. "Changes in Technology." New Zealand Wheat Res. Inst.; Christchurch, New Zealand; 1968.
Meredith, P.W. Private communication. New Zealand Wheat Research Institute, Christchurch, New Zealand, 1968.
Mitchell, T.A. *Baker Miller J.* **1983**, *86*(7), 23.
Webby, B.G. *Baker Miller J.* **1983**, *86*(6), 17.

Eastern Europe

Agababjan, R.J.A. *Baking Pastry Ind. (Moscow)* **1968**, *12*(7), 7–9.
Anon. *Osterreichische Bakerzeitung* **1956**, *6*, 51–52.
Anon. *Backer Konditor* **1961**, *1*, 15–17.
Csaba, J. *Backer Konditor* **1960**, *1*, 13–15, *2*, 44–45; *3*, 76–77; *4*, 109–108.
Czerny, Z. "Polish Cookbook." Polskie Wydawnictwo Gospodarcze; Warsaw; 1961.
Gatilin, N.F. *Osterreichische Bakerzeitung* **1966**, *61*(42), 3; *61*(43), 3.
Glasow, W. *Backer Konditor* **1959**, *12*, 23.
Goroschenko, M.K. *Periscop* **1968**, *3*, 1–20.
Grisin, A.S. *Baking Pastry Ind. (Moscow)* **1968**, *12*(6), 43–46.
Grisin, A.S.; Uchanova, V.A. *Baking Pastry Ind. (Moscow)* **1969**, *13*(2), 2–4.
Gutkowski, H. *Przeglad Pierkarski Cukierniczy* **1965**, *2*, 32–33.
Kosutany,T. "The Hungarian Wheat and Hungarian Flour from the Standpoint of the Farmer, Miller, and Baker." Moluarok Kapja Publishing Co.; Budapest; 1907.
Kowalczuk, M. *Przeglad Pierkarniczy Cukierniczy* **1968**, *16*(3), 55–58.
Kuppers-Sonnenberg, G.A. *Brot Gebaeck* **1961**, *15*, 243–248.
Lubczynska, H. *Przeglad Piekarniczy Cukierniczy* **1969**, *17*(3), 56–58.
Podsiadly, H. *Przeglad Piekarski Cukierniczy* **1967**, *15*, 212–214.
Popova, V. *Arb. Wiss. Forsch. Getreideverarabeitung* **1963**, *1*, 205–210.
Radeff, V. *Brot Gebaeck* **1967**, *21*, 155–159.
Rojter, I.M. "The Modern Technology of Dough Manufacture in Bakeries." Technika Publishing Co.; Moscow; 1968.
Rojter, I.M; Markianova, L.M. *Baking Pastry Ind. (Moscow)* **1969**, *13*(3), 12–15.
Ronnenbeck, G.; Ruffert, H. *Backer Konditor (Leipzig)* **1968**, *16*(22), 38–39.
Tomaszewski, A. *Przeglad Piekarniczy Cukierniczy* **1967**, *15*, 195–196.
Zittlau, G. *Brot Gebaeck* **1958**, *12*, 251–255.

Central Europe

Acker, L. *Brot Gebaeck* **1960**, *14*, 182–183.
Anon. *Muhle* **1935**, *72*(2), 56.
Anon. *Osterreichische Bakerzeitung* **1956**, *39*, 3.
Anon. "Handbook for Bakers." 2d ed. Osterreichischer Gewerbeverlag; Vienna; 1958.
Anon. "Bread Production in Holland." L'Gravenhage, Martinus Nijhoff; The Hague, Netherlands; 1963.
Arndt, J. *Brot Gebaeck* **1967**, *21*, 201–204.
Bailey, C.H. *Cereal Sci. Today* **1959**, *4*, 11–12.
Beccard, E. *Getreide Mehl Brot* **1948**, *2*, 157–159.
Berk, E. *Deutsche Lebensm. Rundschau* **1954**, *50*, 115–116.
Bohnenblust, J.P. *Getreide Mehl Brot* **1980**, *34*, 104–106.
Bure, J. *Brot Gebaeck* **1961**, *15*, 169–179.
Doose, O. "Pumpernickel and Its Modern Production." Gilde Publishing Co.; Alfeld, Germany; 1947.
Doose, O. *Getreide Mehl Brot* **1949**, *3*, 204–207.
Fink, H. "Bread of Our Neighbors." Granum Verlag; Detmold, Germany; **1964**.
Fisher, M.F.K. "The Cooking of Provincial France." Time-Life Books; New York; 1968.
Gerngross, J. *Backer Konditor* **1969**, *17*(23), 196–197.

Huber, H. *Getreide Mehl Brot* **1984**, *38*, 45–49.
Kirchner, O. *Brot Gebaeck* **1961**, *5*, 113–115.
Korber, J.; Tscheinig, M. "The Vienna and Austrian Bakery." 4th Ed. Erwerbs und Wirtschaftsgenossenschaft der Baker Osterreichs; Vienna; 1933.
Kunkel, O. *Brot Gebaeck* **1966**, *20*, 245–252.
Kunkel, O. *Brot Gebaeck* **1967**, *21*, 242–244.
Kunkel, O. *Brot Gebaeck* **1968**, *22*, 37–40.
Landgraf, H.; Weckel, A. "The World of Bread." Fachverlag fur das Osterreichische Backhandwerk; Vienna, 1957.
Lauter, B. *Getreide Mehl Brot* **1980**, *34*, 34–36.
Luddeke, J. *Backer Konditor* **1967**, *15*, 86–87.
Luraschi, A. "The Bread and Its Story." L'Arte Bianca; Torino, Italy; 1953.
Meise, H. *Brot Gebaeck* **1963**, *17*, 151–152.
Menger, A. *Brot Gebaeck* **1961**, *15*, 189–192.
Nergent, H. *Brot Gebaeck* **1961**, *15*, 192–195.
Neumann, M.P. *Z.f.d. ges. Getreidewesen* **1910**, *2*, 75–81, 99–107.
Neumann, M.P. *Mehl Brot* **1937**, *37*(11), 1–4.
Neumann, R. *Urania* **1968**, *7*, 59–63, 82–83.
Orlowski, K. *Brot Gebaeck* **1954**, *8*, 179–180.
Pelshenke, P.F. "Baked Products from German Countries." Gilde Publishing Co.; Alfeld, Germany; 1949.
Pelshenke, P.F. *Steinmetz Nachrichten* **1952**, *15*, 3–4.
Pelshenke, P.F. *Milling* **1963**, *Aug. 9*, 133, 155.
Pelshenke, P.F. *Muhle* **1964**, *101*, 895–896.
Pelshenke, P.F. *Milling* **1968**, *150*(5), 30–31.
Pomeranz, Y. *Cereal Foods World* **1983**, *28*, 387–390.
Rehfeld, G. *Ernahrungsforschung* **1969**, *14*(1), 19.
Richard-Molard, D.; Nago, M.C.; Drapron, R. *Bakers Digest* **1979**, *53*(3), 34–38.
Rotsch, A.; Tessmer, E. "Problems of Panettone Production." Granum Verlag; Detmold, Germany; 1964.
Schroder, F. *Z. gesamt. Getreide Muhlen Backereiwesen* **1935**, *22*,1–13.
Schulz, A. *Getreide Mehl Brot* **1947**, *1*, 69–72.
Schulz, A. *Backer Z.f. Nord-, West-, Mitteldeutschland* **1950**, *4*(24), 537.
Schulz, A. *Brot Gebaeck* **1966**, *20*, 114–117.
Schulz, A.; Stephan, H. *Backer Z.f. Nord-, West-, Mitteldeutschland* **1954**, *8*, 7–8.
Schulz, A.; Stephan, H. *Brot Gebaeck* **1966**, *20*, 61–64.
Schulz, A.; Stephan H. *Brot Gebaeck* **1967**, *21*, 41–45.
Soenen, M. *Brot Gebaeck* **1962**, *16*, 45–52.
Soenen, M.; Pelshenke, P.F. *Brot Gebaeck* **1962**, *17*, 64–66.
Spil, A. *Brot Gebaeck* **1962**, *16*, 176–177.
Steiger, R. Lecture R 8.02, Proc. 4th Int. Bread Congress, Vienna, 1966.
Stroganov, G.V.; Makljukov, V.I. *Baking Pastry Ind. (Moscow)* **1969**, *13*(3), 42–44.

Iberian Peninsula

Anon. *Osterreichische Bakerzeitung* **1956**, *39*, 3.
Beck, H. *Brot Gebaeck* **1964**, *18*, 39–40.
Rinagel, F. *Osterreichische Bakerzeitung* **1956**, *51/52.* 7–8.
Verges Torras, A. *Muehle Mischfuttertechnik* **1983**, *120*(27/28), 380–383.

United Kingdom

Anon. *Milling* **1968**, *150*(12), 23.
Anon. Biscuit Maker Plant Baker **1968**, *19*(6), 436.
Chamberlain, N.; Collins, T.H. *Bakers Digest* **1979**, *53*(1), 18, 19, 22–24.
MacNiven, C.S. "The Highlander's Cookbook. Recipes from Scotland." Ward Ritchie Press; London; 1966.
Muller, H.G. *Getreide Mehl Brot* **1974**, *28*, 85–87.
Sheppard, R.; Newton, E. "The Story of Bread." Routledge & Kegan Paul; London; 1957.

Simmonds, O. "The Book of Bread." MacLaren and Sons; Edinburgh (date unknown).
Spencer, B. *Getreide Mehl Brot* **1981**, *35*, 216–220.
Spencer, B. Proc. 7th World Cereal and Bread Congress, Prague, 1982, pp. 701–706.

Scandinavia

Albertsson, C.E. *Getreide Mehl Brot* **1981**, *35*, 220–221.
Anon. *Backer Konditor* **1962**, *3*, 77–79.
Campbell, A. "The Swedish Bread." A.B. Bennel & Co., Boktryckeri; Stockholm; 1950.
Doerr, R. *Getreide Mehl Brot* **1982**, *36*, 154–158.
Fagerlind, B. *Brot Gebaeck* **1958**, *12*, 97–100.
Freund, O. *Cereal Sci. Today* **1966**, *6*, 320–322, 324, 329.
Henryk, G. *Mlynsko-pekarensky Prum. (Praha)* **1967**, *13*(6), 255–256.
Karp, D. *Brot Gebaeck* **1961**, *15*, 97–100.
Karp, D.; Garping, B. *Brot Gebaeck* **1966**, *20*, 169–176.
Kuukankorpi, P. *Brot Gebaeck* **1966**, *20*, 199–202.
Schulerud, A. "Rye Flour." Moritz Schafer Verlag; Detmold, Germany; 1957.
White, J.E. "Good Food from Denmark and Norway." Fred Muller; London; 1959.

South and Central America

Anon. *Natl. Geographic Mag.* **1908**, *19*, 164–177.
Anon. *Osterreichische Backer Zeitung* **1958**, *53*(18), 3.
Boas, O.; Boas, C.V. *Natl. Geographic Mag.* **1968**, *134*, 425–444.
Day, F. *Cereal Sci. Today* **1974**, *19*, 157–160.
De Andrade, M. "Brazilian Cookery, Traditional and Modern." Charles E. Tuttle Co.; Rutland, VT; 1965.
Dreifuss, M. *Brot Gebaeck* **1963**, *17*, 73–76.
Freund, O. *Cereal Sci. Today* **1961**, *6*, 320–324, 329.
Shellenberger, J.A. *Northwestern Miller* **1944**, *218*(8), 11–16, May 24.
Shellenberger, J.A. "The Milling and Baking Industries of Peru." Institute Inter-American Affairs, Food Supply Div.; Washington, D.C.; 1947.
Staff, F. *Food Eng.* **1969**, *41*(2), 66–67.

North America

Ashley, W. "The Bread of Our Forefathers." Clarendon Press; Oxford, England; 1928.
Belknap, W., Jr. *Natl. Geographic Mag.* **1964**, *125*, 196–212.
Davis, R.E.; Trempler, M. Baking Ind. *129*(1637): 39–58.
Earle, A.M. "Home Life in Colonial Days." Macmillan Co.; New York; 1899.
Hutchinson, R. "The New Pennsylvania Dutch Cook Book." Harper & Brothers; New York; **1958**.
Jacob, H.E. "Six Thousand Years of Bread, its Holy and Unholy History." Doubleday & Co.; New York; 1944.
Moore, M.K. "The Baking Industry." Bellman Publishing Co.; Cambridge, MA; 1946.
Panschar, W.G. "Baking in America." Vols. I and II. Northwestern Univ. Press; Evanston, IL; 1956.
Storck, J.; Teague, W.D. "A History of Milling. Flour for Man's Bread." Univ. Minnesota Press, Minneapolis; 1952.
Wilder, R.M.; Williams, R.R. Bull. 110, Natl. Res. Council, Natl. Acad. Sci., Washington, D.C.; 1944.

U.S. Bread Production—Recent Developments

Alwes, M; Jolly, M.J. *Bakers Digest* **1974**, *48*(4), 28–31, 58.
Anon. *Food Eng.* **1985**, *57*(3), 45.
Anon. *Milling Baking News* **1985**, *63*(47), 35, 37, 39, 45, 46, 51.
Becker, C.A. *Bakers Digest* **1982**, *56*(1), 12, 13, 15.
Chung, O.K.; Pomeranz, Y. In "Advances in Modern Human Nutrition, Vol. III. Roles of

Lipids in Health and Diseases." Kabara, J.J., Ed. Am. Orb. Chem. Lec.; Champaign, IL; 1985, Chapter 28, pp. 315–342.

Cole, M.S. *Bakers Digest* **1973**, *47*(6), 21–23, 64.

Dubois, D.K. *Cereal Foods World* **1981**, *26*, 617–619, 621, 622.

Dubois, D.K. In "Variety Breads in the United States." Miller, B.S., Ed. Am. Assoc. Cereal Chem.; St. Paul, MN; 1981.

Dubois, D.K. *Proc. Int. Symp. Adv. Baking Sci. Technol.*, Dept. Grain Science, Kansas State Univ.; Manhattan, KS; 1984, pp. B1–B14.

French, F.D.; Kemp, D.R. *Cereal Foods World* **1985**, *30*, 344–46.

Jackel, S.S. *Bakers Digest* **1977**, *51*(2), 39–43.

Jackel, S.S. *Cereal Foods World* **1985**, *30*, 524, 526.

Kamman, P.W. *Bakers Digest* **1979**, *53*(1), 26–29.

Kulp, K. *Bakers Digest* **1983**, *57*(6), 20, 22, 23.

Lange, R.L. In Proc. Int. Symp. Adv. Baking Sci. Technol., Dept. Grain Science, Kansas State Univ., Manhattan, KS; 1984, pp. R1–R9.

Matz, S.A. *Sci. Am.* **1984**, *245*, 123–126, 131–134.

Martinez-Amaya, M.A.; Kulp, K. In Proc. Int. Symp. Adv. Baking Sci. Technol., Dept. Grain Science, Kansas State Univ., Manhattan, KS; 1984, pp. D1–D12.

Miller, B.S. (Ed.). "Variety Breads in the United States." Proc. Symp. Am. Assoc. Cereal Chem.; St. Paul, MN; 1981.

Morris, C.E. *Food Eng.* **1985**, *57*(3), 63.

Moss, R.; Stenvert, N.L.; Pointing, G; Worthington, G.; Bond, E.E. *Bakers Digest* **1979**, *53*(2), 10–12, 14, 16, 17.

Olsen, C.M. *Bakers Digest* **1974**, *48*(2), 24–27, 30.

Peppler, H.J. *Cereal Foods World* **1981**, *26*, 609–611.

Ponte, J.G., Jr. In "Variety Breads in the United States." Miller, B.S., Ed. Am. Assoc. Cereal Chem.; St. Paul, MN; 1981.

Pyler, E.J. *Bakers Digest* **1982**, *56*(4), 22–26.

Ranhotra, G.; Gelroth, J.; Novak, F.; Bohannon, F.; Mathews, R. *J. Food Sci.* **1984**, *49*, 642–644, 646.

Rusch, D.T. *Cereal Foods World* **1981**, *26*, 111–113, 115.

Rusch, D.T.; Marnett, L.F. In Proc. Int. Symp. Adv. Baking Sci. Technol., Dept. Grain Science, Kansas State Univ., Manhattan, KS; 1984, pp. W1–W10.

Schnake, L. "White Pan Bread Marketing Spreads and Methodology." ERS staff report No. AGES830224, U.S. Dept. Agric.: Washington, D.C.; 1983.

Sluimer, P. *Bakers Digest* **1981**, *55*(4), 6–8, 10.

Smerak, L. *Bakers Digest* **1973**, *47*(4), 12–15, 18, 20.

Smith, W.M. *Bakers Digest* **1979**, *53*(4), 8–10.

Stenvert, N.L.; Moss, R.; Pointing, G.; Worthington, G.; Bond, E.E. *Bakers Digest* **1979**, *53*(2), 22–27.

Sugihara, T.F. *Bakers Digest* **1977**, *51*(5), 76, 78, 80, 142.

Thompson, D.R. *Bakers Digest* **1980**, *54*(3), 28–30, 32, 34, 36, 37.

Thompson, D.R. *Bakers Digest* **1982**, *56*(3), 28–30, 32, 33.

Thompson, D.R. *Bakers Digest* **1983**, *57*(6), 11, 12, 14–18.

Tipples, K.H.; Kilborn, R.H. *Bakers Digest* **1974**, *48*(5), 34–39.

Ulrich, M.G. *Bakers Digest* **1975**, *49*(2), 43–45, 67.

Vetter, J.L; Dubois, D.K. *Getreide Mehl Brot* **1978**, *32*, 119–121.

Frozen and Refrigerated Baked Goods

Allenson, A. *Bakers Digest* **1982**, *56*(5), 22–24.

Anon. *Cereal Foods World* **1979**, *24*, 50.

Anon. *Milling Baking News* **1985**, *64*(14), 27.

Bamford, R. *Bakers Digest* **1975**, *49*(3), 40–43.

Bruinsma, B.L.; Giesenschlag, J. *Bakers Digest* **1984**, *58*(6), 6, 7, 11.

Chamberlain, N. In Proc. Int. Symp. Adv. Baking Sci. Technol., Dept. Grain Science, Kansas State Univ., Manhattan, KS; 1984, pp. A1–A9.

Chen, R.W. *Cereal Foods World* **1979**, *24*, 46–47.

Davis, E.W. *Bakers Digest* **1981**, *55*(3), 12, 13, 16.

Gajderowicz, L.J. *Cereal Foods World* **1979**, *24*, 44–45.
Hsu, K.H.; Hoseney, R.C.; Seib, P.A. *Cereal Chem.* **1979**, *56*, 419–424.
Hsu, K.H.;Hoseney, R.C.; Seib, P.A. *Cereal Chem.* **1979**,56 424–426.
Ingram, C.E. *Bakers Digest* **1974**, *48*(2), 42, 43, 46, 71.
Lehman, T.A.; Dreese, P. *Am. Inst. Baking Technol. Bull.* **1981**, *III*(7), 1–5.
Lorenz, K. *Bakers Digest* **1974**, *48*(2), 14, 15, 18, 19, 22, 30.
Marston, P.E. *Bakers Digest* **1978**, *52*(5), 18–20, 37.
Oszlanyi, A.G. *Bakers Digest* **1980**, *54*(4), 16–19.
Sanderson, G.W. In Proc. Int. Symp. Adv. Baking Sci. Technol., Dept. Grain Science, Kansas State Univ., Manhattan, KS; 1984, pp. N1–N22.
Varriano-Marston, E.; Hsu, K.H.; Mahdi, J. *Bakers Digest* **1980**, *54*(1), 32–34, 41.
Wolt, M.J.; D'Appolonia, B.L. *Cereal Chem.* **1984**, *61*, 209–212.
Wolt, M.J.; D'Appolonia, B.L. *Cereal Chem.* **1984**, *61*, 213–221.

Chapter 13

Sensory Attributes and Bread Staling

The selection, acceptance, and digestibility of a food are governed by its sensory attributes. Important sensory attributes of bread include appearance (color, size, shape, absence of defects), texture, and flavor (taste and odor). Gloss and color of the crust of a loaf of bread, important appearance factors, depend on dough composition and bread-making procedure. Crumb color (which depends primarily on the color of the flour used), crumb texture, and crumb softness are among criteria most used by the consumer to evaluate acceptability of bread. Undesirable changes that take place in bread after baking (denoted by the term "staling") have increased in significance under modern, large-scale methods of bread production and distribution.

Bread staling has an economic effect both on the baking industry and on the consumer. It has been estimated that in the United States the losses to the industry from stale bread amount to 3–5% of the total production. In addition, there is a loss to the consumer, because some bread becomes stale and unpalatable after its purchase (see also Chapter 11).

STALING

Broadly speaking, staling of bread refers to all changes that take place after baking (Seibel et al, 1968). A phenomenon that indicates decreasing consumer acceptance of bakery products, staling is caused by physicochemical reactions (but not by the action of spoilage microorganisms) that take place as crust and crumb age. These reactions affect redistribution of moisture in the loaf (mainly movement of moisture from crumb to crust), change organoleptic characteristics (including flavor and mouth feel), and cause bread to become hard and to crumble. Changes in flavor are discussed in the second part of this chapter. Crust staling is caused almost entirely by moisture absorption from the atmosphere and the interior of the loaf; as the moisture redistributes, the crust becomes tough and leathery. Crumb staling, however, is much more complex and is the major cause of loss of palatability.

According to D'Appolonia (1984), factors that contribute to bread staling include the biochemical components present in wheat flour, the baking ingredients used in bread production, and the type of processing method employed. Reviews on the phenomenon and the fundamental causes of bread staling, were published by Herz (1965), Knightly (1977), Kulp and Ponte (1981), Lineback (1984), and Zobel (1973).

Part of the firming process in bread staling may be attributed to the migration of water (once part of the protein structure) from the gluten proteins to the starch. Whereas starch crystallinity may be the main factor in staling at room temperature (about 20°C), some additional (yet unidentified) factors may be involved at higher temperatures (30–35°C) at which starch crystallizes up to four times as slowly as at 20°C. Water-soluble pentosans may slow the rate of retrogradation by affecting the amylopectin fraction of starch, and water-insoluble pentosans may do so by affecting both amylose and amylopectin. The pentosans reduce retrogradation by reducing the amount of starch components available for crystallization. Retrogradation of both amylose and amylopectin characterize crystallization of starch gels through the first day of storage; thereafter, amylopectin alone controls retrogradation (Kim and D'Appolonia 1977).

Studies in the author's laboratories have demonstrated that small amounts of wheat flour lipids (polar lipids, about 0.5%, flour basis) are as effective as 3% shortening as antistaling agents and that the effect on staling retardation was highly correlated with the effect on increasing loaf volume. The use of emulsifiers or surfactants as antistaling agents is well recognized; bakers have used monoglycerides for many decades. The monoglycerides act best in combination with surfactants that strengthen the dough and reduce mechanical abuse in processing. The combination provides a balance of dough strengthening and retardation of crumb firming. Monoglycerides complex less with amylopectin than with amylose, and the degree of unsaturation (which determines shape) of monoglycerides relates inversely to the extent of complex formation. Chain length and degree of unsaturation of the fatty acid moiety significantly affect functionality of monoglycerides. Saturated fatty acids of long carbon chain length provide the best results. The amylose-complexing ability and crumb softening index of monoglycerides correlate well, but the degree of complexing between monoglycerides and amylopectin does not correlate with antifirming performance. The evidence points to amylose-complexing as the main mechanism of antistaling. Such reactions are enhanced when monoglycerides are used in certain physical forms, such as the aqueous lamellar phase or a dispersion. Alpha crystalline gel has the highest complexing index with amylose and also the best antifirming effect in bread. The most effective form involves alpha crystallinity, the molecules in the alpha form are postulated to pack such that polar groups are exposed to the water phase.

Researchers have demonstrated a high correlation between the amylose-complexing ability and the crumb softening index of monoglycerides but not necessarily of other surfactants. Studies on model systems seem to indicate that some surfactants such as sucrose esters enter the helical structure of amylose to form insoluble complexes. Other surfactants, such as stearoyl-2-lactylates, have no amylose-binding capacity. Many studies indicate that the main mechanism of the antistaling process involves the formation of a complex between amylose and monoglyceride. To be effective, however, surfactants should interfere with the crystallization of the amylopectins, because the amylopectin fraction is affected 3–4 times more than the amylose fraction.

Crumb Staling

As bread stales, the crumb becomes firmer. Firmness is the characteristic observed first by consumer and chemist and most used by the housewife (who hand-squeezes the bread) to judge freshness. The analyst measures firmness or freshness by determining the force required to compress a slice of bread. In firming, the crumb generally becomes increasingly dry and hard and it finally crumbles. In the process, several additional changes take place: Opacity increases; absorptive capacity, starch solubility, and susceptibility to the action of beta-amylase decrease; and the x-ray diffraction pattern of the starch is altered.

For the past century, scientists have been studying chemical and physicochemical changes that take place in the crumb of aging bread. Crumb staling is frequently confused with drying out of bread crumb. But as early as 1852 Boussingault showed that bread crumb may stale without loss of moisture, that stale bread may be freshened by heating to 70°C, and that this process can be repeated several times (Bechtel, 1955). Many studies have shown that the starch component of bread crumb is partially responsible for the staling phenomenon.

The most comprehensive studies on changes in aging bread have been made by Katz (1928), who attempted to counteract staling to eliminate the need for night baking in Dutch bakeries. Current concepts of bread staling are based on the work of many investigators of the past six decades. Seemingly, the changes that occur in gelatinized starch are responsible for most changes that take place in bread staling. The linear (amylose) fraction of starch retrogrades almost completely during baking and changes little during staling. The branched (amylopectin) part is apparently only partly retrograded in the crumb of a freshly baked loaf; on aging, its solubility and swelling power decrease, and gradually retrogradation takes place. The significance of starch in general and amylopectin in particular in

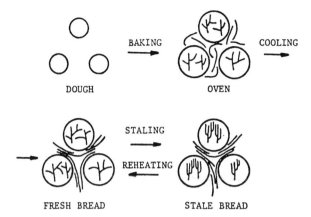

Figure 13-1. Roles of starch fractions during baking and staling of bread. (From Schoch, 1965.)

staling is substantiated by many indirect findings. Thus, the rate that changes occur in gelatinized starch is similar to that in bread starch and accompanying staling. The role of starch in staling has been confirmed by fractionation and reconstitution studies in which wheat starch has been replaced by cereal starches of varying amylopectin concentration or by chemically modified starches.

Measurements of crumb firming indicate that a physical process such as crystallization of starch is involved (Lineback, 1984). X-ray diffraction studies revealed that starch crystallization increased as bread staled. Using breads prepared from doughs containing alpha-amylases, however, showed that firming and crystallization are not synonymous. Although both starch components are involved in staling, amylopectin apparently plays the major role (Lineback, 1984). Amylose has essentially retrograded by the end of the first day after baking and contributes little to staling after that. Refreshing of bread by heating indicates that reversal of the retro-gradation of amylopectin apparently accounts for this behavior (Fig. 13-1). Firming or staling of bread involves, to a large extent, the realignment and association (retrogradation) of amylopectin after baking is completed.

Gluten Proteins

The importance of starch in staling does not detract from that of other components, mainly gluten proteins, in aging of bread. During dough mixing, gluten becomes highly hydrated; rising temperature during baking causes the gluten to coagulate and set to a firm structure. At the same time, the starch gelatinizes and absorbs moisture from the gluten. Microscopic studies have shown that the structure of bread consists of gluten filaments surrounding swollen and partly gelatinized starch granules. During the first few days after baking, the protein component of the bread has little effect on staling. In bread stored for long periods, however, the low-protein loaf stales faster than the high-protein loaf. It is postulated that high gluten content improves resilience and smoothness of the crumb. When gluten content is low, loss of moisture from the gluten to the crust region increases harshness. Increase in gluten, apparently, decreases association among starch granules and retards firming by serving as a moisture reservoir to buffer changes in hydrating capacities of the starch (Bechtel, 1955).

According to Kim and D'Appolonia (1977) there is, in general, an inverse relation between the protein content and staling during storage of bread. The keeping quality of bread, however, is not a linear function of protein content. Willhoft (1973) suggested that the antifirming effect of increased protein is due to dilution of the starch component and an increase in loaf volume affected by gluten enrichment. Banecki (1982) reported a decrease in crumb firmness as a result of increase in gluten contents. He calculated that bread firmness is affected about 6.7 times more by the gluten than by the starch component.

Normal Bread Ingredients

The effects of normal bread ingredients are not well defined. According to Edelmann et al (1950), increasing the sucrose content above 3% or substituting other sugars for sucrose had no significant effect. In the absence of emulsifiers, nonfat milk solids at 3% and 6% decreased firming rate, 2% salt had no effect but 3% increased firmness, and up to 3% lard increased firming but higher levels had no additional effect. In the presence of emulsifiers, milk solids had little effect, there was no difference between 2% and 3% salt, and the use of 3% lard had less effect than the increase from 3% to 8%.

Bread-Making Procedures

Any factor that increases moisture content of bread delays staling. The effects of other variables in bread-making are less clear-cut. Thus, some investigators have reported that length and temperature of fermentation and level of yeast affect staling; others disagree. Generally, staling is slowest when bread is baked properly (in terms of time and temperature) from an optimally processed dough (in terms of mixing, fermentation, and proofing). Bread made from a soft dough (ie, with higher moisture content) has a longer shelf life. Axford et al (1968) measured changes in the elastic modulus of bread crumb to establish the quantitative relationship between loaf specific volume and the rate and extent of staling. They found specific volume to be a major factor in determining both rate and extent of staling; rate and extent decreased linearly as loaf volume increased. Loaf specific volume affected rate of staling more markedly as storage temperature decreased. Bread made from dough mixed in a high-speed mixer staled less rapidly than bread from a conventional bulk-fermentation process.

Retarding Staling

Increasing moisture content of bread increases its softness and retards the firming process, suggesting that bread should have a moisture content as near the U.S. maximum (38%) as possible. Wrapping and sealing bread in moisture-proof materials effectively protects bread against moisture loss from the total loaf, but it cannot prevent changes in moisture distribution between crust and crumb and in the crumb. Most methods of retarding staling are based on preventing rapid redistribution of moisture in the crumb; hence, loss of moisture was for many years considered responsible for firming. Although that assumption was found erroneous when it was demonstrated that stale, firm crumb can contain as much total moisture as fresh crumb, moisture distribution in the crumb is important.

Many researchers have shown that monoglycerides of higher fatty acids are effective softening agents for bread. The compounds apparently alter the surface tension between fat and water and increase the spreading coefficient of the fat, resulting in a finer degree of fat dispersion. The effects of some compounds have been attributed to a decrease in gelatinizing and swelling properties of starch. Presumably, the compounds form complexes with the amylose fraction of starch and decrease starch swelling power (Bechtel, 1955). According to Waldt (1968), certain lipid derivatives and surface-active agents reduce the effective water concentration of the starch phase and assist in increasing the moisture retention of the gluten. It has been contended that monoglycerides basically control the rate of moisture transfer during bread storage and that, during baking, emulsifiers increase the retention of moisture in the heat-coagulated gluten, giving greater initial crumb softness.

According to Krog and Davis (1984), formation of water-soluble starch gels in foods, including bread, can be reduced by the addition of monoacyl surfactants which form insoluble, helical complexes with amylose. The capacity of surfactant-amylose complex formation depends in aqueous systems on the fatty acid chain length and the degree of unsaturation. This capacity is optimal for saturated or transmono unsaturated C16/C18 fatty acid chain length surfactants in the form of lamellar dispersions. The cis-unsaturated monoglycerides in the form of micellar solutions or in solvent systems are effective complexing agents but are inferior to trans monounsaturated or saturated monoglycerides as antifirming agents in bread. The antifirming effect in bread may be related to the thermal stability and the resistance of the amylose complexes toward enzymatic breakdown, which is optimal for complexes with saturated or trans-unsaturated monoglycerides. The thermal stability of saturated

Table 13-1. Crumb Softeners in Yeast-Raised Bakery Products

Conditioner	Typical use levels[b]	Estimated use[a]	
		United States	World[c]
Calcium and sodium stearoyl lactylate	0.25–0.5%	18–27 mlb	30–40 mlb
Diacetyl tartaric acid esters of monoglycerides	0.4–0.7%	0.5–0.8 mlb	4–7 mlb
Distilled monoglycerides		12–14 mlb	22–26 mlbs
Hydrated (25%)	1–1.5%	28–32 mlb	
Hydrated (50%)	0.5–0.75%	1.5–3 mlb	
Powdered (100%)	0.25–0.5%	4–6 mlb	
Mono- and diglycerides	0.4–0.75%	20–25 mlb	no estimate
Succinylated monoglycerides	0.15–0.4%	0.25–0.75 mlb	0.5–1 mlb

[a]In millions of lb/year; range of estimate.
[b]Percent of formula wheat flour.
[c]Non-Communist countries.
Source: Rusch and Marnett, 1984.

monoglyceride complexes is also better than that of complexes with sodium stearoyl lactylate.

In bread dough, the amylose complex formation with surfactants can take place inside the starch granules, reducing gelatinization of starch. This results in reduced crumb firmness and staling rate. Endotherm S is the "staling," and M_2 is the endotherm due to melting of the amylose-lipid complex. The phase behavior of polar lipids, both cereals and added as improvers, on bread volume and bread staling was discussed by Eliasson (1984).

Rusch and Marnett (1984) distinguished between two types of surfactants (fatty acid–based emulsifiers) used in yeast-leavened products: *Dough condtioners* interact with wheat flour protein to provide dough properties similar to those of stronger protein flours. Their effect is to increase the degree of gluten-gluten bonding to provide a stronger, more completely developed film. *Crumb softeners* combine with one or more starch fractions to soften the crumb and slow the staling rate. Uses of those that act mainly as dough conditioners (sodium stearoyl lactylate, calcium stearoyl lactylate, diacetyl tartaric acid esters of monoglycerides, and succinylated monoglycerides) and those that are primarily crumb softeners (distilled monoglycerides and mono- and diglycerides) are listed in Table 13-1.

Use of Enzymes

Emulsifiers prevent hardening of crumb during the first 1–2 days after baking but are less effective after a longer period. Because starch is involved in staling, it would seem that starch might be modified to retard retrogradation, perhaps through the effects of alpha-amylases, which

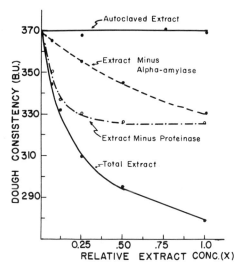

Figure 13-2. Effect of fungal extract, extract minus alpha-amylase, extract minus protease, and autoclaved extract on dough consistency. (From Miller and Johnson, 1955.)

degrade starch. Enzymes that degrade starch have been recognized for many years as being important in bread making. Rheological properties, rate of fermentation, loaf volume, crust color, and crumb characteristics may be affected significantly by enzymatic activity.

In dough, beta-amylase cannot attack raw, mechanically undamaged starch and hydrolyzes damaged starch slowly; alpha-amylase attacks damaged starch much more vigorously and is capable of attacking starch that is not sufficiently damaged to be susceptible to beta-amylase. Sound wheat flour contains abundant saccharifying beta-amylase (which cleaves off a fermentable maltose) and only minute amounts of the dextrinizing alpha-amylase (which cleaves large chains of starch). An adequate level of alpha-amylase activity in a dough containing no added sugar is essential for maintaining a sufficiently high rate of gas production to leaven the dough and give bread desirable crust and flavor. Alpha-amylase activity not only secures adequate levels of fermentable sugars but also influences dough consistency (Figure 13-2), modifies starch during oven baking, and improves the quality of the baked bread. In the past, historical sources of amylase supplements for bread were malted barley and malted wheat flour.

For some time, amylase preparations from molds have gained considerable attention. Certain strains of *Aspergillus oryzae* produce an amylase of the alpha type that is more thermolabile than the alpha-amylases from wheat or barley malt. Fungal amylases are used extensively in bread making because they offer advantages over the cereal amylases, excessive amounts of which may impair bread quality. Bacterial amylases, which because of their high termostability produce a sticky and gummy crumb, have shown little promise as diastatic supplements (see Chapter 11).

The main objection to bacterial alpha-amylase is its high thermostability; used in excessive amounts it produces an undesirable, doughy crumb. On the other hand, combinations of cereal amylases and low levels of bacterial amylase are satisfactory diastatic supplements. The bacterial amylase is so heat-stable that it can survive, to a degree, the oven temperature and continue to degrade starch in the baked loaf, thus offsetting retrogradation. It has been shown to retard crumb firming. In addition to delaying crumb firmness, the enzyme helps retain a moist texture, reduces crumbling, and promotes easy slicing. Thus, if levels of bacterial amylase are carefully controlled, the enzyme treatment can produce a bread with better shelf life than is possible by adding emulsifiers alone (Miller et al, 1953; Pomeranz et al, 1964; Rubenthaler et al, 1965; Shellenberger et al, 1966).

Storage Temperature

The rate of staling increases as the temperature of stored bread decreases to about $-2°C$, but it decreases when the temperature is lowered still further. In liquid air the bread remains fresh. Deep-freezing offers the most satisfactory method of maintaining bread freshness. Bread properly stored for 15–20 days at $-22°C$ is as good as fresh bread, but after longer periods the aroma and flavor deteriorate. Bread sealed in a tin can or vacuum-packed in a can maintains satisfactory flavor and aroma for months (Bechtel, 1955). Freezing maintains bread at the freshness level at the time of freezing, but it cannot improve stale bread. The freezing process is used quite extensively in the United States by the housewife and commercially.

As stated previously, storage below room temperature and above deep freezing increases firming. The firming at low temperatures is not reversed by subsequent storage at room temperature. The first hours after baking are critical, and fresh bread should be protected against undue cold and heat (at temperatures above $43°C$, the crumb is discolored and an off flavor develops). Thus, it is best to store bread at room temperature under conditions that minimize moisture losses. Retrogradation is a basically irreversible physical phenomenon. One cannot "unbake" bread, but one can "unstale" bread to a limited extent by heating it, because much of the amylopectin (though somewhat retrograded) can be made to open again by applying heat. In marketing partially baked rolls, use is made of the freshening effect of the high temperature. The rolls are baked sufficiently to gelatinize the starch, form a rigid structure, and destroy microorganisms. Completing the baking process immediately before consumption assures a fresh product with a crisp crust.

Determining Bread Staling

Several parameters have been used in the analytical control of bread staling: (1) Organoleptic methods—taste panel. (2) Determination of flavor components. (3) X-ray diffraction pattern, differential thermal analysis, and conductance and capacitance. (4) Rheological properties: (a) elasticity-compressibility or penetration; (b) crumbliness; (c) squeezability. (5) Swelling capacity and solubility: (a) absorptive capacity of the crumb by measuring the sedimentation rate of a slurry; (b) resistance to mixing (consistency in a farinograph); (c) consistency in an amylograph; (d) capillary flow of a thin pulverized crumb slurry; (e) changes in starch solubility. (6) Enzymatic tests: (a) susceptibility to beta-amylase; (b) susceptibility to papain.

Determining staleness, basically a matter of consumer judgment, involves several sensory perceptions: firmness of the crumb, the feel of the surface of a cut slice, odor, flavor, and mouth feel. All are used by a taste panel to judge whether bread is fresh or stale. The swelling and solubility and physico-rheological tests are most accepted and useful. The analytical methods, even though they offer convenient and reproducible results, have limited value in that they are not entirely consistent with one another and frequently do not agree with consumer evaluation of freshness. Considering the numerous changes that occur in aging bread, a single and simple chemical or physicochemical test could hardly be expected to be adequate to describe and measure the entire staling process (Bechtel and Meisner, 1951).

FLAVOR

Sensory Evaluation

Two major factors have stimulated extensive research of the chemistry of bread flavor and aroma. The first has been the development of new techniques and instruments capable of separating, isolating, and identifying minute quantities of flavor components. This can be accomplished best by combining gas-liquid chromatography and such techniques as infrared and ultraviolet spectroscopy, nuclear magnetic resonance, and fast scan mass spectroscopy. Classical organic analyses may be used to supplement or confirm information obtained from modern, sophisticated, instrumental methods.

A second factor that has stimulated research on bread flavor and aroma in many countries is the development of continuously mixed bread and the interest in chemically leavened bread. Freshly baked bread has a delectable, appealing flavor. That flavor, however, is not stable, and bread loses much of its appeal in a relatively short time. The trend toward mechanization and wider distribution has made it more difficult to maintain acceptable bread flavor (see Chapter 11).

Flavor responses are subjective and can be measured by recording individual and group responses to pure compounds or foods. The trained expert, the test panel, and the statistical survey of large numbers of consumers are three widely used approaches in determining flavor. Most of us judge food flavor in the same way an amateur judges art: "I don't understand it, but I know what I like" (Olson, 1962). Our flavor preferences are influenced by factors that range from the caprices of fashion to the prevalence of dentures. They are influenced by social, cultural, and religious patterns; psychological factors, variations in climate and in the general physical status of the individual; availability and price; and education (Amerine et al, 1965).

There are regional preferences of flavor and other qualities of bread, and bread type still has status significance. White bread has remained since ancient times a symbol of the upper classes. Manor house records from medieval England indicate using white wheat for the gentry, maslin (a mixture of wheat and rye) for the household staff, and rye and peas for the villeins. Today the nutritional gospel about whole versus white flour has prejudiced many in favor of dark bread from high-extraction flour, in part because of differences in the contents of dietary fiber and regardless of their actual nutritional value and consumer acceptance.

A sentiment widely expressed is that bread flavor today is not as good as it once was. Yet despite complaints about the soft, high-volume, bland white bread characteristic of the modern commercial bakery, a large proportion of consumers in Western countries select that type of bread. It is difficult to assess the effects of preferences and prejudices on bread flavor evaluation, but certain facts are well established. For example, fresh bread is definitely preferred over old bread. But with urbanization, bakeries increased in size and extended their distribution area, which has complicated the problem of getting bakery products to the stores rapidly. So now much of the white bread available to customers is semistale before it is purchased and is even staler by the time it is consumed.

Flavor is influenced by the texture of bread. A dense, compact crumb gives a stronger taste sensation than does a soft bread with a fine, silky crumb. The production of well-leavened, bold loaves from white flour by highly mechanized (especially continuous) processes works to the disadvantage of the commonly distributed white pan bread. Thus, more basic knowledge is needed on the flavor components in freshly baked bread. Research on the chemistry of bread flavor involves isolating—then separating and identifying—flavor components. The basic premise is that once that information is obtained, we may know how to recombine the components to achieve a flavor resembling the original. During the past three decades chemists have catalogued many of the compounds that can be detected in fermented dough, baked bread, and oven vapor (see Chapter 11). But even if all the components were known, additional problems would remain such as methods of introducing and maintaining a synthetic flavor in bread and of reducing loss of flavor and aroma from freshly baked bread. Solving such difficulties could lead to the manufacture of bread with superior natural flavor resistant to staling and to baked products that include mixtures of synthetic flavor components (Collyer, 1964; Coffman, 1967; Johnson et al, 1966).

Chemistry of Bread Flavor

The formation of bread flavor stimuli has been attributed to fermentation and baking in that bread with acceptable flavor cannot be produced without both. Neither a normally fermented dough baked without crust formation nor an improperly fermented dough baked with crust formation has acceptable flavor. However, some of the compounds formed during fermentation are volatilized during baking and do not affect the flavor of the baked bread. Researchers have tended to assume that any compound when isolated from bread that possesses a distinct odor contributes to bread flavor. In practice, the problem of establishing the contribution of individual components to flavor is complicated by the interaction among the components by blending, and by accentuating or suppressing flavor stimuli.

It is generally assumed that "bread" implies a product produced by yeast fermentation and that such a process is essential in developing an acceptable flavor. In many investigations numerous organic compounds have been identified with fermentation of bread dough, but it is not known whether the compounds remain in significant concentrations in the bread to be detected by the consumer or the compounds react during baking to form new products that contribute to flavor. It seems clear, however, that the products of fermentation are essential to producing good bread flavor.

The importance of crust formation and browning in producing acceptable flavor in bread is well established. Browning of baked products can be classified as caramelization and Maillard, or melanoidin, browning (Johnson and Miller, 1961).

The first type, caramelization, occurs when sugars are heated to about 135°C. What compounds are formed in the early stages of caramelization is not well established. In later stages of caramelization, brown unsaturated complex polymers, in some respects similar to products from the Maillard-type browning, are formed. However, caramelization, requires high temperatures of activation, which distinguishes it from Maillard-type browning. The flavors and odors of the compounds produced in the two types of browning are also different.

The second type, Maillard, involves an interaction between a reducing sugar and an amine; the interaction initiates a sequence of reactions the end product of which is a brown pigment. The most important reactions leading to nonenzymic browning of foods are those between reducing sugars and free amino acids or free amino groups of proteins. The reaction has been called Maillard-type browning in honor of the French chemist L.C. Maillard, who did pioneering research to elucidate the basic factors of

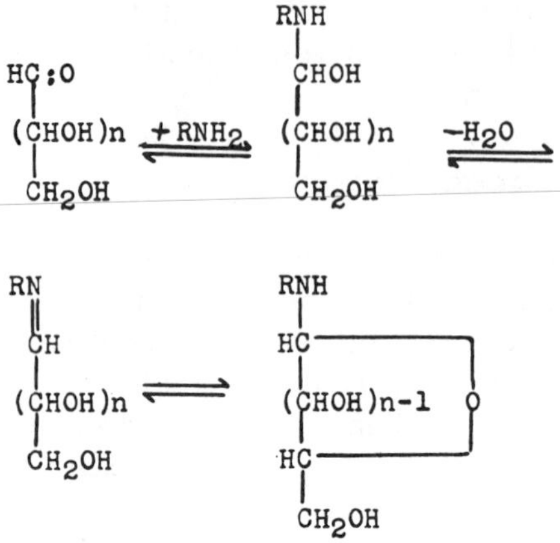

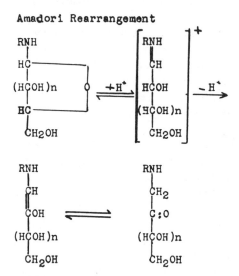

Figure 13-3. First reaction of browning: formation of N-substituted glycosylamine. (From Hodge, 1953.)

Amadori Rearrangement

Figure 13-4. Second reaction of browning: formation of 1-amino-1-deoxy-2-ketose. (From Hodge, 1953.)

browning. The complex chemical reactions involved in Maillard-type browning are summarized in Figures 13-3 through 13-5.

Browning of foods has always been associated with the development of flavors and odors. Some of the changes are desirable because they improve palatability and consumer acceptance of foods—as in roasting coffee or meat, toasting breakfast foods, or baking bread. Crust browning has been recognized as a source of bread flavor. It has been shown that flavor

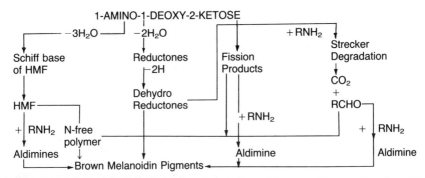

Figure 13-5. Final stages of browning: formation of intermediate and melanoidin pigments. (From Hodge, 1953.)

components produced by browning in the crust diffuse toward the crumb of the bread as it cools; thus, it enhances the flavor of the inside of a loaf. On the other hand, browning (especially at high temperatures) may destroy some of the nutrients in a food. For example, lysine residues of proteins may interact with reducing sugars and be rendered indigestible. If the essential amino acid is nutritionally limiting, nonenzymatic browning may decrease biological value of food.

About 150 organic compounds have been identified as contributors to flavor in preferments, doughs, oven vapors, and bread. These compounds—which include organic acids, alcohols, carbonyls, and esters—arise through a complex series of reactions during fermentation and baking. Compounds responsible for bread flavor sensation appear to be unstable. Many compounds formed during fermentation are volatilized during baking. Bread crust contains larger amounts of carbonyl compounds than crumb, and the crust's losing of carbonyl compounds parallels the staling of bread.

Control of Crust Browning

The extent and rate of browning can be controlled in several ways. Because most browning involves the reaction of free reducing groups of sugars with compounds containing an available amino group, the presence of both groups in a system is a prerequisite to a reaction. Control may be achieved by varying the type of sugar in a formulation. The use of sucrose in cake mixtures eliminates a reactive sugar, but it is ineffective in yeast-leavened systems because the enzymes elaborated by yeast cleave the nonreducing sucrose into two reducing sugars (fructose and glucose). The browning of bread crust can be reduced, although this is seldom desirable, by lowering the sugar level in the dough or by prolonging fermentation. Lowering the pH decreases availability of the reactive aldehyde group of a sugar and the extent of browning. In that browning accelerates at elevated temperatures, crust color can be controlled to some degree by carefully regulating oven temperature. Relative humidity is related to the rate of browning, which increases as the water content of the system decreases. When the relative humidity approaches 30%, browning proceeds rapidly. If evaporation from a surface is high (as in bread), browning may be delayed until the moisture level is reduced and the surface temperature of the crust is increased. Browning at room temperature will not proceed if the materials are dried to low moistures (generally below 7%) for long-term storage.

Kulp and Ponte (1981) concluded in their comprehensive and excellent review that extensive research for over a century has been rather ineffective in reducing staling of white pan bread. Little attention has been paid to flavor changes during staling. Some retardation of staling can be attained by the use of surfactants-emulsifiers; heat-stable alpha-amylase is potentially useful but difficult to control. Retardation of staling by technological means (processing, formulation, storage conditions) has been, thus far, of limited value.

BIBLIOGRAPHY

Amerine, M.A.; Pangborn, R.M.; Roessler, E.B. "Principles of Sensory Evaluation of Food." Academic Press; New York; **1965**.

Axford, D.W.E.; Colwell, K.H.; Cornford, L.J.; Elton, G.A.H. *J. Sci. Food Agric.* **1968**, *19*, 95–101.

Banecki, H. *Getreide Mehl Brot* **1982**, *36*, 272–276.

Bechtel, W.G. *Trans. AACC* **1955**, *13*, 108–121.

Bechtel, W.G.; Meisner, D.F. *Food Technol.* **1951**, *5*, 503–505.

Coffman, J.R. *Bakers Dig.* **1967**, *41*, 50–51, 54–55.

Collyer, D.M. *Bakers Dig.* **1964**, *38*, 43–46, 48, 50, 52, 54.

D'Appolonia, B.L. Proc. Int. Symp. Adv. Baking Sci. Technol. T-1 to T-18. Kansas State University Press; Manhattan; **1984**.

Edelmann, E.C.; Cathcart, W.H.; Bergquist, C.B. *Cereal Chem.* **1950**, *22*, 1–14.

Eliasson, A.-C. Proc. Int. Symp. Adv. Baking Sci. Technol., X-1 to X-14. Kansas State University Press; Manhattan; **1984**.

Herz, K.O. *Food Technol.* **1965**, *19*, 90–103.

Hodge, J.E. *Agric. Food Chem.* **1953**, 1, 928–943.

Johnson, J.A.; Miller, B.S. *Bakers Dig.* **1961**, *35*, 52–59.

Johnson, J.A., Rooney, L.; Salem, A. Adv. Chem. Ser. **1966**, *56*, 153–173.

Katz, J.R. In "A Comprehensive Survey of Starch Chemistry." Walton, R.P., Ed. Chemical Catalog Co.; New York; **1928**.

Kim, S.K.; D'Appolonia, B.L. *Bakers Dig.* **1977**, *51*(1), 38–44, 57.

Knightly, W.H. *Bakers Dig.* **1977**, *51*(5), 52–56, 144–150.

Krog, N.; **1977**, Davis, E.W. Proc. Int. Symp. Adv. Baking Sci. Technol., U1 to U-20. Kansas State University Press: Manhattan; **1984**.

Kulp, K.; Ponte, J.G. *CRC Crit. Rev. Food Sci. Nutr.* **1981**, *15*(1), 1–48.

Lineback, D.R. Proc. Int. Symp. Adv. Baking Sci. Technol., S-1 to S-20. Kansas State University Press: Manhattan; **1984**.

Miller, B.S.; Johnson, J.A. *Bakers Dig.* **1955**, *29*, 95–100, 166–167.

Miller, B.S.; Johnson, J.A.; Palmer, D.L. *Food Technol.* **1953**, 7, 38–42.

Olson, R.L. Proc. First Natl. Conf. Wheat Utilization Res. USDA; Lincoln, NE; **1962**.

Pomeranz, Y.; Rubenthaler, G.L.; Finney, K.F. *Food Technol.* **1964**, *18*, 138–140.

Rubenthaler, G.; Finney, K.F.; Pomeranz, Y. *Food Technol.* **1965**, *19*, 239–241.

Rusch, D.T.; Marnett, L.F. Proc. Int. Symp. Adv. Baking Sci. Technol. Kansas State University Press: Manhattan; **1984**.

Russell, P.L. *J. Cereal Sci.* **1983**, *1*, 297.
Schoch, T.J. *Bakers Dig.* **1965**, *39*, 48–57.
Seibel, W.; Menger, A.; Hampel, G.; Stephan, H. *Brot Gebaeck* **1968**, *22*, 193–203.
Shellenberger, J.A.; MacMasters, M.M.; Pomeranz, Y. *Bakers Dig.* **1966**, *40*, 32–38.
Waldt, L. *Bakers Dig.* **1968**, *42*, 64–66, 73.
Willhoft, E.M.A. *Bakers Dig.* **1973**, *47*(6), 14–16, 18, 20.
Zobel, H.F. *Bakers Dig.* **1973**, *47*(5), 52–61.

Chapter 14

Bread in Health and Disease

Bread supplies a significant portion of the nutrients required for growth and maintenance of health and well-being. Although not an outstandingly good source of any single nutrient, bread is a good source of most nutrients. The potential of wheat and bread for meeting man's nutrients is well established. Enrichment of bread with vitamins and minerals has been instrumental in improving the nutritional adequacy of our diet. Supplementation of bread with protein concentrates and pure amino acids shows great promise for improving the diet of people who depend on bread as the main source of nutrients. This chapter reviews some of our knowledge of the relation of chemical composition of cereals and health.

CHEMICAL COMPOSITION

The chemical composition of the dry matter of wheat, in common with other foods of plant origin, varies widely, as it is influenced by genetic, soil, and climatic factors. Variations are encountered in the relative amounts of proteins, lipids, carbohydrates, pigments, vitamins, and ash; mineral elements present also vary widely. (See also Chapter 4.) Wheat is characterized by relatively low protein and high carbohydrate contents; the carbohydrates consist essentially of starch, dextrins, pentosans, and sugars of which 90% or more is starch. The various components are not uniformly distributed in the different kernel structures. The hulls and bran are high in cellulose, pentosans, and ash; the germ is high in lipid content and rich in proteins, sugars, and ash components. The starchy endosperm contains the starch and is lower in protein content than the germ and in some cereals lower than bran. The starchy endosperm is also low in crude fat and ash components. Wheat is low in nutritionally important calcium, and its calcium and other ash components are greatly reduced during milling. In common with other cereal grains, wheat contains vitamins of the B group but is completely lacking in vitamin C (unless the grain is sprouted) and vitamin D. Wheat contains yellow pigments, but they are almost entirely xanthophylls, which are not precursors of vitamin A. The oils of the embryos of cereal grains, particularly wheat, are rich sources of vitamin E.

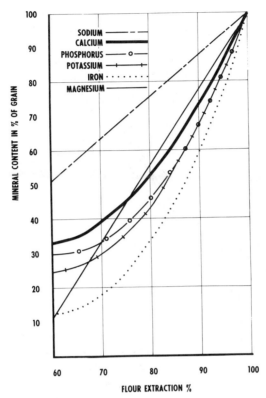

Figure 14-1. Changes in mineral content of wheat flour with percent extraction. (From Thomas, 1968.)

The relative distribution of different vitamins in various kernel structures is not uniform, although the endosperm invariably contains the least.

SUPPLEMENTATION

In progressing from the head to the tail of the mill, the flour streams contain progressively more of the components characteristic of the bran and germ. These structures are richer in protein than the endosperm; also the protein content of the endosperm steadily decreases in progressing from the outer to the central zone. Changes in protein are accompanied by changes in vitamins and minerals (Figures 14-1, 14-2). The ash content of the bran is 20–25 times and that of the germ 10–15 times that of the starchy endosperm. The ash content of a flour is therefore useful to the miller as a measure of the "purity" of the various flour streams. As the purity decreases, the ash content rises much more sharply than the protein content, since the protein content of the aleurone layer exceeds that of the

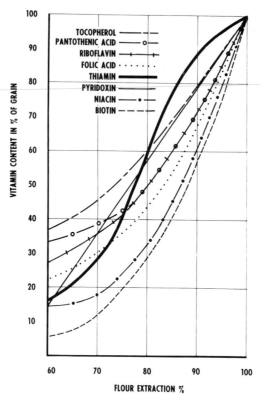

Figure 14-2. Changes in vitamin content of wheat flour with percent extraction. (From Thomas, 1968.)

starchy endosperm by a factor of only about 1.5 times. Thiamine, niacin, biotin, folic acid, and vitamin B_6 apparently are similarly distributed in the wheat kernel, because each is reduced about 80% during milling of wheat into white flour. The reduction of inositol and p-aminobenzoic acid is even bigger, but it is smaller in pantothenic acid. Choline is apparently distributed throughout the wheat kernel. The substantial increases in thiamine, riboflavin, and niacin in bread are from enrichment factors added in manufacture (discussed later in this chapter). Table 14-1 shows the vitamin content of wheat, flour, and milling by-products. The vitamin content of flour is markedly affected by the degree of refinement; the more highly refined flours lose the most vitamins in milling. Although removing wheat germ decreases the vitamin content of flour, the factor that accounts for the greatest loss of vitamins is removing the aleurone layer. This is to be expected in view of the high concentration of vitamins in the aleurone layer and the fact that much more bran than germ is removed in milling. The bran fraction contains these proportions of the vitamins in the whole wheat: thiamine 23%, riboflavin 40%, niacin 73%, pantothenic acid 53%, folic acid 26%, biotin 57%, p-aminobenzoic acid 58%, inositol 63%, and vitamin B_6 51% (Bradley, 1962).

Table 14-1. Vitamin Content in Various Milling Streams (mg/100 g dry weight)

Vitamin	Patent flour	First clear flour	Low-grade flour	Red dog	Shorts	Bran	Germ	Whole wheat
Thiamine	0.076	0.245	1.080	2.80	1.34	0.629	1.350	0.393
Riboflavin	0.032	0.048	0.124	0.322	0.347	0.334	0.487	0.107
Niacin	1.01	2.09	3.86	8.01	16.00	26.60	4.53	5.45
Pantothenic acid	0.483	0.675	0.915	1.82	2.66	3.91	1.04	1.09
Folic acid	0.011	0.018	0.042	0.120	0.135	0.088	0.205	0.05
Biotin	0.0014	0.0042	0.0108	0.0250	0.350	0.0440	0.0174	0.0114
p-Aminobenzoic acid	0.033	0.126	0.295	0.781	1.26	1.48	0.370	0.383
Choline	161	151	148	174	176	154	265	163
Inositol	33	113	341	808	1080	1340	852	315

Source: Bradley, 1962.

Supplementation with Vitamins and Minerals

Investigations of the diets of American and other people, before World War II, established that serious and widespread deficiencies existed in all age groups in intake of thiamine, riboflavin, nicotinic acid, and iron. A deficiency of calcium was found in certain population groups, and vitamin D deficiency was quite prevalent among infants and children. U.S. millers, bakers, and many nutrition authorities realized that a rapid improvement in general nutrition could not be achieved by consuming long-extraction flours or whole wheat flours because the public prefers white flour and white bread. It was recognized that controlled addition of commercially prepared vitamins and minerals to white flour and other types of refined food products, which had long met with public acceptance, would assure the best distribution of thiamine, riboflavin, niacin, and iron. Three basic approaches are possible in dealing with the improvement of the nutritional value of cereals: (1) restoration—addition of one or more nutrients to a processed food to restore it to a preprocessed (or natural) level; (2) enrichment—adding specific amounts of selected nutrients to a processed food in accordance with official regulations; and (3) fortification—adding nutrients to foods and food products at levels that may exceed natural levels.

Enrichment of white flour and bread in the United States is so designed that six slices of enriched bread in the average daily diet suffice to protect against defficiencies in thiamine, riboflavin, niacin, and iron. Flour standards in 1941 under the U.S. Food, Drug, and Cosmetic Act of 1938 provided for enriching flours. In a series of publications, the Committee on Cereals, Food, and Nutrition of the National Academy of Sciences kept pace with developments in the cereal enrichment program from its inception in 1941. New information led to revised standards. The present standards are given in Table 14-2. Flour is enriched at the mill by metering a vitamin-mineral premix mixture with a mechanical feeder into a flowing flour stream or by adding a weighed quantity of enrichment mixture to flour in a batch mixing operation. The miller is relieved of the troublesome task of making premixes and assured of more uniform enrichment by using commercially available, accurately compounded, finely milled enrichment mixtures with wheat or corn starch as the carrier of vitamins and iron. In the United States, relatively little interest has been shown in adding the optional ingredients, calcium and vitamin D, to flour at the mill.

Many bakers use regular flour and add wafers or tablets of the necessary ingredients. The tablets consisting of amounts of ingredients required for units of 50 or 100 lb of flour with a rapid-swelling starch as carrier are allowed to disintegrate in a portion of the water used to make the dough, and the resulting suspension is added during dough mixing.

The cost of enriching 100 lb of bread is less than 1¢. The enriched products are nutritionally improved; yet the color, texture, aroma, and

Table 14-2. U.S. Cereal Enrichment Standards

Product	Thiamine	Ribo-flavin	Niacin	Iron	Calcium	Vitamin A
Enriched flour	2.9	1.8	24	20	(960)[a]	
Enriched self-rising flour	2.9	1.8	24	20	960	
Enriched farina	2.0–2.5	1.2–1.5	16–20	13	(500)[a]	
ASCS domestic wheat flour	2.9	1.8	24	20		
ASCS export wheat flour	2.9	1.8	24	20	500–625	4,000–6,000
ASCS soy-fortified flour	2.9	1.8	24	20	500–625	4,000–6,000
Enriched corn meal and grits	2.0–3.0	1.2–1.8	16–24	13–26	(500–750)[a]	
Enriched self-rising corn meal	2.0–3.0	1.2–1.8	16–24	13–26	(500–1,750)[a]	
ASCS domestic corn meal	2.0–3.0	1.2–1.8	16–24	13–26		
ASCS domestic corn grits	2.0–3.0	1.2–1.8	16–24	21–26		
ASCS corn masa flour	2.0	1.2	16	13–26		
ASCS export corn meal and soy-fortified corn meal	2.0–3.0	1.2–1.8	16–24	13–26	500–750	4,000–6,000
Enriched rice	2.0–4.0	1.2–2.4[b]	16–32	13–26	(500–1,000)[a]	
Enriched macaroni and noodle products[c]	4.0–5.0	1.7–2.2	27–34	13.0–16.5	(500–625)[a]	
Enriched bread[d]	1.8	1.1	15	12.5	(600)[a]	
U.S. RDA (mg or IU/day)	1.5	1.7	20	18	1,000	5,000

This table shows the final nutrient levels required for enriched cereal products as specified by the U.S. Food and Drug Administration and the USDA Agriculture Stabilization and Conservation Service (ASCS). All figures are in milligrams per pound of product except for vitamin A, which is in International Units (IU) per pound. When two figures are shown it indicates a minimum-maximum range. Where one number is shown it indicates the minimum level with overages left to good manufacturing practice. Standards in parentheses are optional.

[a]No claim of calcium enrichment can be made when calcium is present for technological reasons at levels less than the minimum value shown except as required by nutritional labeling (21 CFR 101.9).

[b]The riboflavin standard for rice has been stayed for many years.

[c]Enriched pasta products are normally made from semolina, durum flour, or wheat flour enriched to these same levels. There are, however, no official standards for enriched cereals used in pasta production.

[d]Bread containing 62% or more of flour enriched to the standards for enriched flour will meet the standards for enriched bread. Bread can be additionally fortified to meet the standards proposed by the Food and Nutrition Board of the National Academy of Sciences. In addition to the above, the flour would contain, in mg/lb: 0.3 folic acid, 2.0 pyridoxine, 900 calcium, 200 magnesium, 10 zinc, and 5,000 IU/lb vitamin A. Such bread is to be labeled "special formula enriched bread." The above is a condensation of regulations in effect as of July 1983.

overall appearance are unaffected. Nearly 95% of the white bread in the United States is enriched. Enrichment regulations do not apply to cakes, pies, pastries, or similar sweet goods.

CHANGES DURING FERMENTATION AND BAKING

Data on time-temperature relations during baking indicate that the crust temperature rises to 100°C during the first 10 minutes of heating and may reach 160–180°C or occasionally 200°C at the end of baking. The temperature at the center of the crumb rises more slowly, varying with size and shape of loaf, and usually barely reaches 100°C at the end of baking. This large difference in temperatures of crumb and crust is a primary factor in the stability of nutrients in the two parts of the bread. It has great nutritional significance because desirable modifications during the baking stage in digestibility, leavening, production of a porous and well-texturized bread, and development of delectable taste and flavor are accompanied by undesirable (though basically unavoidable) destruction of heat-labile nutrients or by the inactivation of nutrients through the Maillard reaction. On the other hand, bread that has not been baked properly has a poor, inelastic crumb, unappetizing crust, decreased consumer acceptance, and lowered digestibility.

Losses of 20–30% of thiamine have been reported in European-type breads. There is significantly less thiamine in the crust than in the crumb of whole wheat bread; the maximum difference is about 35%. Auerman et al (1954) reported that flour of 72–85% extraction retained 64–88% less original riboflavin than did whole wheat bread during bread making. It is generally agreed that niacin is fairly stable and that 95–100% of it is retained under normal baking conditions.

Thomas (1968) reported that losses of thiamine were up to 5% during fermentation; further losses occurred during baking (about 17% and 13% in bread made of flours milled to 80% and 60% extraction, respectively). The higher destruction in bread from dark flours apparently results from longer baking time. Losses of vitamin E in baking were 5.5–15% and 20–30% in white bread and whole wheat or dark wheat bread and pumpernickel, respectively.

According to Kennedy and Joslyn (1966), data on the thiamine content of wheat, dough, crumb, and crust of whole wheat bread show no change from wheat to dough but an increase in crumb and a 30% decrease in crust. Other studies indicate that thiamine destruction occurs mainly in the crust, and averages 15%. Up to 75% of phytic acid in bread dough may be destroyed in converting dough to bread. The extent of destruction is affected by the pH of the dough, fermentation time, and baking conditions.

Maleki and Daghir (1967) have studied the effects of high baking temperatures (400–500°C) and baking time on retention of thiamine, riboflavin, and niacin in flat, circular Arabic bread baked from white or brown wheat flour. Loss of niacin under all conditions was negligible. Destruction of thiamine was greater in brown (34%) than in white (25%) Arabic bread, and was positively related to intensity in baking. About 10–25% of riboflavin was destroyed, with retention higher in white than in brown bread.

In view of the significance of bread proteins in nutrition, it is interesting to review changes occurring in the nitrogenous components. Gorbach and Regula (1964) found that baking destroyed 26% and 84% tryptophan, 23% and 73% lysine, 14% and 57% methionine, and 13% and 66% threonine in the crumb and crust, respectively. The decrease in the limiting amino acids lysine, threonine, and methionine, especially in the crust, substantially decreased the biological value of the proteins. The decrease was accompanied by substantial destruction of polyunsaturated fatty acids of fat in the crust. Kennedy and Joslyn (1966) concluded that the biological value of proteins in the crust is considerably below that in the crumb. Apparently, less impairment in biological value of cereal proteins and less loss in available lysine content occurs in bread baked by high frequency than in conventional ovens. Factors that influence the percentage of amino acids destroyed include the form of the added amino acid (D or L), the temperature and time of baking, the amount of reducing sugars in the dough, and whether the amino acids are in the protein or added as pure amino acids.

Amino Acids in Wheat and Milled Products

Differences in the amino acid composition of wheat and flour and destruction of certain amino acids during baking have prompted extensive investigations. The essential amino acids of various milled products are given in Table 14-3. The low-grade flours, particularly red dog, have a completely different pattern of amino acid composition from that of white flour. Thus, for instance, the red dog proteins contain twice as much lysine as white flour proteins. Similarly, the lysine content of shorts, bran, germ, and whole wheat proteins is much higher than of white flour proteins. Threonine and, to a lesser extent, tryptophan follow a similar pattern. The high concentration of essential amino acids in the proteins of by-products of white flour manufacture makes attractive the proposition to use them in producing nutritionally enriched products (see also Chapter 9).

PROTEIN-ENRICHED BREAD

The need to enrich wheat flour used in bread making with proteins and amino acids is not nearly as clearly established as the need to enrich white

Table 14-3. Concentration of Essential Amino Acids in Various Milling Streams (g/16 g N)

Amino Acid	Patent flour	First clear flour	Low-grade flour	Red dog	Shorts	Bran	Germ	Whole wheat
Arginine	3.73	3.87	4.68	6.84	6.85	6.60	6.88	4.71
Histidine	1.92	2.06	2.14	2.22	2.20	2.22	2.26	2.12
Isoleucine	3.91	4.02	3.72	3.42	3.31	3.29	3.48	3.78
Leucine	6.63	6.59	6.33	5.77	5.64	5.51	5.75	6.52
Lysine	1.97	1.94	2.54	4.13	4.18	3.77	5.28	2.67
Methionine	1.73	1.71	1.67	1.70	1.62	1.48	1.91	1.74
Phenylalanine	4.77	5.04	4.64	3.55	3.44	3.58	3.38	4.43
Threonine	2.64	2.73	2.76	3.11	3.03	2.86	3.42	2.76
Tryptophan	0.92	1.01	1.01	1.25	1.29	1.58	0.98	1.13
Valine	4.32	4.44	4.45	4.91	4.84	4.69	4.90	4.69

Source: Bradley, 1962.

flour with minerals and vitamins. Over 70 years ago, Osborne and Mendel (1914) showed in studies with young rats that wheat protein is deficient in lysine. During the milling of wheat to white flour, the concentration of lysine in the protein is substantially decreased. Considerable research has been carried out to increase by plant breeding and agrotechnical practices the quantity and quality of protein in cereal grains. Generally, however, as the protein content increases, the concentration of lysine and other limiting amino acids in the protein decreases. Adding lysine, threonine, and methionine to wheat flour fed to rats improves the nutritional value of the flour. The growth-promoting effects of germ, soy flour, or milk supplement in bread are primarily due to the amounts of lysine contributed by these supplements. Such supplementation makes the combined proteins in the enriched bread nearly comparable in quality to the proteins in milk and meat.

On the other hand, numerous arguments have been raised against indiscriminate supplementation of bread with proteins and amino acids. The protein level in wheat ranges from 8% to as much as 20%; most of the flour used in bread making in the United States has about 11% protein, at least. In the manufacture of bread, the formula includes two good sources of lysine—about 2% compressed yeast and about 4% of soy, milk, or soy-whey solids. The adult requirement for lysine is about 0.8 g per day; that requirement can be met by about 14 oz of bread. As that amount of bread supplies only 40% of the caloric requirements, proteins from additional foods would safely provide an adult's lysine needs (Bradley, 1967).

Data in Table 14-4 compare amounts of essential amino acids supplied by wheat diets with minimum estimates of amino acids needed. A diet composed of practically all wheat or 80% extraction flour meets the theoretical needs for many of the nutrients, including protein, which is

Table 14-4. Comparison of FAO Minimal Amino Acid Requirements and
Content of Wheat Diets

Amino acid	7-yr child (24 kg)		14-yr boy (49 kg)		Adult woman (58 kg)		Adult man (70 kg)	
	FAO min	80% Flour 1,800 cal	FAO min	80% Flour 3,000 cal	FAO min	80% Flour 2,300 cal	FAO min	80% Flour 3,200 cal
Isoleucine	0.71	3.1	1.64	5.1	0.97	3.9	1.18	5.5
Leucine	0.82	4.5	1.87	7.6	1.10	5.8	1.34	8.2
Lysine	0.71	1.7	1.64	2.9	0.97	2.3	1.18	3.2
Phenylalanine	0.48	3.6	1.10	6.0	0.64	4.6	0.78	6.5
Phenylalanine + tyrosine	0.98	5.2	2.20	8.8	1.28	6.7	1.56	9.5
Methionine	0.37	1.0	0.86	1.6	0.51	1.2	0.62	1.8
Total sulfur containing amino acids	0.71	2.5	1.64	4.3	0.97	3.3	1.18	4.6
Threonine	0.48	1.9	1.10	3.2	0.64	2.5	0.78	3.5
Tryptophan	0.24	0.67	0.55	1.1	0.32	0.85	0.39	1.2
Valine	0.71	2.9	1.64	4.8	0.97	3.7	1.18	5.1

Source: Hegsted, 1962.

often considered unsatisfactory. The diet is, of course, deficient in vitamins A, C, and D; in calcium; and probably in riboflavin. Needs for these nutrients can easily be met by fruits, vegetables and exposure to sunlight. Although such a diet appears to be nutritionally adequate, its acceptability is admittedly questionable.

A useful method of assessing the protein value of foods and diets customarily eaten was derived by Platt et al (1961). They state that in terms of net dietary protein calories percent (NDP cal%), it is possible to represent the usable protein in a foodstuff or diet. Provided the total caloric requirements are met, NDP cal% of 4.6 represents a satisfactory protein allowance for adults other than pregnant or nursing women. For children up to 3 years old the desirable figure is about 8, and for pregnant or nursing women, about 9.5. Wheat has a value of 6.0—higher than rice (5.1) or corn (4.7). Diets containing wheat as the staple foodstuff together with appropriate amounts of protein-rich foods can satisfy protein requirements in terms of NDP cals% at all stages of development, and wheat alone would meet the requirements of most adults. This is confirmed by studies showing that the final stages of growth and maintenance of adult rats can be attained by feeding bread as the only protein source (Moran and Pace, 1967). Feeding trials with humans have shown that wheat by itself can meet the protein requirement of the adult male.

Thus, it comes as no surprise that the U.S. National Research Council of the National Academy of Sciences found no clear indication that adding lysine to white bread would improve the diets of the population in the

United States (Jansen, 1969). Yet the committee on amino acids recognized that supplementation with proteins or specific amino acids can be nutritionally effective and that such supplementation can be justified for people who depend on cereals as the main source of proteins.

PROTEIN CONCENTRATES FROM WHEAT PRODUCTS

Protein concentrates can be produced from milled products by using aqueous systems or by separating air-dry products from conventional milling. Gluten containing 80% protein on a dry-matter basis can be separated from starch of wheat flour by a gentle stream of water. The viscoelastic mass thus obtained consists of the insoluble proteins of wheat (gliadin and glutenin), lipids, small amounts of carbohydrates, and trapped soluble proteins. If the gluten is dried carefully, it may retain (after hydration) some of its functional properties in bread making and justly be called "vital" gluten. Dry, purified gluten can be prepared by dispersing a washed gluten in dilute acetic acid, centrifuging, and lyophilizing. The purified gluten contains, on a dry-matter basis, 90% protein, 8% fat, and 2% other components (carbohydrates, minerals etc). Although vital gluten produced under controlled conditions from strong flours can be an excellent supplement to improve the bread-making quality of a flour, its use for nutritional enrichment has several disadvantages. They include high cost of manufacture and amino acid composition less balanced than in unfractionated flour.

Concentrates containing up to 90% protein can be obtained from wheat feeds by extracting with dilute acids or alkali and then drying the neutralized extracts under mild conditions. The price of such concentrates for large-scale supplementation, however, is generally prohibitive.

Protein concentrates can be manufactured from wheat flours produced by conventional roller milling by three basic methods: fractionating feed streams, air-classifying flours (and combinations of the two methods), and separating the wheat germ. Protein fractions (up to 30% protein) from air classification can be used to increase the bread-making quality of low-protein or functionally weak flours. The air-classified protein concentrates are comparable in amino acid balance to the original flours or are slightly poorer. The subaleurone layer in the starchy endosperm, one-cell deep and situated adjacent to the aleurone cells, is very rich in protein and can be obtained by dry milling and air-classifying a third-break flour. The product may contain about 40% protein.

GERM BREAD

Germ is not an intentional component of white wheat flour, though some germ remains in flour. The amount of germ in a flour increases with increased flour extraction rate. Because even the most refined flour contains some germ, the effects of germ in bread making are of interest. Furthermore, the high protein content and excellent amino acid balance make it attractive to supplement wheat flour with germ. About 0.5% fairly pure germ can be isolated during conventional flour milling. Meaningful improvement of nutritional value of white bread requires about 10% germ addition. The germ, including scutellum, comprises 1.5–3.3% (average 3.0%) of the wheat kernel. Purified germ contains (on an as is basis) about 27% protein, 9% fat, 46% carbohydrates, 2% crude fiber, and 4% minerals. The minerals include 5 mg sodium, 837 mg potassium, 69 mg calcium, 8 mg iron, and 1,100 mg phosphorus. Among the vitamins are 0.2 mg carotene, 27.5 mg vitamin E, 2.0 mg thiamine, 0.7 mg riboflavin, 4.5 mg nicotinamide, 1.0 mg pantothenic acid, 3.0 mg vitamin B_6, and 0.5 mg folic acid; vitamin C is absent. The embryo and scutellum, which comprise only 1.2% and 1.5% of the kernel weight, contain respectively 3% and 59% of the kernel's thiamine.

The germ is the main source of tocopherol in the wheat kernel, and wheat germ contains more tocopherol than is found in other grains. Germ fat is rich in polyunsaturated fatty acids. Wheat germ proteins are rich in essential amino acids. The proteins are particularly rich in lysine, arginine, aspartic acid, threonine, alanine, and valine. The concentration of glutamic acid and proline in the germ proteins is much lower than in proteins of wheat flour or wheat gluten. The amounts of sulfur-containing amino acids in proteins of wheat germ and wheat flour are comparable.

Freshly prepared germ becomes rancid in a few days at room temperature. Action of wheat lipases that results in forming oxidizable free fatty acids is most important in deterioration of germ in storage. Shelf life can be increased by storing at low temperatures (around 4°C). Most commonly, the germ is stabilized by wet or dry heat treatment to inactivate enzymes responsible for rancidity. Similar results can be obtained by drying at low temperatures to 5% moisture and storing the dried product in waterproof containers.

High levels of raw germ have a deleterious effect in bread making. The deleterious effects of glutathione were identified, and several methods were proposed to counteract the effects. Heat treatment, increasing fermentation length, and increasing oxidant levels all decrease the harmful effect.

SUPPLEMENTATION WITH NONWHEAT PROTEINS

The effects on bread quality of supplementing wheat flour with proteins from plant and animal sources have been studied by many investigators.

The plant proteins included protein-rich residues from oilseed processing (mainly soybean, cottonseed, peanuts, sesame, and sunflower). In addition, food-grade yeast and fish meal were suggested as promising and inexpensive sources of high-quality proteins. Most promising with regard to both their compatibility in bread making and their high nutritional value is supplementation with soy proteins. Excellent bread can be made from blends containing up to 8% soy flour provided that a rich formula, a high-protein wheat flour, and optimum oxidant levels are used. Unfortunately, in countries where enrichment with soy flour would be most beneficial, lean formulas are generally used, high-protein wheat flours are too expensive or unavailable, and the addition of oxidizing agents is prohibited by law.

It has been suggested that soy flour protein isolates be used as a wheat flour supplement. These isolates contain twice as much protein as ordinary soy flour, have a bland taste and flavor, and have a light color. Such protein isolates have found a market in the United States as supplements in breakfast cereals, biscuits, baby foods, milk substitutes, and texturized foods. Though less expensive than animal proteins, soy protein isolates are about three times as expensive as soy flour on a comparative protein basis.

Studies of Pomeranz et al (1969a,b) have shown that wheat flour enriched with plant or animal proteins and supplemented with natural or synthetic glycolipids could well be the answer to producing low-price, protein-enriched bread. Without added glycolipids, relatively low levels of 3–6% lysine-rich protein supplements impaired bread quality. The amount of nutritionally limiting amino acids was approximately tripled in bread that was protein enriched with 16% soy flour. At the same time, consumer acceptance of the enriched bread was maintained by adding glycolipids (Figure 14-3). The use of glycolipids in producing acceptable, nutritionally improved bread seems particularly promising as it can be carried out by existing bread-making processes and requires virtually no change in dough formulation, bread-making schedules, or baking equipment. Best results were obtained, especially from the standpoint of taste and flavor, by the use of soy flour from germinated and defatted beans (see also Chapter 9).

PROTEIN SUPPLEMENTATION OF NONWHEAT BREADS

Rotsch (1954) reported that bread can be prepared from doughs in which gluten is replaced by other gel-forming substances. Jongh (1961) found that a starch-water dough offered great resistance to mixing and had a firm appearance; as soon as mixing was stopped, the dough became liquid and could be poured from a bowl. Adding 0.05–1.0% glyceryl monostearate reduced resistance during mixing and lowered the dough's tendency to

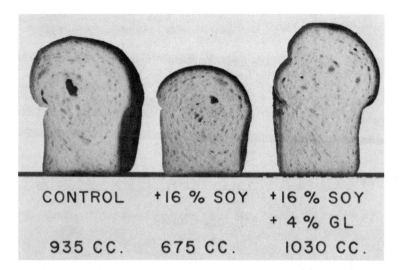

Figure 14-3. Bread baked from 100 g of wheat flour showing reduction in loaf volume with addition of soy flour and the restored loaf volume by the addtion of sucrose tallowate. (From Pomeranz et al, 1969b.)

become fluid at rest. At the same time, gas retention capacity of the dough increased and bread quality of the baked starch dough was improved considerably. The effect of glyceryl monostearate was attributed to enhanced aggregation of starch granules, leading to formation of a plastic-like structure. It was suggested that any compound capable of improving coherence between starch granules and of imparting adequate gas retention during fermentation and oven spring would be a suitable binder in bread making. These requirements were met by adding 5% lard, 5% egg albumin, or 11% gliadin. Gluten was considered also as a binder which imparts considerable elasticity to the dough.

The investigations of Jongh (1961) led to significant developments in which nonwheat baked products were produced from mixtures of tuber flours and defatted oilseed flours (De Ruiter, 1978). The potential of these contributions was utilized in the production of so-called composite flours in the Chorleywood Bread Process. Bread made with 60% wheat flour, 30% cassava starch, and 10% soy flour was only slightly inferior to a regular wheat flour bread.

AMINO ACID SUPPLEMENTATION

Because agricultural plant breeding programs are slow and developing new protein sources is difficult, amino acid supplementation offers a possibility of indirectly increasing the protein supply by improving protein quality. Theoretically, optimum amino acid supplementation could double the biological value of wheat proteins. Optimization, however, is difficult to attain. Supplementing bread with proteins in combination with amino acids has several advantages. The protein supplement raises the protein level and generally supplies enough of the second limiting amino acid so that a better response of the first limiting amino acid (generally lysine) can be obtained. At the same time, adding lysine makes it possible to produce a nutritionally improved bread at a level of protein supplementation less likely to impair the functional properties of the dough or consumer acceptance of the bread.

MISCELLANEOUS HEALTH ASPECTS

Dietary Goals

It is common to hear that nutrition cannot be sold—that people buy foods primarily for their taste and appearance and for the enjoyment they provide rather than for nutrition. It is probably true that in the order of food priorities caloric requirements are first, food aestetics and enjoyment are second, and the need to be adequately fed is third. But the situation is changing.

In January 1977, the Select Committee on Nutrition and Human Needs of the U.S. Senate called on Americans to increase their consumption of fruits, vegetables, whole grains, poultry, fish, skim milk, and vegetable oils and to reduce their consumption of whole milk, meat, eggs, butterfat, and foods high in sugar, salt, and fat. Specifically, it asked consumers (1) to increase carbohydrate consumption to account for 55–60% of caloric intake; (2) to reduce overall fat consumption from approximately 40% to 30% of energy intake; (3) to reduce saturated fat consumption to account for about 10% of total energy intake and balance that reduction by consuming polyunsaturated and monounsaturated fats, which should account for about 10% of energy intake; (4) to reduce cholesterol consumption to about 300 mg a day; (5) to reduce sugar consumption by about 40% to account for about 15% of total energy intake, and (6) to reduce salt consumption by about 50–85% to approximately 3 g a day.

In December 1977, a second edition of the report on dietary goals appeared. The new report recommended that people reduce their intake of animal fat and choose meats, poultry, and fish that would reduce saturated fat intake. The new edition did not contain advice about reducing the whole milk and egg consumption by young children. For adults, the new edition raised the suggested limits for salt to 5 g. A release by the U.S. Department of Agriculture and the Department of Health, Education, and Welfare entitled "Dietary Guidelines for Americans" largely follows the recommendations published by the U.S. Senate Select Committee, including the suggestion that Americans increase their intake of complex carbohydrates present in cereal grains and their products.

Similar plans and targets were developed for other countries, including Norway, Sweden, Canada, and the United Kingdom. Norwegians were encouraged to adopt a diet composed of more vegetables, cereals, fish, and poultry and less red meat, saturated fats, and sugar. The French Health Ministry is urging citizens to cut down on fatty meat, pastries, and alcoholic beverages and is promoting milk and starchy foods, particularly bread. Similar efforts are under way in Canada and the United Kingdom (Chou, 1979).

Obesity

Consumption of bread has been traditionally linked with obesity. It is well established that the amount of food consumed and its caloric content, rather than food consumption, are responsible for most cases of overweight. According to Mickelsen (1975), contrary to what most people think, bread in large amounts is an ideal food in a weight-reducing regimen.

Atherosclerosis

Atherosclerosis and coronary heart disease are the primary fatal diseases of man in the United States. Atherosclerosis is derived from the Greek *athero*, meaning mushy, and *sclerosis*, meaning hardening. The word was coined to describe the progression of changes in the human aorta, starting with deposition of fat and proceeding to calcification. The disease is not peculiar to our civilization, having been found in Egyptian mummies dating from 1500 BC. Atherosclerosis is frequently accompanied by abnormally high blood beta-lipoproteins and associated cholesterol. Saturated fats and cholesterol have been suggested as possible contributors to these degenerative diseases; unsaturated fats seem to have a protective effect. Wheat flour lipids are highly unsaturated, but the shortening employed in baking contains more saturated fats. High cereal intakes are generally considered to have a protective effect. Unlike sugar, starch does not increase the cholesterol level in blood. Extensive investigations have shown that incidence of atherosclerosis and coronary heart disease was markedly reduced in cases where bread supplied up to 80% of the caloric intake.

Constipation

Constipation is largely due to nervous tension or lack of crude fiber in the food. Indigestible crude fiber enhances secretion and peristaltic reflex of the intestine. Consumption of dark bread that has a relatively high fiber content decreases the incidence of constipation. A large intake of whole meal bread, on the other hand, may lead to diarrhea.

Fiber in Bread Making

Epidemiological observations that several diseases of "civilization" such as coronary heart disease, diabetes, and some colon diseases are more prevalent in Western countries have heightened interest in the inclusion in our diets of nutritive fiber that resists human digestive secretions and intestinal flora. Burkitt's report (1971) suggested that the incidence of colon cancer was associated with low levels of dietary fiber. If low fiber content of diets causes such diseases, they might be reduced by the addition of fiber to diets. The importance of dietary fiber in food was reviewed by Inglett and Falkenhag (1979), James and Theander (1981), Spiller and Amen (1977), Spiller and Kay (1980), and Wisker et al (1985).

Insofar as effects on functional properties in bread making are concerned, three approaches have been used: (1) production of fiber-enriched bread that resembles conventional bread as much as possible; (2) production of bread that differs substantially in overall characteristics (loaf volume, crumb grain, color, etc) from conventional bread and that should be considered a specialty product; and (3) production of fiber-enriched extrusion types of products.

Bread was baked from wheat flour with up to 15% of the flour replaced by seven celluloses, four wheat brans, or two oat hulls (Pomeranz et al, 1977). Adding 15% oat hulls somewhat reduced water absorption; bran increased absorption about 4%; celluloses increased absorption about 10%. Oat hulls increased mixing times little, celluloses increased them considerably, and wheat bran had no consistent effect. Adding up to 5% fiber materials decreased loaf volume to an extent expected from dilution of functional proteins. At levels above 7%, fiber materials decreased loaf volume much more than expected from dilution of gluten. The large decrease resulted from lowered gas retention rather than gas formation. Effects of fiber materials on bread crumb texture were confirmed by visual observations, light microscopy, and scanning electron microscopy. Oat hulls imparted to bread an objectionable gritty texture; the celluloses modified bread taste and mouth feel little; bran modified the taste and mouth feel somewhat, but the modification was not objectionable. Overall effects on color from added fiber materials were smallest for the celluloses and largest for bran. Bran decreased bread softness more than celluloses did; oat hulls softened bread somewhat.

To counteract the deleterious effects of fiber in bread making, blends of wheat flour and wheat, corn, or soybean bran or coconut residue were baked into bread. The deleterious effects of up to 15 parts of wheat bran per 85 parts of wheat flour could be largely counteracted with the addition of vital gluten and one of several surfactants alone or in combination with shortening. Surfactants included diacetyl tartaric acid esters (DAT), ethoxylated monoglycerides (EMG), lecithin, sucrose monopalmitate (SMP), and sodium stearoyl-2-lactylate (SSL) (Shogren et al, 1981). In the absence of shortening, 1 g each of DAT, EMG, or SMP per 100 g of mixture materially increased loaf volume beyond the level produced by 3 g shortening. In the presence of 3 g shortening per 100 g of mixture, adding 0.5 g of DAT, EMG, lecithin, SMP, or SSL produced only small improvements above the levels produced when shortening alone was added.

Allergies

The old proverb that "one person's food is another person's poison" is most appropriate with allergies. Certain protein-rich foods contain nutrients that may cause disease or even threaten life for some individuals. Such disorders are of an immunological nature. They are manifested by various clinical forms, the most important of which are skin, respiratory, and gastrointestinal diseases. Up to 10% of people in the United States show symptoms of food allergy; about one- third of the 10% are allergic to wheat flour but normally not to bread because of modifications in the proteins during baking. Identifying the allergies is rather difficult and even after a painstaking and tedious procedure is not always clear-cut. Consequently, statistics regarding allergic persons must be considered with caution. Baldo and Wrigley (1984) described the general categories of cereal allergies as those caused by inhalation or ingestion. Inhalation may involve pollinosis (caused by pollen), bakers' asthma (from flour or bran), and grain fever (in grain handlers from bran or straw). Ingestion allergies include food allergies, celiac condition, and food intolerances.

Celiac Condition

Faulty absorption from the small intestine may be due to defective digestion or incompetence of the intestine. If the incompetence results from intolerance to wheat gluten, the disease is called celiac condition. The symptoms of this gluten-induced condition are increased fecal fat and protein, damaged internal mucosa, folic acid deficiency, inhibition of the peristaltic reflex, loss of appetite, and eventually "gliadin" shock. The gliadin fractions of gluten is responsible for the disease. The gliadin fractions of wheat and rye are most deleterious; other cereals and dietary proteins have less or little effect. Acid hydrolysis or digestion with crude papain, but not crystalline papain, removes the toxic effect.

The nature of malabsorption in celiac has not been fully established. Frazer (1962) favors the theory that the deleterious effect is related to faulty proteolysis. According to Berger and Freudenberg (1961), the effects of gliadin are immunological. That theory seems to be supported by subjects with celiac producing increased amounts of antibodies to food proteins, and especially to gliadin. Biserte (1967) suggested that the celiac infant shows no deficiency of gastric or pancreatic enzymes but has no demonstrable intestinal mucosal enzymes. Consequently, in the breakdown of gliadin, proline-glutamine peptides that the infant cannot utilize are formed.

Cariogenicity

Dental decay is the most prevalent of chronic diseases. It is recognized that dental caries occurs by the interaction of three factors: the susceptible host, oral bacteria, and diet. Inherent differences in caries susceptibility in human beings range from high caries susceptibility to resistance. Differences may depend on hereditary characteristics or may result from nutritional influences during tooth development.

According to Lorenz (1984), human studies have not consistently implicated cereals and cereal products as bein cariogenic, with the exception of high-sugar cakes and cookies. White bread causes little dental caries, and consumption of whole wheat bread or inclusion of cereal bran in baked products has not been shown to have consistent anticariogenic effect. Several factors, in addition to the sugar contents, determine the cariogenicity of cereals. They include mineral content, amino acid composition, phytate content, degree of oral retention, pH buffering capacity, and acidogenic potential.

Adventitious Toxic Factors

Sound wheat and baked wheat products contain no significant heat-labile antinutritional factors or toxicants naturally occurring in many foods. Associated with cereal grains is phytic acid (inositol hexaphosphate), which forms insoluble compounds with calcium and iron and which makes the two minerals largely unavailable. In diet limited in dairy products, high levels of phytic acid (such as in high-extraction flour) can be conducive to rickets. A major part of phytic acid in dough is likely to be hydrolyzed to phosphoric acid and inositol by the enzyme phytase.

Wheat can be contaminated during cultivation and processing by adventitious toxic contaminants. Wheat contaminated with certain noxious seeds may cause bread poisoning. Cereal grains may be attacked by a fungus *Claviceps purpurea* (Fr.) Tul., which results in ergoty grain that is toxic at 1% in bread. Much research has centered around aflatoxins elaborated by the mold *Aspergillus flavus* Lk. ex Fr. The fungus grows under certain conditions also on wheat. *A. flavus* is not the only mold to cause concern. Toxin-producing fungi are numerous and widespread; they proliferate on a number of foods and are able to grow under a wide range of conditions.

Certain carcinogenic substances may increase three- to tenfold while grain is being dried with combustion gases. Grain grown near industrial zones contained four times more of such compounds than grain grown far from industrial areas.

Bleaching flour destroys 50–90% of its vitamin E content. A similar decrease is observed when flour is stored 6 months. Since vitamin E deficiency seems to be common in some countries, vitamin E losses may be nutritionally significant. Oxidation of essential fatty acids by bleaching agents is not significant.

Mental and Physical Capacity

Experiments conducted in schools and factories have shown that the rapid fall in performance during the later part of the morning can be prevented by a small meal of bread and fruit. Glatzel (1966) studied the effects of various types of bread and crumb and crust on blood sugar contents and on hunger sensation. Crust improved performance better than crumb, and dark rye bread was better than white bread or buns. The reason for the superiority of crust over crumb in preserving an adequate sugar level in blood for extended periods is not known. In practice, crust-rich bread is recommended by physicians both to athletes before a major physical effort and to improve capacity of mental workers.

Criteria of Nutritional Quality

The nutritional quality of our bread is one of the most important factors in deciding the nutritional adequacy of our diet. Nutritional quality is generally computed in terms of meeting needs of growth, reproduction, and maintenance of adequate health. For many years, decreased infant mortality and the taller statures of our young were sources of pride, traceable in part to improved nutrition. Improved nutrition has contributed to improved mental capacity and earlier adolescence, so it is associated with socioeconomic problems. However, advantages of additional increase in height by future generations are highly questionable. Admittedly, vitamin supplementation has been—and is—highly important. Yet it is possible that excessive levels might enhance metabolic rates and lead to premature aging.

Lengthening the life-span of our older citizens is an important medical and scientific achievement. Yet physical survival alone is not enough; it must not be accompanied by too prolonged emasculation or by impaired physical and mental capacity. Providing a properly balanced diet for best overall performance at all ages is a challenging task that requires a coordinated effort of food manufacturers, nutritionists, biochemists, physiologists, and sociologists around the world.

BIBLIOGRAPHY

Auerman, L.J., et al. *Biokhim Zerna* **1954**, *2*, 193–201.

Baldo, B.A.; Wrigley, C.W. *Adv. Cereal Sci. Technol.* **1984**, *6*, 289–356.

Berger, E.; Freundenberg, S. *Ann. Paediatr.* **1961**, *196*, 234.

Biserte, G. *Ann. Biol. Anim. Biochim. Biophys.* **1967**, *3*, 121.

Bradley, W.B. Proc. U.S. Dept. Agric. Conf., Role of Wheat in World's Food Supply, Albany, California; **1962**.

Bradley, W.B. *Bakers Dig.* **1967**, *41*, 66–67, 70–71.

Burkitt, D.P. Cancer **1971**, *28*, 3–13.

Chou, M. "Critical Food Issues of the Eighties." Chou, M.; Harmon, D.P., Ed; Pergamon Press; New York; **1979**.

De Ruiter, D. *Adv. Cereal Sci. Technol.* **1978**, 2, 349–385.

Frazer, A.C. *Adv. Clin. Chem.* **1962**, *5*, 69.

Glatzel, H. *Ernahr. Umschau* **1966**, *13*, 169.

Gorbach, G.; Regula, E. *Fette Seifen Anstrichmittel* **1964**, *66*, 920–925.

Hegsted, D.M. Proc. U.S. Dept Agric. Conf., Role of Wheat in World's Food Supply, Albany, California, **1962**.

Inglett, G.E.; Falkenhag, S.I. (Ed.). "Dietary Fiber: Chemistry and Nutrition." Academic Press; New York; **1979**.

James, W.P.T.; Theander, O. (Ed.). "Analysis of Dietary Fiber in Food." Marcel Dekker; New York; **1981**.

Jansen, G.R. *Am. J. Clin. Nutr.* **1969**, *22*, 38–43.

Jongh, G. *Cereal Chem.* **1961**, *38*. 140–152.

Kennedy, B.M.; Joslyn, M.A. *Bakers Dig.* **1966**, *40*, 60–62, 64,87.

Lorenz, K. *Adv. Cereal Sci. Technol.* **1984**, *6*, 83–137.

Maleki, M.; Daghir, S. *Cereal Chem.* **1967**, *44*, 483–487.

Mickelsen, O. *Cereal Foods World* **1975**, *20*, 308–310.

Moran, T.; Pace, J. *Milling* **1967**, Aug. *4*, 90–92,96.

Osborne, T.B.; Mendell, L.B. *J. Biol. Chem.* **1914**, *17*, 325–349.

Platt, B.S.; Miller, D.S.; Payne, P.R. In "Recent Advances in Human Nutrition." Brock, J.F., (Ed.). J. & A. Churchill; London; **1961**.

Pomeranz, Y. *Adv. Cereal Sci. Technol.* **1980**, *3*, 1–40.

Pomeranz, Y.; Shogren, M.D.; Finney, K.F. *Cereal Chem.* **1969a**, *46*, 503–511.

Pomeranz, Y.; Shogren, M.D.; Finney, K.F. *Cereal Chem.* **1969b**, *46*, 512–518.

Pomeranz, Y.; Shogren, M.D.; Finney, K.F.; and Bechtel, D.B. *Cereal Chem.* **1977**, *54*, 25–41.

Rotsch, A. *Brot Gebaeck* **1954**, *8*, 129–130.

Shogren, M.D.; Pomeranz, Y.; Finney, K.F. *Cereal Chem.* **1981**, *58*, 142–144.

Spiller, G.A.; Amen, R.J. (Ed.). "Fiber in Human Nutrition." Plenum Press; New York; **1977**.

Spiller, G.A.; Kay, R.M. (Ed.). "Medical Aspects of Dietary Fiber." Plenum Medical Book Co.; New York; **1980**.

Thomas, B. *Qualitas Plant. Mater. Vegetabiles* **1968**, *15*, 350–371.

Wisker, E.; Feldheim, W.; Pomeranz, Y.; Meuser, F. *Adv. Cereal Sci. Technol..* **1985**, *7*, 169–238.

Chapter 15

Soft Wheat Products

Soft wheat flours (generally low extraction, low protein) are uniquely suited for the production of cookies (biscuits), most cakes, wafers, cake doughnuts, and similar baked products (Yamazaki and Greenwood, 1981).

COOKIES AND CRACKERS

A high ratio of spread to thickness (W/T) is used as criterion of adequacy of cookie flour. A low-hydration flour permits more syrup formation in a flour-sugar dough and makes the dough more slack at elevated temperatures. This in turn enables the leavening to expand the dough to a greater extent before it sets (Yamazaki and Lord, 1971). According to Tamili (1976), whereas weak, low-protein soft wheat flours have a W/T ratio of 8.5–10.0, strong, high-protein, hard wheat flours have a ratio of 6.5–8.0. The main types of cookies, depending on their production methods, are wire cut, rotary, and deposit.

A comparison of changes in properties of cookie dough recipes as affected by the ratio of fat to sugar is given in Table 15-1. The relationship of sugar and fat enrichment in cookie (biscuit) recipes is shown in Figure 15-1. The ratios of fat, sugar, and water in biscuit doughs are given in Figure 15-2. Note that in Figure 15-1 the sugar and fat are calculated per 100 parts of flour (bakers' percentages); in Figure 15-2 the totals of water, fat, and sugar are adjusted to 100 and their ratios are plotted independent of flour. The 20°C and 40°C lines indicate saturated levels at the two temperatures. Dough recipes below those lines contain some crystalline sugar prior to the temperature rise in the oven.

CAKES

Special cakes are prepared from batters rather than doughs. The distinguishing differences between the two are summarized in Table 15-2. High-quality cakes should have a large volume, fine grain, and a moist,

Table 15-1. Changes in Properties of Cookie Dough Recipes as Affected by
Fat:Sugar Ratio

| | Crackers | Semisweet | Short | | Soft |
			High fat	High sugar	
Moisture in dough	30%	22%	9%	15%	11%
Moisture in biscuit	1–2%	1–2%	2–3%	2–3%	3 + %
Temperature of dough	30–38°C	40–42°C	20°C	21°C	21°C
Critical ingredients	Flour	Flour	Fat	Fat and sugar	Fat and sugar
Baking time	3 min	5.5 min	15–25 min	7 min	12 + min
Oven band type	Wire	Wire	Steel	Steel	Steel

Source: Manley, 1983.

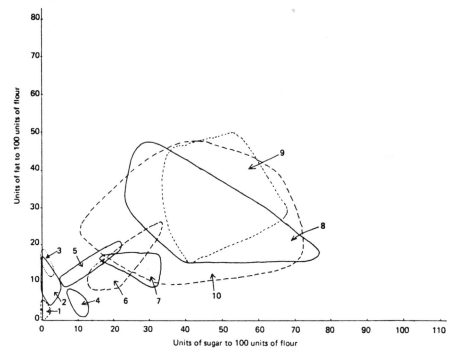

Figure 15-1. Relationship of sugar and fat enrichment in biscuit recipes. By areas:
1, bread, pizza, and crispbread; 2, water biscuits and soda crackers; 3, cream
crackers; 4, cabin biscuits; 5, savory crackers; 6, semisweet/hard–sweet; 7, "Continental" semisweet; 8, short doughs (molded); 9, wire-cut types; 10, short dough
(sheeted). (From Manley, 1983.)

tender crumb. The flour used for their production is milled from low-
protein soft wheats. Chlorine treatment makes it possible to produce
sponge cakes with evener crumb texture, increased volume, and greater
symmetry (Gough et al, 1978). Chlorinated flours are used in layer,
genoese, yellow, madeira, and fruit cakes made with greater proportions of
sugar and liquor (so-called "high-ratio" cakes). Ingredients used in layer

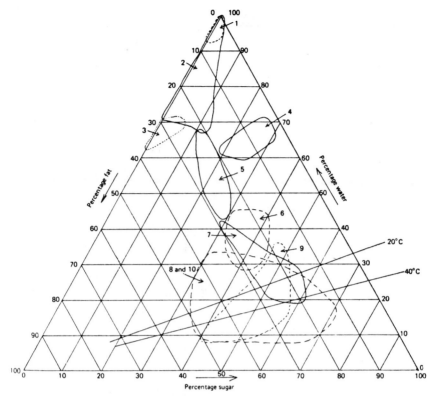

Figure 15-2. Ratios of fat, sugar, and water in biscuit doughs. By areas: 1, bread, pizza, and crispbread; 2, water biscuits and soda crackers; 3, cream crackers; 4, cabin biscuits; 5, savory crackers; 6, semisweet/hard–sweet; 7, "Continental" semisweet; 8, short doughs (molded); 9, wirecut types; 10, short dough (sheeted). (From Manley, 1983.)

Table 15-2. Distinguishing Characteristics of Doughs and Batters

Characteristics	Dough	Batter
Basic ingredients in recipe	Flour, sugar, fat	Sugar, eggs, fat, flour
Processing	Kneading, mixing	Beating, stirring, mixing, short heating
Raising	Biological, chemical, physical	Chemical, physical
Factors affecting binding of water or consistency	Wheat gluten, pentosans, damaged starch, swelling agents	Eggs, fat, sugar, damaged starch, and in part wheat gluten and swelling agents
Consistency	Elastic to plastic	Foamy, soft-plastic, pastelike to semifluid

Source: Menger and Bretschneider, 1972.

Table 15-3. Ingredients Used in Layer Cakes

Ingredients	Source	Essential	Non-essential
Proteins (soluble and insoluble)	Flour	+	
	Egg	+	
	Milk	+	
Carbohydrates (soluble and insoluble)	Sucrose		+
	Dextrose		+
	Lactose		+
	Starch (from flour)	+	
	Thickeners		+
Lipids	Shortening		+
	Emulsifiers	+	
	Flour lipids		+
	Milk lipids		+
	Egg yolk lipids	+ (in yellow cakes)	
Other	Sodium chloride		+
	Leavening acids	+	
	Sodium bicarbonate	+	
	Flavor		+
	Color		+
	Water	+	

Source: Howard, 1972.

cakes are listed in Table 15-3; development of cakes (yellow, white, and devil's food) is described in Table 15.4. Small cake items were described by Ash (1979).

DOUGHNUTS

Production of cake doughnuts is probably one of the most critical processes in the wholesale bakery operation (Dixon, 1983). The cake doughnut is the only item produced in the bakery that does not have some mechanical means of forming the dough piece into the desired product. The fluid-viscous batter is deposited into the frying fat. We depend on the flow and characteristics of this batter in the fluid medium to flow, fry, and set in the desired size and shape. Consistency and uniformity of cake doughnut mixes and bases are primary objectives of manufacturers (Gorton, 1984). Careful maintenance of equipment is vital to making quality doughnuts (Anon., 1984).

In general, a bakery mix contains 55–65% flour (mix weight basis) of 9–10% protein to yield the proper tenderness-structuring profile (Gorton, 1984). Sugar in the range of 22–30% is added to sweeten, tenderize, aid moisture retention, accelerate crust formation, and effect spreading in the

Table 15-4. Cake Development and Formula Balance

Steps	Yellow		White		Devil's food	
	Plastic mono-glyceride short-ening	Fluid cake short-ening	Plastic mono-glyceride short-ening	Fluid cake short-ening	Plastic mono-glyceride short-ening	Fluid cake short-ening
1. Flour: Assume a basic level	100	100	100	100	100	100
2. Shortening: Assume a level to give desired tenderness and shelf life	40–45	35–40	40–45	35–40	50–55	45–50
3. Sugar: Assume a level to meet tenderness/flavor requirements	125	125	125	125	140	140
4. Cocoa: Assume a level to meet color/flavor requirements	—	—	—	—	20	20
5. Leavening (good average figures)						
a. Baking powder level (oz/lb of flour)	1	¾	1	¾	¾	¾
b. Soda (oz/lb of flour)	—	—	—	—	¼	¼
6. Calculate a liquid level						
a. For mono shortening—1.4 × flour for yellow and white; 1.8 × flour for devil's food	140		140		180	
b. For fluid shortening—1.5 × flour for yellow and white; 1.9 × flour for devil's food		150		150		190
c. Distribute the liquids—						
⅔ water	95	100	95	100	120	125
⅓ eggs	45	50	45	50	60	65
7. Salt and flavors	To taste		To taste		To taste	

Production Note: If a trial bake with a newly balanced formula produces cakes that dip slightly, the liquid level may be a little low.

Source: Lawson, 1970.

fryer. Some shortening, 3–9%, is also included to aid tenderness, increase shelf life, and lubricate the protein structure for proper flour performance. Dried egg yolks (0.5–3.0%) provide richness and tenderness, and 3–5% nonfat milk solids act as a binder and structure builder and contribute to crust color, shelf life, gas retention, and "crowning." Leaveners (1.75–3%) are usually blends of fast-acting sodium pyrophosphate and slower-acting sodium aluminum phosphate, monocalcium phosphate, and sodium bicarbonate. Sometimes, glucono delta lactone may be used (Gorton, 1984).

Optional ingredients include 1–3% soy flour to bind water, control fat absorption, improve gas retention, and control crust coloring and volume; 0.75–1.5% salt to enhance the flavor profile; up to 2% potato starch to aid water absorption, reduce staling and coating breakdown, and reduce fat absorption during frying; up to 0.5% lecithin to control fat absorption, batter flow, and symmetry; surface active agents to improve shelf life and eating quality; and 0.10–0.25% gums to reduce fat absorption and aid moisture retention. After mixing, the cake batter is allowed to rest, cut to

Table 15-5. Characteristics of Leavening Acids during Cake Production

| Leavening acid | Neutral-izing value | CO$_2$ released (%) | | | Uses |
		2 min after mixing	Bench action	During baking	
Monocalcium phosphate monohydrate	80	60	0	40	Commercial cakes, baking powders
Monocalcium phosphate anhydrous	83	15	35	50	—
Sodium acid pyrophosphate	72	28	8	64	Doughnuts, refrigerated doughs
Sodium aluminum phosphate (SALP)	100	22	9	69	Cakes, breading and batter mixes
Sodium aluminum sulfate	100	0	0	100	Pancakes, waffle mixes, muffins
Dicalcium phosphate dihydrate (DCP)	33	0	0	100	Only in mixed formulas
Potassium acid tartarate	50	70	Trace	30	—
Gluconodeltalactone	45	25	40	35	Cake doughnuts, "instant" bread mixes

Adapted from La Baw, 1982.

size and weight, and fried. According to Dixon (1983), the temperature of the water added to make the mix regulates the temperature of the batter and in turn the rate of hydration of the mix, batter viscosity, and rate of leavening. The floor time is needed to hydrate the ingredients and allow initial leavening retention; it is important for the production of a doughnut with uniform symmetry. Overlap on the cutter regulates the size and shape of the doughnut. Volume, spread, crust color, and fat absorption of the doughnut are affected by the frying temperature and frying time. Shape and fat absorption are affected by fat levels. Typically, cake doughnuts contain 20–25% moisture and 20–25% fat, of which 80–85% is absorbed frying fat.

LEAVENING AGENTS

Soft wheat products are generally leavened (made light by aeration) through the use of chemical leavening agents (Conn, 1981; La Baw, 1982). In the first chemical leavens, carbon dioxide was released from sodium bicarbonate by the action of an organic acid (generally sour milk or buttermilk containing small amounts of lactic acid). The first record of a complete chemical leavening dates back to about 1835, when cream of tartar (potassium tartarate) from wine making was combined with bicarbonate of soda.

The leavening of cake batters occurs in two steps. In the mixing step, air

Table 15-6. Typical Baking Powder Compositions (%)

Ingredients	Household baking powders			Commercial baking powders							
	Single-acting			Double-acting							
Soda, granular	28	28	27	30	30	30	30	30	30	30	30
Monocalcium phosphate monohydrate	35			8.7	12	5.0	5.0		5.0	5.0	10
Monocalcium phosphate anhydrous		34									
Corn starch (redried)	37	38	20	26.6	37.0	19	24.5	26	27		38
Sodium aluminum sulfate				21.0	21.0	26					
Sodium acid pyrophosphate							38	44	38		
Sodium aluminum phosphate										38	22
Cream of tartar			47								
Tartaric acid			6								
Calcium sulfate				13.7		20					
Calcium carbonate											
Calcium lactate							2.5				

Source: La Baw, 1982.

bubbles are incorporated in the liquid batter. In the subsequent step, carbon dioxide from the reaction of the leavening components produces additional bubbles that are finely dispersed and retained in the batter (generally containing suitable emulsifiers). The system is then "set" by heat in the baking stage.

The action of the commercial leavening agents depends on their particle size but also on their composition and associated acidity-neutralizing power (Table 15-5). The amount of acid leavener is calculated from the ratio of soda ($\times 100$) divided by the neutralizing value of the leavener. Thus 100 kg of SALP but 303 kg DCP is required to neutralize 100 kg $NaHCO_3$.

The timing of carbon dioxide release is critical for various types of products. Double-acting preparations can be obtained by combinations of powders to achieve CO_2 release at the mixing-bench stage and during the baking stage. Typical baking powder compositions are listed in Table 15-6. The reactions between sodium bicarbonate and an acid phosphate salt, monocalcium phosphate, sodium acid pyrophosphate (SAPP), and acidic sodium aluminum phosphate (SALP) are given in equations I, II, III, and IV, respectively (from Reiman 1977):

$NaHCO_3 + H^+ \rightarrow Na^+ + CO_2 + H_2O$	Equation I
$3Ca(H_2PO_4)_2 + 8NaHCO_3 \rightarrow$	
$\qquad Ca_3(PO_4)_2 + 4Na_2HPO_4 + 8CO_2 + 8H_2O$	Equation II
$Na_2H_2P_2O_7 + 2NaHCO_3 \rightarrow Na_4P_2O_7 + 2CO_2 + 2H_2O$	Equation III
$NaAl(SO_4)_2 + 3NaHCO_3 \rightarrow 3CO_2 + Al(OH)_3 + 2Na_2SO_4$	Equation IV

BIBLIOGRAPHY

Anon. Careful maintenance of equipment is vital to making quality donuts. *Milling Baking News* **1984**, *63*(6), BE-38.

Ash, D.J. *Bakers Digest* **1979**, *53*(4), 30–33, 39.

Conn, J.F. *Cereal Foods World* **1981** 26, 119–123.

Dixon, J. *Bakers Digest* **1983**, *57*(5), 26, 31, 34, 36.

Gorton, L. *Bakers Digest* **1984**, *58*(2), 8, 11.

Gough, M.M.; Whitehouse, M.E.; Greenwood, C.T. *CRC Crit. Rev. Food Sci. Nutr.* **1978**, *12*, 91–113.

Howard, H.B. *Bakers Digest* **1972**, *46*(5), 28–30, 32, 34, 36, 37, 64.

La Baw, G.D. *Bakers Digest* **1982**, *56*(1), 16–18, 20, 21.

Lawson, H.W. *Bakers Digest* **1970**, *44*(6), 36–39, 66.

Manley, D.J.R. "Technology of Biscuits, Crackers, and Cookies." Ellis Horwood, Ltd.: Chichester, England; 1983.

Menger, A.; Bretschneider, F. *Getreide Mehl Brot* **1972**, *26*, 120.

Reiman, H.M. *Bakers Digest* **1977**, *51*(4), 36, 42.

Tamili, V.H. *Cereal Foods World* **1976**, *21*, 624, 625, 627, 628, 644.

Yamazaki, W.T.; Lord, D.D. "Wheat Chemistry and Technology." Pomeranz, Y., Ed. Am. Assoc. Cereal Chem.; St. Paul, Mn; **1971**, pp. 749–750.

Yamazaki, W.T.; Greenwood, C.T. (Ed.). "Soft Wheat, Production, Breeding, Milling and Uses." Am. Assoc. Cereal Chem.; St. Paul, MN; **1981**.

Chapter 16

Rice

Rice is a covered cereal. In the threshed grain (or rough rice), the kernel is enclosed in a tough siliceous hull, which renders it unsuitable for human consumption (Bechtel and Pomeranz 1977, 1978 a, b, 1980). When this hull is removed, the kernel or caryopsis, comprising the pericarp (outer bran) and the seed proper (inner bran, endosperm, and germ), is known as brown rice. Brown rice is little in demand as a food. It tends to become rancid and is subject to insect infestation. When brown rice is subjected to further milling processes, the bran, aleurone layer, and germ are removed, and the purified endosperms are marketed as white rice or polished rice, which is classified according to size as head rice (at least three-fourths of the whole endosperm) and various classes of broken rice, known as second-hand, screenings, and brewers' rice, in decreasing size. (Figure 16-1).

MILLING

The objective of rice milling is to remove the hull, bran, and germ with minimum breakage of the starchy endosperm (White, 1970). The rough rice, or paddy, is cleaned and conveyed to shelling machines that loosen the hulls. Conventional shellers consist of two steel plates, 4×5 feet in diameter, mounted horizontally. The inner surfaces are coated with a mixture of cement and carborundum. One plate is stationary and the other is rotated. As the plate revolves, the pressure on the ends of the upturned grains disengages the hulls. The hulls are removed by aspiration, and the remaining hulled and unhulled grains are separated in a paddy machine that consists of a large box shaker fitted with vertical, smooth steel plates set on a slight incline to form zigzag ducts. The plates and the shaking action cause the less dense paddy grains to move upward while the heavier hulled grains move downward. Rough rice may also be shelled with rubber rolls or with a rubber belt operating against a ribbed steel roll. The process causes less mechanical damage and improves stability against rancidity. Hulled rice is sent to machines that consist of grooved tapering cylinders that revolve rapidly in stationary, uniformly perforated cylinders. The entire

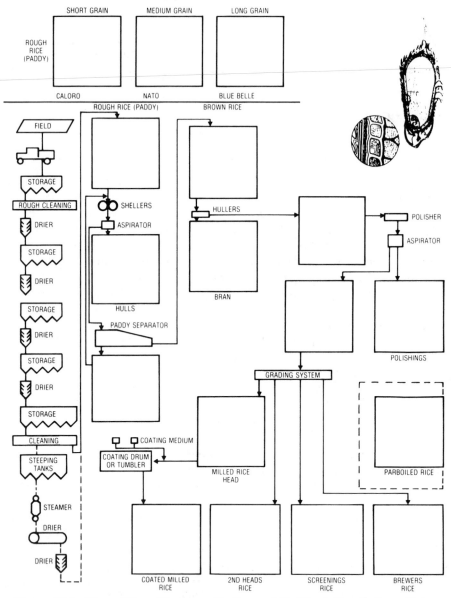

Figure 16-1. Rice milling. (Courtesy, Professor A.B. Ward, Kansas State University.)

machine is filled with grain, and the packing force is regulated by a blade that protrudes between the upper and lower halves of the perforated cylinder. The outside bran layers and the germ are removed by the scouring action of the rice grains moving against themselves near the surface or the perforated cylinder. After passing through a succession of

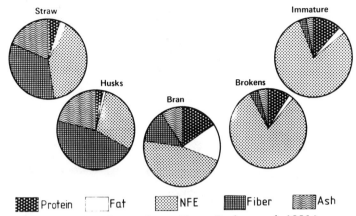

Figure 16-2. Rice milling by-products. (From Barber et al, 1981.)

hullers, the rice is practically free from germ and outer bran. Scouring is usually completed by polishing in a brush machine. The polished rice contains whole endosperms and broken particles of various sizes. They are separated by grading reels of disc separators.

The yield of white rice normally varies between 66% and 70%, based on the weight of rough rice. As head rice is the most valuable product, its yield determines the milling quality of rough rice. The price obtained for the various classes of broken rice decreases with size.

A solvent-extraction process was developed to increase the yield of whole grain rice. Dehulled brown rice is softened with rice oil, to improve bran removal. Fully milled rice is sometimes treated with a talc-and-glucose solution to improve its appearance. After the coating is evenly distributed on the kernels and dried with warm air, the rice emerges from the equipment with a smooth, glistening luster and is known as coated rice.

The amounts of rice milling by-products and their chemical composition are illustrated in Figures 16-2 and 16-3. The annual production of bran has a potential for 5 million tons of food protein and 6 million tons of edible oil; the husks, for 256,000 billion kcal as fuel; and the straw, for 30,000 billion kcal as metabolizable energy for cattle.

CHEMICAL COMPOSITION

Starch, the major component of rice, is present in the starchy endosperm as compound granules that are 3–10 μm in size. Protein, the second major component, is present in the endosperm in the form of discrete protein bodies that are 1–4 μm in size. The concentration of nonstarchy carbohydrates is higher in the bran and germ fractions than in the starchy endosperm. Brown rice contains about 8% protein, 75% carbohydrates, and small amounts of fat, fiber, and ash. After milling, the protein content

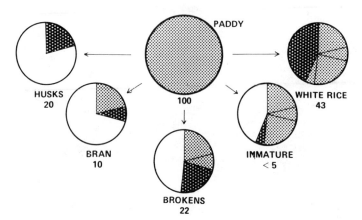

Figure 16-3. Gross chemical composition of rice by-products. (From Barber et al, 1981.)

of rice is about 7% and the carbohydrate content (mainly starch) about 78%. Starch is found primarily in the endosperm; fat, fiber, minerals, and vitamins are concentrated in the aleurone layers and in the germ (Houston, 1972; Luh, 1980; Bechtel and Pomeranz, 1980).

Starch, the main carbohydrate of rice, comprises up to 90% of the rice solids. In common rice, amylose amounts ot 12–35% of the total starch; in waxy (glutinous) rice, the amylose content is much lower. *Indica* rices generally contain more amylose than *japonica* rices. Brown rice contains 0.6–1.4% and milled rice contains 0.3–0.5% sugars (mainly sucrose) (Houston and Kohler, 1970).

Protein composition of milled rice is unique among cereals. The rice proteins are rich (at least 80%) in glutelins and have a relatively good amino acid balance. Among the protein fractions, albumin has the highest lysine content, followed by glutelin, globulin, and prolamin. The high lysine content of rice protein is primarily due to their low prolamin content. Proteins in milled rice are generally lower in lysine than proteins in brown rice. The proportions of albumin and globulin and the total protein are highest in the outer layers of the milled rice kernel and decrease toward the center; proportions of glutelin have an inverse distribution. In rice, as in other cereal grains, the proteins differ considerably in their amino acid composition and biological value. The most notable differences are in the high concentration of lysine in albumins and of cystine in globulins, and in the very low lysine and cystine concentrations in the prolamines. Rice protein is not ideally balanced; it is relatively low in lysine concentration when compared with the FAO Reference Pattern; supplementation with lysine and threonine significantly increases the biological value of rice protein.

The subaleurone region, which is rich in protein, is only several cell layers thick, lies directly beneath the aleurone, and is removed rather easily

Table 16-1. Levels of Essential Amino Acids (g/16.8 g N) in Parts of IR 8 Rice

Amino acid	Hull	Brown rice	Milled rice	Pericarp	Aleurone	Embryo	LSD (5%)
Isoleucine	4.2	4.5	4.7	4.5	4.3	3.8	0.2
Leucine	8.6	8.3	8.5	8.2	7.8	6.8	0.2
Lysine	5.7	4.4	4.0	5.7	5.1	6.8	0.5
Methionine + cystine	0.9	3.6	3.5	3.0	3.2	2.9	0.6
Phenylalanine + tyrosine	6.9	9.6	9.8	8.4	8.6	7.4	0.7
Threonine	5.3	3.9	3.9	4.6	4.0	4.5	0.2
Tryptophan	0.8	1.2	1.3	1.0	1.3	1.4	0.09
Valine	7.8	6.6	6.8	6.9	6.3	6.3	0.4
Amino acid score (%)	24[a]	76[b]	69[b]	76[a]	87[a]	79[a]	—
Protein (N × 5.95) (%)	1.5	7.8	7.2	16.0	15.8	25.3	—
Relative distribution of weight	(26.2)	100	92.2	2.2	3.4	2.3	—
Relative distribution of protein	(4.9)	100	82.0	4.3	6.5	7.2	—

[a]100% = 3.76 g met + cys/16.8 g N.
[b]100% = 5.78 g lysine/16.8 g N.
Source: IRRI Annual Rep., 1974, p. 102.

during milling. From a nutritional standpoint, it is therefore desirable to mill rice as lightly as possible and retain some of the protein in the subaleurone or to breed cultivars that have either an increased number of aleurone layers or have the protein more evenly distributed throughout the endosperm. Protein content of the grain determines the protein distribution between bran polish and milled rice. Protein distribution is more uniform throughout the grain as the grain increases in total protein content. Also, high-protein milled rices usually have more thiamine. The increase in protein content is related mainly to an increase in the number of protein bodies and a slight increase in their size.

Aminograms of grain parts were determined in a study of changes in amino acids in protein fractions of developing rice (Table 16-1). Lysine was the first limiting amino acid in brown rice and milled rice protein. In other fractions, particularly the hull, sulfur amino acids were limiting. The aminogram of the milling fractions was consistent with the lowered lysine content of the milled rice protein because the protein of the bran fractions is richer in lysine. Glutamic acid was high in proteins of the milling fractions that were low in lysine.

Brown rice contains 2.4–3.95% lipids. The lipid content depends on the variety, degree of maturity, growth conditions, and lipid extraction method. The lipid content of bran and polished rice is affected by the degree of milling and the milling procedure. Polishing gradually removes the pericarp, tegmen, aleurone layer, embryo, and parts of the endosperm, but parts of the lipid-rich germ may remain attached to the endosperm even after advanced polishing and removal of up to 20% of the rice kernel (Fujino, 1978). The major proportion of the lipid in rice is removed with the bran (containing the germ) and the polish (Houston and Kohler, 1970).

Commercial bran contains 10.1–23.5%, polish 9.1–11.5%, brown rice 1.5–2.5%, and milled rice 0.3–0.7% oil, respectively.

In the rice kernel, as in other cereals, lipid content is highest in the embryo and in the aleurone layer, and the lipid is present as droplets or spherosomes. The spherosomes are submicroscopic—about 0.5μm or less in the coleoptile cells. Much higher quantities of lipids are present outside the aleurone granules than inside them. The testa contains a fatty material, and a sheath of fat-staining material encloses the aleurone granules. Rice lipids are mainly triglycerides, with smaller amounts of phospholipids, glycolipids, and waxes. The three main fatty acids are oleic, linoleic, and palmitic. The main glycolipids are acyl sterol glycosides and sterol glycosides, and either diglycosyl diglyceride and ceramide monohexoside.

The distribution of lipid types is not uniform in the rice kernel. Approximate ratios of neutral and polar lipids are 90:10 in bran, 50:50 in the starchy endosperm, and 33:67 in the starch. Thus the bran is rich mainly in neutral lipids; the endosperm contains relatively high concentrations of polar lipids.

There is a considerably higher concentration of ash and of individual minerals in outer layers of the milled rice kernel than toward the center. P, K, Mg, Fe, and Mn are concentrated in the aleurone layer; P, K, and Mg are particularly high in the subcellular particles of the aleurone layer; Ca is abundant in the pericarp. The phytin-P constitutes almost 90% of the total bran-P and 40% of the milled rice-P.

Rice and its by-products contain little or no vitamin A, ascorbic acid, or vitamin D. Thiamine, riboflavin, niacin, pyridoxin, panthotenic acid, folic acid, inositol, choline, and biotin are lower in milled rice than in brown rice and substantially lower than in rice bran, polish, or germ.

NUTRITIONAL IMPLICATIONS OF PROCESSING

Production of brown rice from rough rice increases protein, fat, and starch contents, since the hulls are low in those constituents. Conversely, there is a decrease in the crude fiber and ash contents. Conversion of brown rice to white or polished rice removes about 15% of the protein, 65% of the fat and fiber, and 55% of the minerals (Table 16-2). Rough rice and brown rice differ little in vitamin content, but conversion of brown rice to white rice decreases the vitamin values considerably. Thus head rice contains only 20% as much thiamine, 45% as much riboflavin, and 35% as much niacin as brown rice. The losses have created much interest in the development of practical methods to retain more of the B vitamins in the milled rice kernel. The problem of improving the vitamin content of milled rice has been approached by processing the rice before milling to diffuse the vitamins

Table 16-2. Composition of Rice (%)

Material	Moisture	Protein	Lipid	Fiber	Ash	Degree of polishing
Brown rice	15.5	7.4	2.3	1.0	1.3	0
Rice bran	13.5	13.2	18.3	7.8	8.9	—
Polished rice	15.5	6.2	0.8	0.3	0.6	8–10

Source: Fujino, 1978.

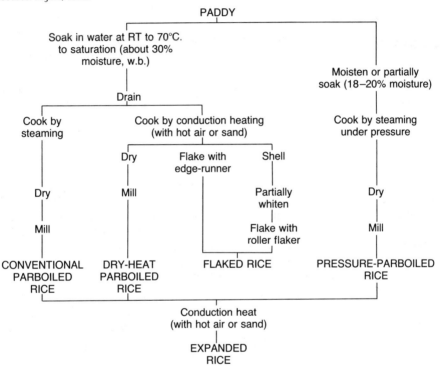

Figure 16-4. Flow chart of steps for making different kinds of parboiled rice and expanded and flaked rice. RT = room temperature. (From Bhattacharya and Ali, 1985.)

and other water-soluble nutrients in the outer portion of the grain into the endosperm. Processing of rough rice to increase vitamin retention involves parboiling or some modification thereof (Bhattacharya and Ali, 1985) (Figure 16-4). For parboiling, rough rice is soaked in water, drained, steamed, and dried. In 1940, a process for the manufacture of "converted rice" was developed and patented in England. The cleaned rough rice is exposed to a vacuum, treated with hot water under pressure, and then steamed, dried, and milled. The converted rice process is particularly effective for the retention of vitamins.

Parboiling is performed to improve the nutritional and also the storage and cooking attributes of rice. The main modifications are transfer of some vitamins and minerals from the aleurone and germ into the starchy endosperm, dispersion of lipids from the aleurone layer and germ, inactivation of enzymes, and destruction of molds and insects (Gariboldi, 1974). Those changes are accompanied by reduced chalkiness and increased vitreousness and translucence of the milled rice, and improved digestibility and cooking properties. Parboiling strengthens the attachment of the germ and aleurone to the starchy endosperm and prevents the separation of the germ during husking. However, the strengthening of these attachments and hardening of the endosperm increase the difficulty of milling the husked grains of parboiled rice (Bechtel and Pomeranz, 1978). Compared with nonparboiled rices, parboiled rices disintegrate less during cooking and remain better separated and less sticky after cooking. Parboiling reduces the amount of solids leached into the cooking water and the extent to which the kernels solubilize during cooking.

Rice must have acceptable market and eating qualities and good nutritional value. Grain quality is related mainly to the amylose/amylopectin ratio, which governs water absorption and volume expansion during cooking, and to cohesiveness, color, gloss, and tenderness of cooked rice (Juliano, 1977). Long-grain types generally cook to dry, fluffy products that harden on keeping and are preferred by some. Short-grain types tend to be more cohesive and moist and to remain relatively tender when kept and consumed cold. Waxy (1–2% amylose) rices, in contrast to high-amylose (over 25%) rices, are glossy and sticky when cooked. Rices with intermediate amylose contents and intermediate gelatinization temperatures are preferred in the tropics.

The modern trend in processed foods is toward convenience items. Quick-cooking rices may be prepared by precooking in water and drying under controlled conditions or by application of dry heat. Other convenience items include canned and frozen cooked rice.

WILD RICE

Wild Rice (*Zizania aquatica*), a grass indigenous to North America, grows extensively in shallow lakes and sluggish streams from southern Canada to the Gulf of Mexico and from the Atlantic Ocean to the Rocky Mountains (Anderson, 1976). The limited supply and its unique flavor and color (brown) have made wild rice a favorite of the gourmet chef. Traditional processing for wild rice includes (1) short-term wet curing that involves active fermentation, (2) parching from 35–50% to 7–12% moisture, (3) hulling, and (4) cleaning, separating, and/or grading (Lund et al, 1975). Wild rice contains, on an as-is basis, 12–15% protein, 0.5–1.0% fat, and 0.5–1.2% crude fiber. It is a fairly good source of protein (second only to oats, among the cereals), and the proteins are rich in lysine (about 4.5%) and methionine (about 3.0%).

BIBLIOGRAPHY

Anderson, R.A. *Cereal Chem.* **1976**, *53*, 949–955.

Barber, S.; Benedito de Barber, C.; Tortosa, E. In "Cereals—A Renewable Resource; Theory and Practice." Pomeranz, Y.; Munck, L., Am. Assoc. Cereal Chem.; St. Paul, MN; **1981**, pp. 471–488.

Bechtel, D.B.; Pomeranz, Y. *Am. J. Bot.* **1977**, *64*, 966–973.

Bechtel, D.B.; Pomeranz, Y. *Am. J. Bot.* **1978a**, *65*, 684–691.

Bechtel, D.B.; Pomeranz, Y. *Am. J. Bot.* **1978b**, *65*, 75–85.

Bechtel, D.B.; Pomeranz, Y. *J. Food Sci.* **1978c**, *43*, 1538–1542, 1552.

Bechtel, D.B.; Pomeranz, Y. *Adv. Cereal Sci. Technol.* **1980**, *3*, 73–113.

Bhattacharya, K.R.; Ali, S.Z. *Adv. Cereal Sci. Technol.* **1985**, *7*, 105–167.

Fujino, Y. *Cereal Chem.* **1978**, *55*, 559–571.

Gariboldi, F. "Rice Milling Equipment; Operation and Maintenance." Agric. Serv. Bull. No. 22; FAO; Rome; **1974**.

Houston, D.F. (Ed.). "Rice, Chemistry and Technology." Am. Assoc. Cereal Chem.; St. Paul, MN; **1972**.

Houston, D.F.; Kohler, G.O. "Nutritional Properties of Rice." Natl. Acad. Sciences; Washington, D.C.; **1970**.

Juliano, B.O. *Cereal Foods World* **1977**, *22*, 284–287.

Luh, B.S. (Ed.). "Rice: Production and Utilization." Avi Publ. Co.; Westport, CT; **1980**.

Lund, D.B.; Lindsay, R.C.; Stuiber, D.A.; Johnson, C.E.; Marth, E.H. *Cereal Foods World* **1975**, *20*, 150–154.

White, G.C. *Assoc. Operative Millers* **1970**, 3147–3160.

Chapter 17

Corn

Component parts of mature dent corn kernels and their chemical composition are given in Table 17-1. About three-fifths of the processed corn (or maize; the terms are used interchangeably) is used in the United States to produce corn starch, sweeteners, corn oil, and various feed by-products. The remainder is used to prepare various food products and alcoholic beverages. Corn is prepared in several ways as human food: (1) parched to be eaten whole; (2) ground to make hominy, corn meal, or corn flour; (3) treated with alkali to remove the pericarp and germ to make lye hominy; and (4) converted to a variety of breakfast foods.

DRY MILLING

Dry milling of corn is carried out by the old-process milling from nondegermed grain and by the new-process milling from degermed grain.

Old-Process Milling

In the old process, corn is ground to a coarse meal between millstones run slowly at a low temperature, with the meal frequently not being sifted. In the larger mills, about 5% of the coarser particles of the hull are sifted out. The meal is essentially a whole corn product and has a rich, oily flavor, as it contains much of the germ. The product stores poorly. The meal is soft and flourlike. In some larger mills, the corn is dried to 10–12% moisture before grinding. Kiln drying facilitates rapid grinding and improves the keeping qualities of the meal.

New-Process Milling

In this process, steel rolls are used as in the milling of wheat. The objective is to remove the bran and germ and to recover the endosperm in the form of hominy or corn grits, coarse meal, fine meal, and corn flour. A schematic outline of new dry corn milling is given in Figure 17-1. Corn grits and

Table 17-1. Component Parts of Mature Dent Corn Kernels and Their Chemical Compositions

Part	Dry wt of whole kernel (%)		Composition of kernel parts (%)[a]				
			Starch	Fat	Protein	Ash	Sugar
Germ	range	10.5–13.1	5.1–10.0	31.1–38.9	17.3–20.0	9.38–11.3	10.0–12.5
	mean	11.5	8.3	34.4	18.5	10.3	11.0
Endosperm	range	80.3–83.5	83.9–88.9	0.7–1.1	6.7–11.1	0.22–0.46	0.47–0.82
	mean	82.3	86.6	0.86	8.6	0.31	0.61
Tip cap	range	0.8–1.1		3.7–3.9	9.1–10.7	1.4–2.0	
	mean	0.8	5.3[b]	3.8	9.7	1.7	1.5
Pericarp[c]	range	4.4–6.2	3.5–10.4	0.7–1.2	2.9–3.9	0.29–1.0	0.19–0.52
	mean	5.3	7.3	0.98	3.5	0.67	0.34
Whole kernels	range		67.8–74.0	3.9–5.8	8.1–11.5	1.27–1.52	1.61–2.22
	mean	100	72.4	4.7	9.6	1.43	1.94

[a]Dry basis.
[b]Composite from nine different corn-belt hybrids.
[c]Also known as hull or bran.
Source: Earle et al, 1946.

coarse meal consist largely of particles of flinty endosperm, and the fine meal and corn flour are obtained mainly from the soft and starchy endosperm. Flint varieties of corn are considered too "sharp" for grinding to a meal. Dent corn is almost invariably used. Since the grits used in the manufacture of corn flakes are made mainly from white corn, large quantities are used to make grits. Meal and flour are considered as by-products.

The corn is cleaned and passed through a scourer to remove the tip cap from the germ end of the kernel. The hilar layer under the tip is frequently black, and it causes black specks in the meal. The corn is tempered by two additions of water to a moisture content of 21–24%. Subsequently, it is passed through a corn degerminator, which frees the bran and germ and breaks the endosperm to two or more pieces (Figure 17-2). Stock from the degerminator is dried to 14–16% moisture and cooled in revolving or gravity-type coolers.

The largest endosperm pieces are used for making corn flakes. The stocks are passed through a hominy separator. It first separates the fine particles and then grades the larger fragments to four sizes and "polishes" them. The various grades of broken corn are passed through aspirators to remove loose bran from the endosperm fragments. The corn fragments are reduced to coarse, medium, and fine grits by gradual reduction between corrugated rolls and subsequent sifting of the stock. The coarsest stock from the aspirators goes to the first-break rolls. The rolls are spaced further apart, have coarser corrugations, and operate at lower speed differential than the subsequent breaks. The coarsest grade of hominy is

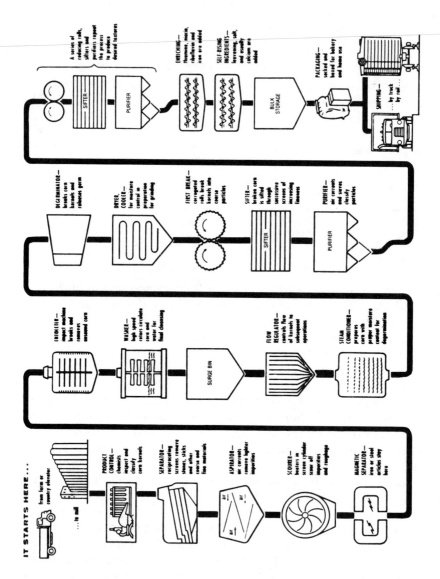

Figure 17-1. How corn meal is milled; a simplified diagram. (Courtesy Wheat Flour Institute.)

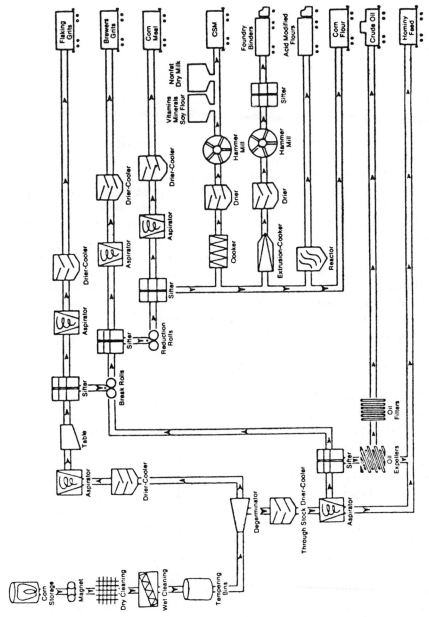

Figure 17-2. Simplified corn tempering-degerming process flow (Courtesy Krouse Milling Co., Milwaukee, WI).

Table 17-2. Yield and Particle-Size Range of Milled Corn Products

Product	Particle size		Yield (%)
	Mesh	Inches	
Grits	14–28	0.054–0.028	40
Coarse meal	28–50	0.028–0.0145	20
Fine meal	50–75	0.0145–0.0095	10
Flour	through 75	below 0.0095	5
Germ	3–30	0.292–0.0268	14
Hominy feed			11

Source: Stiver, 1955.

highly contaminated with germ. The germ is flattened between the break rolls with minimum endosperm grinding and separated by sieving. The successive steps in the gradual reduction for corn are similar to those described for wheat. Modern corn mills can produce a variety of grits, meals, and flours. They are dried at 65°C and cooled before packing. The flattened germs are used to produce corn oil. The germ is dried to about 2–3% moisture, ground, tempered with steam, and passed through expellers. The germ cake from which most of the oil has been expelled is frequently reground and may be solvent extracted before packaging.

Hominy or grits for industrial uses, such as brewing and manufacture of wallpaper paste, are flaked. The grits are steamed and passed between heavy-duty heated iron rolls, and the flakes are dried but not toasted. The heating process gelatinizes the starch.

Yields

Relative yields of mill products depend on whether the main objective is to produce grits or meal and whether the corn was degermed before grinding. Typical yields and particle sizes of milled corn products are given in Table 17-2. In milling corn for grits and meal by the degerminating process, the following average yields are obtained: grits 52%, meal and flour 8%, hominy feed 35%, and crude corn oil 1%. When corn is not degermed before grinding, about 72% corn meal and 20% feed are produced. Of the total meal produced, about two-thirds contain about 1.4% fat and one-third about 4.7% fat. Typical composition of dry milled products from degermed maize is listed in Table 17-3. Grits and meal are largely produced from the horny or vitreous endosperm; they contain less than 1.0% and 1.5% fat, respectively. Flour produced by grinding the starchy endosperm contains 2–3% fat from broken germ during processing. The large surface area and relatively high fat content of corn flour lower its shelf life.

The Food and Drug Administration (FDA) has established standards of

Table 17-3. Typical Yields and Analyses of Products From a Degerming Type
of Maize Dry Mill

Products	Yield (%)	Particle size range[a]	Moisture (% wb)[b]	Fat (% db)[c]	Crude fiber (% db)	Ash (% db)	Crude protein (% db)
Maize	100		15.5	4.5	2.5	1.3	9.0
Primary products							
Cereal flaking							
(hominy grits)	12	−3.5 + 6	14.0	0.7	0.4	0.4	8.4
Coarse grits	15	−10 + 14	13.0	0.7	0.5	0.4	8.4
Regular grits	23	−14 + 28	13.0	0.8	0.5	0.5	8.0
Coarse meal	3	−28 + 50	12.0	1.2	0.5	0.6	7.6
Dusted meal	3	−50 + 75	12.0	1.0	0.5	0.6	7.5
Flour[d]	4	−75 + pan	12.0	2.0	0.7	0.7	6.6
Oil	1						
Hominy feed	35		13.0	6.3	5.4	3.3	12.5
Alternative products							
Brewers' grits	30	−12 + 30	13.0	0.7	0.5	0.5	8.3
100% meal	10	−28 + pan	12.0	1.5	0.6	0.6	7.2
Fine meal	7	−50 + pan	12.0	1.6	0.6	0.7	7.0
Germ fraction[e]	10	−3.5 + 20	15.0	18.0	4.6	4.7	14.9

[a]U.S. standard sieve.
[b]Wet basis.
[c]Dry basis.
[d]Break flour.
[e]Yield is distributed between maize oil and hominy feed.
Source: Brekke, 1970.

identity for dry-milled corn products used for food. According to those standards (Code of Federal Regulations, Title 21, part 15), the fat content of corn meal may not differ more than 0.3% from that of cleaned corn; that of bolted corn meal should not be less than 2.25% or more than 0.3% greater than the fat of cleaned corn; that of degerminated corn meal should be less than 2.25%; that of corn grits should be not more than 2.25%; and that of corn flour may not exceed that of cleaned corn.

Composition of commercial maize products is listed in Table 17-4. The meals and flours are produced from maize ground to typical granulations. The cooked flour hydrates readily in cold water to form a stable paste. The toasted germ is a food-grade product in flake form; the stabilized product contains all the original oil of the germ. The germ cake is a feed product from maize germ from which most of the oil has been removed. It is used as a carrier for vitamins and antibiotics in animal feed formulations. *Massa harina* (yellow regular grind, or yellow coarse grind, or white) is a food-grade product. It is milled from maize that has been steeped, ground, and dried to produce a stable flour for the production of Mexican foods.

Table 17-4. Composition in Percent of Typical Maize Products

Maize product	Moisture	Protein[a]	Fat[a]	Fiber[a]	Ash[a]
Yellow maize					
Meal[b]	8.2–12.6	8.0–9.0	0.9–2.3	0.3–0.7	0.3–0.7
Flour[b]	8.5–12.1	7.6–8.8	1.4–3.0	0.6–1.0	0.6–1.0
Fine flour	7.1–11.3	6.4–7.4	2.3–3.3	0.04–0.60	0.66–1.12
Cooked flour	6.7–11.5	9.0–9.2	0.6–0.7[c]	—	0.55–0.73
White maize					
Meal[b]	11.0–13.0	8.0–9.0	1.7–2.2	0.5–1.0	0.3–0.7
Flour[b]	8.0–12.0	7.0–8.0	2.0–2.7	0.5–1.0	0.4–0.9
Fine flour	10.0–13.0	8.0–8.8	4.0–5.2	0.5–1.2	0.8–1.0
Toasted corn germ	4.2	17.0	25.4[c]	4.2[d]	7.2
Corn germ cake	5.0 max	14.0 min	3.5 min	8.5 max	—
Massa harina	10.0–12.0	7.0–9.0	3.5–4.5	1.8–2.6	1.2–1.7

[a]Dry-matter basis.
[b]From degermed maize.
[c]Ether extract.
[d]Dietary fiber = 20.8%.

Courtesy Quaker Oats Co., Chicago.

WET MILLING

The main products of wet milling are starch (unmodified and modified, including syrups, and dextrose) and several coproducts. The coproducts, used mainly as feed ingredients, include gluten meal, gluten feed, corn germ meal, and condensed, fermented corn extractives (about 50% solids). Processing maize germ yields refined oil (along with fatty acids from crude oil refining) and corn germ meal (Harness, 1978) (Figure 17-3).

The cleaned corn is first softened by steeping in a very dilute solution of sulfur dioxide at 48–52°C for 30–50 hours. For optimum milling and separation of corn components, at the end of the steeping period the corn should have absorbed about 45% water, released about 6.0–6.5% of its dry substance as solubles into the steepwater, absorbed about 0.2–0.4g sulfur dioxide per kilogram, and become quite soft. When corn has been optimally steeped, the germ can be removed easily and intact; the starch can be separated from fiber by milling and screening and can easily be removed from the gluten by centrifuging (Watson, 1967).

According to Meyer (1984), for the past 10 years the corn wet-milling industry in the United States has grown at a rate of about 10% per year and has increased the grinding capacity from 800,000 bushels per day to over 2 million bushels per day. This growth has been fueled by the development of the high-fructose corn syrup as a lower-priced industrial sweetener and of ethanol as an octane booster with built-in tax relief. The corn wet-milling grind in 1983 was 33% high-fructose corn syrup, 21% starch, 21% ethanol, 19% corn syrup, and 6% dextrose.

Coproducts of maize starch wet milling amount to about one-third of the

Table 17-5. Composition of Feeds from Maize Wet Milling[a]

Component	Corn gluten feed	Corn gluten meal		Corn germ meal	Condensed fermented corn extractives (about 50% solids)
Guaranteed analysis (%)					
Protein (min)	21.0	60.0	41.0	20.0	23.0
Fat (min)	1.0	1.0	1.0	1.0	0.0
Fiber (max)	10.0	3.0	6.5	12.0	0.0
Typical analysis (%)					
Fat (average)	2.5	2.5	2.5	1.9	0.0
AOAC (range)	1.4–3.5	1.0–5.2	1.2–4.4	1.0–2.9	0.0
Total fat[b] (average)	3.8	5.7	4.8	4.6	0.0
Total fat (range)	2.7–4.7	4.4–7.9	3.6–6.4	4.1–5.3	0.0

[a]Adapted from Corn Wet-Milled Feed Products; Corn Refiners Association Inc., Washington, D.C. 1975, and S.A. Watson, AACC Short Course, April 1980.
[b]As determined by extraction with a chloroform-methanol mixture 4:1; widely used in Europe.

total output. Except for maize oil and steep liquor (used in industrial fermentations), the coproducts are mainly sold as feed ingredients. In decreasing value they are corn gluten meal, corn gluten feed, spent germ meal, corn starch molasses or hydrol, steep liquor (condensed corn fermentation extractives), corn bran, and hydrolyzed fatty acids. Composition of the main maize wet-milling feeds is summarized in Table 17-5.

Corn gluten meal is a high-protein product, used as a protein-balancing ingredient in feed formulations. It is used widely in broiler and layer rations because of its high content of carotenoid pigments. Among the three carotene isomers (beta, zeta, and beta-zeta), only beta-carotene has significant vitamin A activity. The dihydroxy xanthophylls are potent pigments for coloring poultry skin and egg yolks. The major isomer, lutein, is slightly superior to zeaxanthin in producing color. The monohydroxy pigments, zeinoxanthin and cryptoxanthin, have less than half the pigmenting value of the dihydroxy pigments. Xanthophyll levels in gluten meal are highest in winter months and drop gradually to half the original value by the end of summer.

The linoleic acid content, on as-is basis, is 3.2% in corn gluten meal, 2.2% in corn gluten feed, and about 0.5% in corn germ meal (Rapp, 1978). Corn gluten meal is relatively rich in xanthophylls (100–225 mg/lb); 10 mg/lb is present in corn gluten feed, and practically none is present in corn germ meal and concentrated steepwater. Corn gluten meal contains 30–65 vitamin A equivalents as retinol (0.15 mg retinol = 5,000 IU vitamin A) and 20–30 mg of beta-carotene per pound.

Maize contains about 4.5% oil, of which 85% is present in the germ (Reiners, 1978). The germ fraction separated from maize by the wet-milling process contains about 50% oil and by the dry-milling process about 25% oil. Germ oil can be extracted by a continuous screw press (expeller) to yield a meal with a residual oil content of 7–10%; solvent extraction

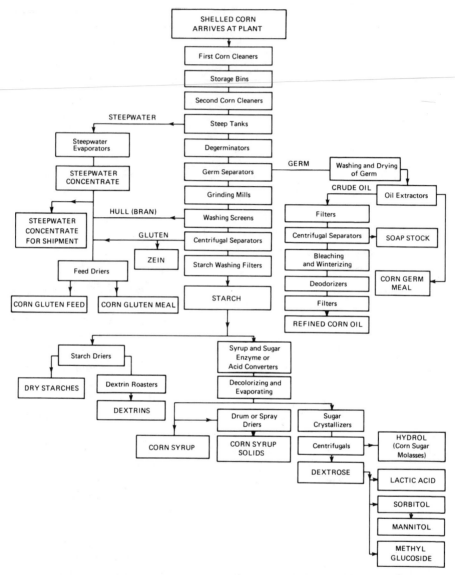

Figure 17-3. The corn refining process. (Courtesy, Corn Refiners Assoc., Washington.)

(directly or following expeller extraction) produces a meal with a residual oil content of 1–3%. About 1.75 lb of oil can be recovered from a bushel of maize by solvent extraction of the germ. The products are crude oil and corn germ meal. Crude and refined maize oils are compared in Table 17-6.

Refining maize oil removes free fatty acids, phospholipids, waxes, and carotenoids. The main triglyceride fatty acids are about 60% linoleic acid

Table 17-6. Composition of Corn Oil (%)

Component	Crude	Refined
Triglycerides	95.6	98.8
Free fatty acids	1.7	0.03
Phospholipids	1.5	—
Phytosterols	1.2	1.1
Tocopherols	0.06	0.05
Waxes	0.05	—
Carotenoids	0.0008	—

Source: Reiners, 1978.

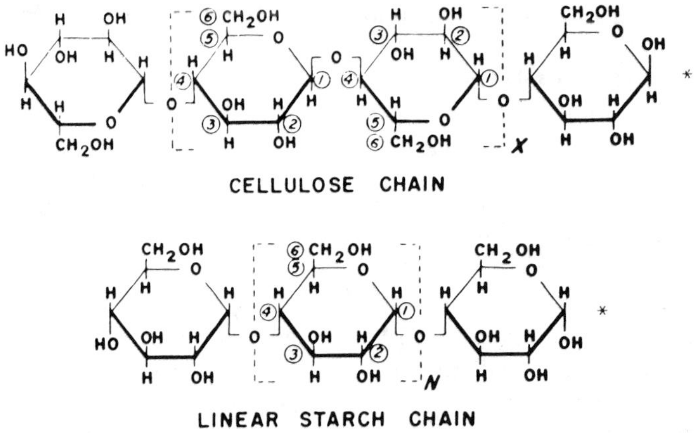

CELLULOSE CHAIN

LINEAR STARCH CHAIN

Figure 17-4. Polymeric chain structures of cellulose and amylose. (From T.J. Schoch, Corn Products Co., Inc., Argo, IL.)

(all in the cis-cis configuration), 25% oleic acid, and 13.5% palmitic + stearic acid. The iodine value of refined maize oil produced in the United States shows little variation (125.4–127.6). Oils from African maize may have iodine values as low as 110 and correspondingly reduced ratios of polyunsaturates to saturates (Reiners, 1978).

Whereas maize starch contains only 0.04% fat as determined by ether extraction, total lipids amount to about 0.54%. The lipids are almost entirely associated with amylose, the linear starch fraction, and they are predominantly free fatty acids. Commercial maize starch products contain less than 0.1% ether extractives, typically 0.03% (D.W. Harris, Clinton Corn Processing Co., private communication, Sept. 14, 1981).

STARCHES

Occurrence and Appearance

Next to cellulose, starch is the most abundant and widely distributed substance of vegetable life (Figure 17-4). Starch occurs in form of granules

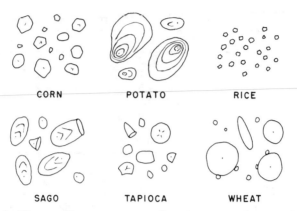

Figure 17-5. Microscopic appearance of various granular starches. (From T.J. Schoch, Corn Products Co., Inc., Argo, IL.)

as reserve food in various parts of most plants: in the seeds of cereal grains, the roots of tapioca, the tubers of potatoes, and—in small amounts—in the stem pith of sago and fruit of the banana. The shape and size of the starch granules are characteristic for each plant. The appearance of the granules under microscopic examination (Figure 17-5) is used to identify starches from various sources. Potato starch has a relatively large, oval granule with pronounced "oyster shell' radiations. Corn starch granules are medium-size and polygonal or spherical. Tapioca granules are often truncated and rounded; rice and oat starches occur in compound granules that break up into small, polygonal individual granules during processing. Extremes in size of starch granules are shown by taro or dasheen root (1 μm in diameter) and by root of the edible type of canna (150–200 μm in diameter).

Under the polarizing microscope, the granular starches have a characteristic birefringence, spherical granules giving a "Maltese cross" interference pattern and others a differently formed cross, indicating an organized spherocrystalline structure. The birefringence of the starch granule implies that there is a high degree of molecular orientation within the granule. The crystalline form is due to the amylopectin content of the granules. This is indicated by two facts: (1) Waxy starches containing no amylose give an x-ray pattern similar to that of common-regular starch, and (2) if the amylose is leached selectively from the granule, the structure remains intact. It has been suggested that some degree of double helix in amylopectin may explain the apparent anomaly that a branched polymer is the source of structural order.

Crystallinity varies with starch type, as shown by the different x-ray spectra of root and cereal starches. In the potato starch granule the relatively high degree of crystallinity is contributed solely by the amylopectin; the amylose is in the amorphous state. Below the gelatinization temperature, the crystallinity contributes to a compact structure that

Table 17-7. Properties of Starch Components

Property	Amylose	Amylopectin
General structure	Essentially linear	Branched
Average chain length	$\sim 10^3$	20–25
Degree of polymerization	$\sim 10^3$	10^4 to 10^5
Iodine complex	blue (~ 650 nm)	purple (~ 550 nm)
Iodine affinity	19–20%	1%
Blue value	1.4	0.05
Stability of aqueous solution	Retrogrades readily	Stable
Conversion to maltose (%)		
with β-amylase	100	55–60
with limit dextrinase and β-amylases	100	100

Source: Pomeranz, 1984.

reduces, eliminates penetration by alpha-amylase. In addition, the strong radial orientation of the molecules provides a surface that is not receptive to enzymic attachment. In consequence, alpha-amylolysis is low. On gelatinization, the granule swells until it bursts. The effects of heat treatment are change of the x-ray diffraction pattern to the A type as a result of dehydration, and conversion of part of the amorphous amylose to a helical form. The helical form is less soluble than the amorphous form, and when the granule gelatinizes, the helical regions act as weak centers of crystallinity and prevent the granule from bursting.

In the maize starch granule, the amylopectin may also constitute the crystalline skeleton, but the overall degree of crystallinity and orientation is lower than in the potato. The surface of the granule is more susceptible to the action of alpha-amylase, and the smaller degree of orientation of the molecular chains allows the enzyme easier access to the interior of the molecule. Much of the amylose seems to be present as a complex with fat and is probably in the V form. The limited solubility of this form is sufficient to set up a weak crystalline structure that retards granule swelling.

In the amylomaize (high-amylose) starch granules, amylose is in the retrograded form and constitutes the crystalline structure. Hydrolysis of the granule is very slow, as the retrograded amylose is resistant to alpha-amylase. Only a small fraction may be involved in the long regions of the double helix. Still, the stability of such a structure may prevent considerable swelling.

Chemical Composition

Starch is a natural high polymer of D-glucose units; the polymer is formed through successive condensation of glucose units involving enzyme systems of a developing plant. Complete chemical or enzymatic hydrolysis of starch yields only glucose units.

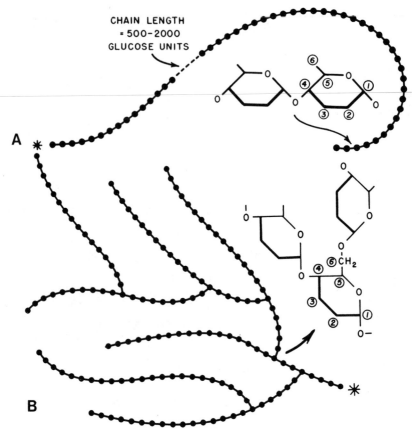

Figure 17-6. Structures of starch fractions. (a). Structure of the linear fraction. (b). Structure of the branched starch molecule. (From T.J. Schoch, Corn Products Co., Inc., Argo, IL.)

Ordinary starches contain two types of glucose polymers. The linear fraction, called amylose, is a linear chain of some 500–2,000 glucose units (Figure 17-6a). Corn starch amylose contains an average of about 500 glucose units and has a molecular weight of about 80,000. The individual units in amylose are connected by alpha-1,4 linkages—ie, from the aldehydic carbon 1 of each glucose to carbon 4 of the preceding glucose molecule.

The branched fraction, amylopectin (Figure 17-6b) contains several hundred short linear branches, with an average branch length of 25 glucose units. The average molecular weight of amylopectin is at least 1 million. The interglucose linkages in each linear portion are alpha-1,4 (as in amylose); the branch points are through alpha-1,6 linkages.

Common starches (corn, wheat, tapioca, and potato) contain amylose and amylopectin in (generally) fixed proportions. Starches in certain varieties of corn, sorghum, and rice (but not of wheat or common root starches) are

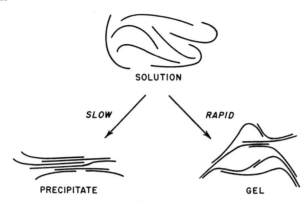

SOLUTION

SLOW   RAPID

PRECIPITATE   GEL

Figure 17-7. Mechanism of retrogradation of the linear starch fraction. (From T.J. Schoch, Corn Products Co., Inc., Argo, IL.)

composed entirely of amylopectin and are known as waxy starches. On the other hand, wrinkled-seeded, garden-type varieties of pea and some corn varieties ("amylomaize") have starches that contain 70% to over 80% amylose. Waxy starches have certain unique properties that make them preferred for various uses. The linear fraction, in cooked starch, tends to deposit in the form of discrete insoluble particles, as a result of starch, retrogradation; amylopectin shows little tendency to retrograde (Figure 17-7). Properties of starch components are compared in Table 17-7.

Starches can be fractionated into amylose and amylopectin by gelatinization in water at elevated temperatures and pressures. The dissolved starch granules are cooled in the presence of polar agents such as butyl alcohol, amyl alcohol, thymol, and nitroparaffins. A complex precipitate of the polar components with amylose can be isolated by ultracentrifugation. Amylose similarly treated with fatty acids forms a complex that separates at 90 +5°C as a pulverulent precipitate.

Commercially, corn and potato starch can be fractionated into amylose and amylopectin by two processes: (1) The starch is gelatinized under pressure at 140–150°C in a magnesium sulfate solution, and cooled to 90–100°C to precipitate the linear fraction. Further cooling to room temperature precipitates the branched fraction. The fractionation is effective, and the fractions are relatively pure though somewhat degraded. (2) The starch is gelatinized under pressure in a water medium by steam injection; the linear fraction precipitates after slow cooling. The degradation effect is reduced, but the separation is not as effective as in the magnesium sulfate procedure.

Amylose fractions are available commercially. Their unique character makes them useful in manufacture of edible food- packaging films—ie, casings for sausages or containers for dried soups. The amylose fraction can form inclusion complexes with a variety of substances. Compounds susceptible to oxidation can be protected by complex formation. Such complexes may be useful for the protection of vitamins A and D, flavors, or drugs and pharmaceutical products. The ease of retrogradation of the amylose fraction makes it possible to use them in semipermanent finishing of textiles or as beater size in paper manufacture to impart wet strength and stiffness. As pastes of amylopectin do not ordinarily retrograde except after long standing, they remain fluid and noncongealing over useful periods of time. Amylopectin fractions can therefore be used in products in which minimum retrogradation is desired. In addition, a variety of physically and chemically modified amylopectins are available commercially for specific food and industrial uses.

Physical Properties

Certain physical and chemical properties of some starches are summarized in Table 17-8. The intense blue color that starches give with iodine is due solely to the amylose fraction; amylopectin gives a red or violet-red color. Starch granules immersed in water or exposed to a humid atmosphere readily take up moisture. The hydration is a process of simple absorption that causes the granule to increase slightly in size. Under those conditions, undamaged granules retain their birefringence. Granules that have been damaged by shearing lose birefringence and swell more.

The phenomenon exhibited by a granule subjected to shear and placed in cold water is apparently the same as that which a granule undergoes on heating in water during gelatinization. Starches from various species swell and gelatinize in water at different rates and temperatures, indicating varietal differences in the molecular organization within the granule. Thus, tapioca and particularly potato starch gelatinize at relatively low temperatures and swell freely and enormously. Corn starch and sorghum starch gelatinize at relatively higher temperatures and swell more slowly; they show a restricted two-stage swelling. Characteristics of gelatinized and cooled starch dispersions are compared in Table 17-9.

During gelatinization, the starches undergo certain changes. Thus, when starch is heated in excess water, birefringence is first lost around the hilum and then progressively outward, as heating continues, until none is left. Some wheat starch granules, for example, start to lose birefringence at 58°C, whereas others may retain all or much of their refringence at 70°C. All birefringence of all granules in a wheat starch sample is lost at around 70–78°C. Although granule size is not clearly related to the gelatinization temperature, the last granules to loose birefringence are usually very small. Gelatinization temperature curves of unmodified starches are given in Figure 17-8.

As the pasting process progresses with increase in temperature, the granules swell and viscosity increases. Viscosity of starches can be measured with the amylograph (see Chapter 7). Modified starches in the food industry are listed in Table 17-10. Digestibility of practically all commercial

Table 17-8. Physical and Chemical Properties of Common Starches

| Starch | Granule size (μm) | | Amylose (%) | Swelling power (at 95°C) | Solubility at 95°C (%) | Gelatinization range (°C) | Source | Taste | General description of granules |
	Range	Average							
Barley	2–35	20	22	—	—	59–64	Cereal	Low	Round, eliptical
Corn									
regular	5–25	15	26	24	25	62–72	Cereal	Low	Round, polygonal
waxy	5–25	15	~1	64	23	63–72	Cereal	Low	Round, oval indentations
high amylose	—	15	up to 80	6	12	85–87	Cereal	Low	Round
Potato	15–100	33	24	1000	82	56–69	Tuber	Slight	Egg-like, oyster indentations
Rice	3–8	5	17	19	—	61–78	Cereal	Low	Polygonal clusters
Rye	2–35	—	23	—	—	57–70	Cereal	Low	Eliptical, lenticular
Sago	20–60	—	27	97	—	60–72	Pith	Low	Egg-like, some truncate forms
Sorghum	5–25	15	26	22	22	68–75	Cereal	Low	Round, polygonal
Tapioca (cassava)	5–35	20	17	71	48	52–64	Root	Fruity	Round-oval, truncated on side
Wheat	2–35	—	25	21	41	62–75	Cereal	Low	Round, eliptical
Oats	—	25	27	—	—	—	Cereal	Low	Round

Source: Pomeranz, 1984.

Table 17-9. Characteristics of Gelatinized and Cooled Starch Dispersions

Starch	Hot-paste viscosity	Texture	Clarity	Stability to retrograde	Freeze-thaw stability	Resistance to shear
Corn	Medium	Short-stiff	Opaque	Low	Low	Medium
Waxy maize	Medium-high	Soft-cohesive	Clear	High	Medium	Low
Potato	Very high	Long-cohesive	Clear	Medium	Low	Low
Rice	Medium-low	Short-stiff	Slightly opaque	Low	Low	Medium
Sago	Medium-high	Long-cohesive	Clear-translucent	Medium	Low	Low
Tapioca	High	Long-cohesive	Clear-translucent	Medium	Low	Low
Wheat	Medium-low	Soft-short	Slightly opaque	Low	Low	Medium

Source: Pomeranz, 1984.

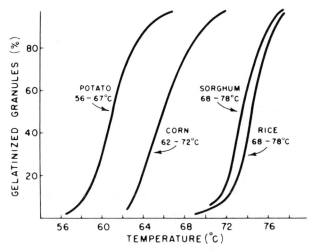

Figure 17-8. Gelatinization temperature curves of various unmodified starches. (From T.J. Schoch, Corn Products Co., Inc., Argo, IL.)

Table 17-10. Modified Starches in the Food Industry

Modified starch	Major properties	Examples of uses
Pregelatinized	Cold-water-soluble	Cake mixes
		Instant desserts
Cross-linked	Delayed thickening; stable to wide pH range and high temperature	Pie fillings, soups, baby foods, sauces, and gravies
Cross-linked etherified	Clarity, delayed thickening	Frozen foods, canned foods
esterified	Freeze-thaw stability	
	Stable to pH range and high temperature	Pie fillings
Oxidized	High clarity, thin-boiling	Jellies, lemon curd
Acid-thinned	Gel strength	Jellies, gums, pastilles
Dextrins	High solubility	Toffees, glazes

Source: Selby, 1977.

starches is high (96–98%). Potential uses of raw starches in foods depend largely on the characteristics of gelatinized dispersions. Some of those characteristics can be modified by chemical treatment of the starches. The treatment is responsible for new types of processed starches (specialty, modified, and pregelatinized) that are tailored to meet special requirements of the food manufacturer. Viscosity ranges of different types of starches are compared in Figure 17-9.

Starch thickeners represent the major part of starch sales to the food industry. Traditionally, thick-bodied consistency was imparted to foods by thick-cooking unmodified cereal starches: corn, wheat, and rice. A disadvantage of those starches was the tendency to give rigid gels because of the retrogradation of the amylose fraction. Waxy starches show no immediate tendency to retrogradation, but they are mechanically unstable. Dispersing

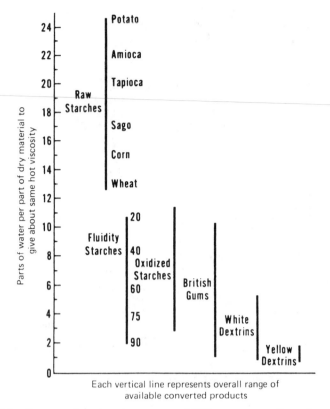

Figure 17-9. Comparative viscosity ranges of different types of starches. (Courtesy, Corn Refiners Assoc., Washington.)

the starch in dilute alkali or adding a small amount (less than 0.1%) of sodium trimetaphosphate or epichlorhydrin introduces ester or ether cross-linkages into the amylopectin molecule. The treatment reduces swelling and solubility and at the same time improves gel stability. To modify texture of potato starch gels from a highly extensible and cohesive viscoelastic to a short gel, stearic acid or monoglycerides are added. The lipids form complexes with the linear fraction of the starch. The complexes show reduced swelling and solubility during gelatinization. Cross-bonding of waxy starches is also important in increasing stability of starch thickeners to high temperatures employed in sterilization of canned foods; to mechanical agitation and shear during pumping, mixing, and homogenization; and to thinning effects of highly acid food ingredients.

Other functions of starches in foods can similarly be modified advantageously by chemical treatment. The treatments include, in addition to cross-bonding for improvement of paste clarity, introduction of ionic phosphate groups into cross-bonded waxy starch (to maintain high water-holding capacity and improve freeze-thaw stability of frozen foods).

$$2STOH + CH_3-\overset{\overset{O}{\|}}{C}-O-\overset{\overset{O}{\|}}{C}-(CH_2)_n-\overset{\overset{O}{\|}}{C}-O-\overset{\overset{O}{\|}}{C}-CH_3 \rightarrow STOC-(CH_2)_n-\overset{\overset{O}{\|}}{C}-O-ST$$

$$+ 2CH_3-\overset{\overset{O}{\|}}{C}-O^- -Na^+$$

$$POCl_3 + 3STOH \rightarrow STO-\overset{\overset{O}{\|}}{\underset{\underset{OST}{|}}{P}}-OST + NaCl$$

Figure 17-10. Cross-linking by ester formation. (From Wurzburg and Szymanski, 1970.)

For industrial uses starch is generally gelatinized and dried as a penetrating and coating film (in paper or textile sizing) or as an adhesive film. Numerous physically—or chemically—modified starches are manufactured for industrial uses. They include starches of various particle size; pregelatinized starches (sizing agents for home laundry, adhesives for metal casting, bodying agents for oil-well drilling); acid- or enzyme-modified thinned starches; white and yellow dextrins (obtained by heating of dry acidified starch, and used as adhesives); starches oxidized with alkaline hypochlorite to give water-soluble products with practically no retrogradation and low viscosity (for use in paper sizing); hydroxyethylether and acetate ester starches prepared by treating alkaline starch suspensions with ethylene oxide or acetic anhydride (giving strong clear films for size finishing of cotton textiles and for clay-coating of printing papers); starches cross-bonded with formaldehyde (for use in paste adhesives); and some industrial starch derivatives (cationic starch ethers in which the starch has a positive charge from introduction of quarternary ammonium groups, or dialdehyde starches obtained by oxidizing granular starch with periodate to introduce two aldehyde groups per glucose unit). Enzymes are also sometimes used in making specific types of dextrins.

Cross-linking by ester formation is depicted in Figure 17-10, and preparation of acetates is shown in Figure 17-11. Figure 17-12 illustrates the effect of cross-linking on viscosity of waxy maize starch, and Figure 17-13 shows the effect of shear on viscosity of unmodified and lightly cross-linked starch. The effect of cross-linking on resistance to shear is quite dramatic.

CORN SWEETENERS

Production of Corn Syrup

The manufacture of corn sweeteners is a multistep continuous process. (see Figure 17-3). To convert the starch granules in the slurry to corn syrup, the

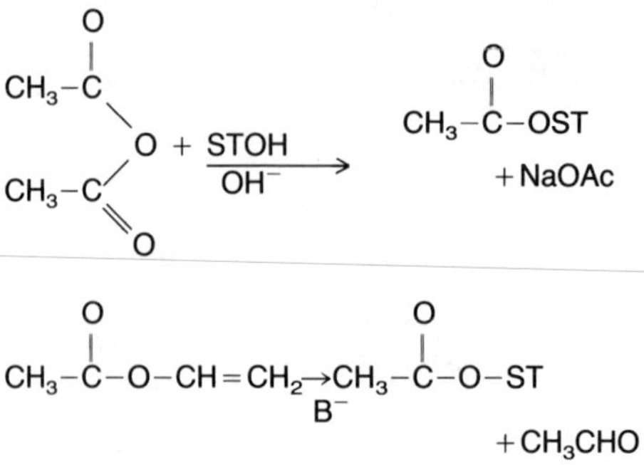

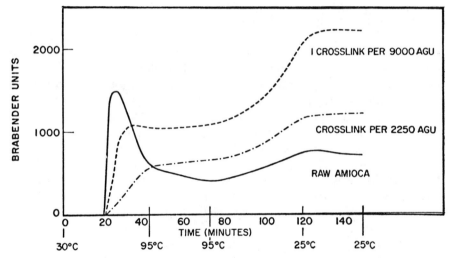

Figure 17-11. Preparation of starch acetates. (From Wurzburg and Szymanski, 1970.)

Figure 17-12. Effect of cross-linking on viscosity of waxy maize starch 5% solids. (From Wurzburg and Szymanski, 1970.)

granules must be gelatinized and the starch depolymerized in a conversion process that is halted as soon as the desired composition is reached. Two or more interrelated processes may be involved. Thus, for instance, an acid primary conversion may be followed by an enzymic conversion. The products of the conversion may be used in the production of isomerized corn syrups. The starch conversion products are classified by their dextrose equivalent DE. This is a measure of the reducing sugar content calculated as anhydrous dextrose and expressed as a percentage of total dry substance. DE is a useful parameter in classifying corn syrups, but it does not provide full information on actual composition.

While the most common methods used in the production of corn syrup

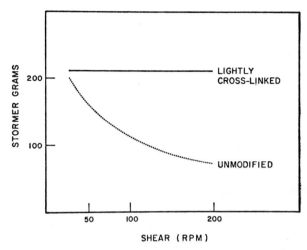

Figure 17-13. The effect of shear on viscosity of lightly cross-linked and unmodified starch. (From Wurzburg and Szymanski, 1970.)

are the acid and acid-enzyme processes, some syrups are produced by multiple enzyme processes. In the acid conversion process, a starch slurry of about 35–40% dry matter is acidified with hydrochloric acid to pH of about 2 and pumped to a converter (Figure 17-14). In the converter, the steam pressure is adjusted to about 30 lb/sq in. and the starch is gelatinized and depolymerized to a predetermined level. The process is terminated by adjusting the pH to 4–5 with an alkali. The liquor is clarified by filtration and/or centrifugation and is concentrated by evaporation until it contains about 60% dry matter. The syrup is further clarified and decolorized by treatment with powdered and/or granular carbon, refined by ion exchange to remove soluble minerals and proteins and to deodorize and decolorize, and further concentrated in large vacuum pans or continuous evaporators.

In the acid-enzyme process, the liquor containing a partially converted product is treated with an appropriate enzyme or combination of enzymes to complete the conversion. Thus, in the production of a 42-DE high-maltose syrup, the acid conversion is carried through until dextrose production is negligible. At this point, beta-amylase (a maltose-producing enzyme) is added and the conversion is continued. The enzyme is deactivated, and the purification and concentration are continued as in the acid process.

In enzyme-enzyme processes, the starch granules are cooked, preliminary starch depolymerization is done by starch-liquefying alpha-amylase, and the final depolymerization is done by a single enzyme or a combination of enzymes (Figure 17-15). Combinations of enzymes make possible the production of syrups with specific composition and/or properties, such as high-maltose or high fermentable syrups.

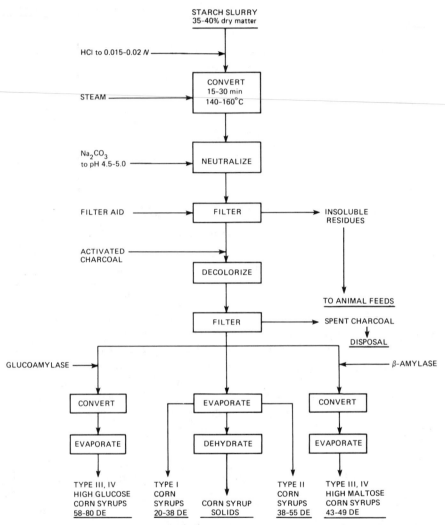

Figure 17-14. Acid-converted corn syrup process. (Courtesy, Corn Refiners Association, Washington.)

Types of Corn Sweeteners

There are five types of corn sweeteners.

Corn syrup (glucose syrup) is the purified concentrated aqueous solution of saccharides obtained from edible starch. It has a DE of 20 or more.

Dried corn syrup (dried glucose syrup) is corn syrup from which the water has been partially removed. Refined corn syrups are spray or vacuum drum-dried to low moisture content, and they form granular, crystalline, or powdery amorphous products. They are mildly sweet and moderately

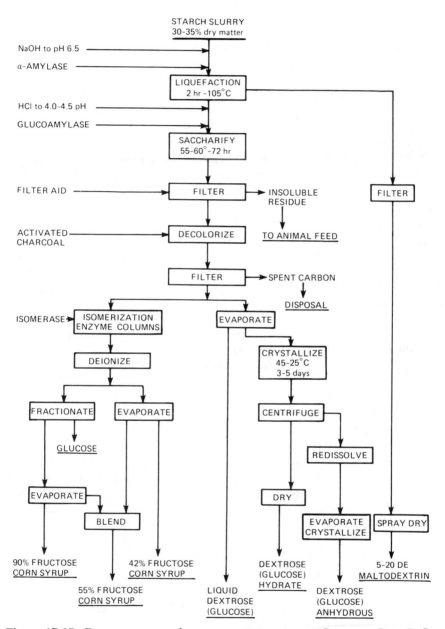

Figure 17-15. Enzyme-converted corn sweetener process. (Courtesy, Corn Refiners Association, Washington.)

hygroscopic. Because of their hygroscopicity, they are packed in multiwall, moisture-proof paper bags. The products are comparable in chemical composition to their liquid corn syrups, except for lower moisture contents.

Maltodextrin is a purified, concentrated, aqueous solution of saccharides or the dried product derived from the solution obtained from starch. It has a DE of less than 20. Maltodextrins are produced in the same manner as corn syrups except that the conversion process is stopped at an early stage to keep the DE below 20. Both acid and enzyme processes can be used. Maltodextrins are usually dried to white, free-flowing powders and are packed in multiwall bags. However, in some cases, moderately concentrated solutions of maltodextrins are sold.

Dextrose monohydrate is purified and crystallized D-glucose containing one molecule of water of crystallization per molecule of D-glucose. For the manufacture of dextrose, complete depolymerization of the starch substrate and recovery of the product by crystallization are required.

The starch slurry is gelatinized as in the manufacture of corn syrup and is partially converted by acid or alpha-amylase. Then a purified glucoamylase enzyme, free of transglucosylase activity, is added to the intermediate substrate. When the dextrose conversion is complete, the enzyme is deactivated and the dextrose liquor is filtered to remove residual suspended materials and purified and decolorized with granular or powdered carbon.

The liquor is concentrated to about 75% solids, cooled, and pumped into crystallizers. The temperature is slowly lowered to about 25°C. Crystallization is induced by seed crystals left in the crystallizer from the previous batch. Dextrose monohydrate crystallizes from the mother liquor, is separated by centrifugation, and is washed in the centrifuges with a spray of water. The wet crystals are dried in warm air to about 8.5% moisture. The mother liquor is reconverted, refined again, concentrated, and crystallized to produce a second crop of dextrose hydrate.

Dextrose anhydrous is purified and crystallized D-glucose without water of crystallization. Anhydrous dextrose is obtained by redissolving dextrose hydrate and refining the solution to a highly purified and clear filtrate. The solution is evaporated to a high solids content, and anhydrous alpha-D-glucose is precipitated by crystallizing at an elevated temperature. The anhydrous crystals are separated by centrifugation, washed with a warm water spray, and dried. Anhydrous dextrose also can be made by direct crystallization from a high-DE liquor.

Syrup Characteristics and Properties

The characteristics and functional properties of corn syrups vary according to their composition. Corn syrups are classified into four types on the basis of dextrose equivalents (DE):

Type I   20–38 DE
Type II   38–58 DE
Type III   58–73 DE
Type IV   73 DE and above

To evaluate a syrup adequately, its actual carbohydrate composition must be known. Such information may be obtained by chromatographic analysis. Results of such analyses are given in Table 17-11.

In general, type I syrups and maltodextrins have a relatively small concentration of low-molecular-weight sugars such as dextrose, maltose, and maltotriose. Types III and IV syrups have relatively small concentrations of oligosaccharides above maltoheptaose. Compositions of type II syrups depend on the process and the extent of conversion.

It is frequently possible to describe the composition of a syrup in terms of DE and one or more of the saccharide fractions. For example, 43-DE high-maltose syrup is made by an acid-enzyme process, in which a partially converted corn syrup is treated with beta-amylase to produce a syrup high in maltose and low in dextrose. A regular type II 42-DE acid-converted syrup contains about 20% dextrose and 14% maltose; a 43-DE high-maltose type II syrup contains about 8% dextrose and 40% maltose.

Corn syrups are also characterized according to solids content. Most commercial corn syrups are sold on a Baumé (Bé) basis, which is a measure of the dry-matter content and specific gravity. However, the high fructose corn syrups (see later in this chapter) are sold on the dry-matter basis. Since all corn syrups, except high-fructose syrups and high-dextrose syrups, are very viscous at room temperature, the Baumé determination is made at 60°C and an arbitrary correction of 1.00 Bé is added to the observed reading. This provides a value called "commercial Baumé." Most corn syrups are available in the range of 41–45 Bé, corresponding to a dry-matter content of about 77–85%. High-fructose syrups and high-dextrose syrups are available at about 71% dry-substance content.

Corn syrups and dextrose are sweet but otherwise essentially tasteless. The sweetening power of corn sweeteners depends on the percentages of sweetener solids and the combination of sweeteners. A 2% water solution of dextrose is about two-thirds as sweet as a 2% sucrose solution. As the concentration is increased, the difference decreases.

When corn syrup or dextrose is used in combination with sucrose, the sweetness is usually greater than expected. For example, at 45% solids, a mixture of 25% 42-DE corn syrup and 75% sucrose is as sweet as a sucrose solution of 45% solids. Consequently, when corn sweeteners are used with sucrose in products high in total sweetener concentration, there is no noticeable loss of sweetness. In less-sweet products, or when only one sweetener is used, a decrease in sweetness may be expected if sucrose is replaced by some corn sweeteners.

Sweetness is also influenced by temperature and the presence of nonsugar substances. It is therefore difficult to state accurately the relative sweetness of syrups and sugars for use in the food industries. Each product should be considered individually.

Corn syrup, corn syrup solids, and dextrose are compatible with other sweeteners and with food flavors. They are widely used with other sweeteners, mainly sucrose. In canned fruits, for example, a liquid packing medium containing a blend of several corn sweeteners gives optimum sweetness combined with superior gloss and mouth feel. In ice cream, body and texture are improved through the use of corn sweeteners and the product is smoother, has a better "melt-down," and is more resistant to "heat shock". In sherbets and ices, corn syrup sweeteners tend to eliminate crystallization and promote smoothness.

Physical characteristics of sweeteners are summarized in Table 17-12. Those and other physical properties of the sweeteners are important to food and beverage products. Practically all corn sweeteners are readily soluble in water; aqueous solutions containing 70–80% dissolved solids can be obtained at room temperature. Dextrose has a negative heat of solution; ie, it cools the mixture in which it is dissolved. Anhydrous dextrose melts at 146°C, and dextrose hydrate melts at 82°C. Dried corn syrups or commercial corn syrup solids are granular, crystalline, or amorphous powders which, when heated gradually, soften or dissolve in their own trace moisture.

Corn sweeteners are hygroscopic; the degree of hygroscopicity increases as the DE increases. Corn syrups and dextrose are employed as moisture conditioners and stabilizers. The higher saccharides impart to corn syrup cohesive and adhesive properties, and they exhibit some attributes of vegetable gums. The higher saccharides also contribute a chewy texture to confections and chewing gum.

Corn sweeteners, particularly the corn syrups, control crystallization of sucrose. This property is advantageous in confections, ice cream, frozen desserts, jams, jellies, and preserves.

Dextrose and fructose have a relatively high osmotic pressure. This enhances their effectiveness in inhibiting microbial spoilage. Corn syrup of 55 DE has about the same average molecular weight as sucrose or lactose and hence about the same osmotic properties as those sugars. Corn syrups of lower DE have higher molecular weights and correspondingly lower osmotic pressures.

The effect of corn syrups and sugars on the freezing point of a solution is significant in the manufacture of ice cream and frozen desserts. The lowering of a solution's freezing point is inversely proportional to the molecular weight of the dissolved solids. Generally, type I and type II syrups depress the freezing point somewhat less than an equal weight of sucrose. Type III corn syrups have about the same effect as sucrose on the freezing point. Dextrose and type IV corn syrups contain substantial proportions of monosaccharides and lower the freezing point to a greater

Table 17-11. Examples of Carbohydrate Composition of Commercially Available Corn Syrups

Type of conversion	Dextrose equivalent	Saccharides							
		Mono	Di	Tri	Tetra	Penta	Hexa	Hepta	Higher
Acid	30	10.4	9.3	8.6	8.2	7.2	6.0	5.2	45.1
Acid	42	18.5	13.9	11.6	9.9	8.4	6.6	5.7	25.4
Acid-enzyme[a]	43	5.5	46.2	12.3	3.2	1.8	1.5	—	29.5[a]
Acid	54	29.7	17.8	13.2	9.6	7.3	5.3	4.3	12.8
Acid	60	36.2	19.5	13.2	8.7	6.3	4.4	3.2	9.5
Acid-enzyme[a]	63	38.8	28.1	13.7	4.1	4.5	2.6	—	8.2
Acid-enzyme[a]	71	43.7	36.7	3.7	3.2	0.8	4.3	—	7.6[b]

[a]The carbohydrate composition of acid-enzyme syrups will vary as a result of different processes used. The values given here are to be considered only as examples of ranges of values that are available commercially.
[b]Includes heptasaccharides.
Source: Nesetril, 1967.

Table 17-12. Physical Characteristics of Sweeteners (100 lb of product)

Sweetener	Solids	Fermentable solids (%)	Water (%)	Viscosity CPS (100°F)	Storage temp (°F)	Percent dry basis composition		
						Dextrose	Levulose	Maltose
Dry sugar	100	100	—	—	—	—	—	—
Dry dextrose	91.5	91.5	8.5	—	—	100	—	—
Corn syrup (70 DE)	82.5	82	17.5	5,200	95	50	—	29
30/70 blend[a]	77.9	86.6	22.1	1,550	85	39	10.1	18.4
Liquid sugar	67	67	33	90	Room temp	—	—	—
Liquid dextrose	70	70	30	80	130	100	—	—

[a]30% Liquid sugar/70% corn syrup—70 DE.
Source: Piekarz, 1968.

extent. Corn sweeteners are used by the ice cream industry to control sweetness and improve body and texture.

Fermentability is an important property of corn syrups and sugars, particularly in the baking and brewing industries. The lower-molecular-weight sugars, mainly the mono- and disaccharides, glucose, fructose, maltose, and sucrose, are readily fermentable by yeasts. The total fermentability of corn syrups is roughly proportional to their content of mono-, di-, and trisaccharides—the higher the DE, the higher the fermentability.

Dextrose and fructose combine with nitrogenous compounds at elevated temperatures to produce brown coloration resulting from the Maillard reaction. This makes corn sweeteners useful in the manufacture of caramel color, promotes crust color in baking, and produces desirable color and caramel flavor in other food products and confections.

Isomerized Corn Syrups

In recent years, isomerized corn syrups have become a sizable part of the total production of syrups by the U.S. corn wet-milling industry. Commercial isomerized corn syrups are clear, sweet, bland, low-viscosity sweeteners high in dextrose (glucose) and levulose (fructose). High-levulose corn syrups are functionally equivalent to liquid invert sugar in most foods and beverages and can be substituted for it with little or no change in formulation, processing, or final product (Wardrip, 1971).

The high-levulose syrups (also called high-fructose corn syrups) are prepared by the enzymatic action of dextrose isomerase on dextrose. That converts a portion of the dextrose to levulose. The substrate can be dextrose or a high-conversion corn syrup composed mostly of dextrose. The levulose content in the syrup may be 50% or more, depending on the substrate, method of preparation, etc. The principal high-levulose corn syrup marketed in the United States today contains 42% levulose.

The high-levulose corn syrups currently on the market are composed mainly of the simple sugars dextrose and levulose. The composition of a 42% levulose syrup is similar to that of commercial invert sugar (Table 17-13). The high-levulose corn syrups are low in viscosity and are easy to ship, store, and blend.

According to Fruin and Scallet (1975), attributes of high-levulose corn syrup that are of particular interest in foods are their ability to do the following: (1) retain moisture and/or prevent drying out; (2) control crystallization; (3) produce an osmotic pressure that is higher than for sucrose or medium invert sugar and thereby help control microbiological growth or help in penetration of cell membranes; (4) provide a ready yeast-fermentable substrate; (5) blend easily with sweeteners, acids, and flavors; (6) provide a controllable substrate for browning and Maillard reactions; and (7) impart a degree of sweetness that is essentially the same as in invert liquid sugar.

Table 17-13. Carbohydrate Profile of Various Nutritive Sweeteners (in percent)

Sweetener	DP[a]		DP₂		DP₃	DP₄ and higher
	Fructose	Dextrose	Sucrose	Maltose	Triose	
Sucrose	—	—	100	—	—	—
Medium invert	25	27	46	2[b]	—	—
Total invert	45	48	3	4[b]	—	—
Dextrose	—	100	—	—	—	—
42 DE, corn syrup	—	20	—	14	12	54
42 DE, high maltose	—	8	—	40	15	37
62 DE, corn syrup	—	39	—	31	7	23
42% high fructose corn syrup	42	52	—	6[b]	—	—
55% high fructose corn syrup	55	40	—	5[b]	—	—
90% high fructose corn syrup	90	9	—	1[b]	—	—

[a]DP = degree of polymerization.
[b]DP₂ and higher saccharides.
Source: Young, 1981.

High-levulose corn syrups have many applications in foods and beverages. They are equivalent to invert sugar in providing humectancy and sweetness and in controlling crystal size in some candy. In some candy products, however, there may be a problem of excessive moisture pickup. Large amounts of high-levulose corn syrup may unduly darken, especially at high temperatures, and soften some candies by inhibiting sugar crystallization. In bread, sucrose or invert sugar can be replaced on a pound-per-pound basis by high-levulose corn syrups. Similarly, blends of high-DE corn syrups with sucrose or with high-levulose syrups are functionally equivalent on a solids basis in bread, sweet-goods dough, Danish, pie fillings, soft-moist cookies, and bakers' jellies and soft fillings. The high-levulose syrups must be used sparingly in icings that depend on sucrose crystal development for body and consistency and in baking of light-colored and dense cakes.

According to Fruin and Scallet (1975), high-levulose corn syrups can replace up to 100% sucrose in bread, pie fillings, and jellies and fillings, up to 70% in soft cookies, 25–75% in boiled and marshmallow icings, 20–50% in dark-colored cakes, up to 30% in chiffon or angel food cake, up to 20% in hard cookies or white icings, and 10–15% in flat icings.

Corn sweeteners in combination with sucrose or invert sugars are available commercially. In using the combinations, one must consider the moisture content and percent of fermentables of each blend component.

BIBLIOGRAPHY

Anderson, R.A.; Watson, S.A. In "Handbook of Processing and Utilization in Agriculture." Wolf I.A., Ed. CRC Press; Boca Raton, FL; **1982**, Vol. 2, Part I, pp. 31–61.

Brekke, O.L. In "Corn: Culture, Processing, Products." Inglett, G.E., Ed. Avi Publ. Co.; Westport, CT; **1970**.

Earle, F.R.; Curtis, J.J.; Hubbard, J.E. *Cereal Chem.* **1946**, *23*, 504–511.

Fruin, J.C.; Scallet, B.L. *Food Technol.* **1975**, *29*(11), 40, 42, 44, 45.

Harness, J. In "Seminar Proceedings Products Corn Refining Industry in Food." Corn Refiners Assoc.; Washington, D.C.; **1978**, pp. 7–10.

Horn, H.E. *Cereal Foods World* **1981**, *26*, 219–223.

Meyer, P.A. *Milling Baking News* **1984**, *62*(53), 19–20.

Nesetril, D.M. *Bakers Digest* **1967**, *41*(3), 28–30, 32.

Piekarz, E.R. *Bakers Digest* **1968**, *42*(5), 67–69.

Pomeranz, Y. "Functional Properties of Food Components." Academic Press: Orlando, FL; **1984**.

Rapp, W. In "Seminar Proceedings Products Corn Refining Industry in Food." Corn Refiners Assoc.; Washington, D.C.; **1978**, pp. 11–17.

Reiners, R.A. In "Seminar Proceedings Corn Refining Industry in Food." Corn Refiners Assoc.; Washington, D.C.; **1978**, pp. 18–21.

Selby, K. *Chem. Ind. (Lond.* **1977**, 494–498.

Stiver, T.S. *Bull. Assoc. Operative Millers* **1955**, 2168–2179.

Wardrip, E.K. *Food Technol.* **1971**, *25*(5), 47.

Watson, S.A. In "Starch Chemistry and Technology," Vol. II; "Industrial Aspects." Whistler, R.L.; Paschall, E.P., Eds. Academic Press; New York; **1967**.

Wurzburg, O.B.; Szymanski, C.D. *Agric. Food Chem.* **1970**, *18*, 997–1001.

Young, L.S. In "Conference on Formulated Foods and Their Ingredients." Agric. Food Chem., Div. Am. Chem. Soc.; Anaheim, CA; **1981**.

Chapter 18

Barley

FOOD USES

Uses of barley and barley products are listed in Table 18-1. Most barley that goes into human food is consumed as pot barley or pearl barley (Wiebe, 1968). Pot and pearl barley are both manufactured by gradually removing the hull and outer portions of the barley kernel by abrasive action. The pearling or decortication process used to produce pot barley is merely carried out further to produce pearl barley; 100 lb of barley normally yields 65 lb of pot barley or 35 lb of pearl barley. Barley flour is a secondary product, and polishings are a by-product of the pearling process. On the basis of decreases in weight and changes in chemical composition, it has been computed that six pearlings remove 74% of the protein, 85% of the fat, 97% of the fiber, and 88% of the mineral ingredients contained in the original barley. Table 18-2 compares the gross composition of barley products obtained in pearling barley (Rohrlich and Bruckner, 1966). Composition of commercial barley products is listed in Table 18-3. Chester and Portage are creamy white, pearled barley products; both are used in food and pet food applications as thickeners and fillers. About 90–95% and 80–90% is retained on U.S. sieve No. 8 for Chester and Portage barley products, respectively. The quick-cooking product is a creamy white, pearled barley that has been steamed and rolled into a thick flake for quick cooking (100% of the thick flakes is retained on U.S. sieve No. 8). The quick-cooking barley is used as a major ingredient in dry soups and as a thickener when a quick-cooking product is required. The flakes are a creamy white, pearled barley that has been steamed and rolled into thin flakes. Barley flakes provide a less chewy texture than oat flakes. Barley flakes can be used as an ingredient in granola products. They can also be used in specialty breads to provide texture. Barley flour is milled from barley grain that has been pearled, steamed, and ground to produce a stable product in which enzyme activity has been minimized. Only 0–2% is retained on U.S. sieve No. 2, and 70–80% passes U.S. sieve No. 100. It can

Table 18-1. Present Uses of Barley and Barley Products

Type	Use
Feed	Livestock Poultry
Pearling	Pot barley for soups and dressings Pearled barley for soups and dressings Flour Feed
Milling	Flour for baby foods and food specialties Grits Feed

Malting	Brewers' beverages
	Brewers' grains for dairy feeds
	Brewers' yeast for animal feed, human food, and fine chemicals
	Distillers' alcohol
	Distillers' spirits
	Distillers' solubles for livestock and poultry feeds
	Distillers' grains for livestock and poultry feeds
	Specialty malts
	  High dried
	  Dextrin for breakfast cereals, sugar colorings, dark beers, and coffee
	    substitutes
	  Caramel for breakfast cereals, sugar colorings, dark beers, and coffee
	    substitutes
	  Black for breakfast cereals, sugar colorings, dark beers, and coffee
	    substitutes
	Export
	Malt flour for wheat flour supplements, human and animal food production
	Malted milk concentrates for malted milk, malted milk beverages, and infant
	  food
	Malted syrups for medicinal, textile, baking uses, and for breakfast cereals
	  and candies
	Malted sprouts for dairy feeds, vinegar manufacture, and industrial
	  fermentations

Source: Phillips and Boerner, 1935.

be used as a thickener, stabilizer, binder, or protein source for baby foods, malt beverages, prepared meats, and pet foods (Pomeranz, 1974).

MALTING AND BREWING

Beer is a fermented, hopped, malt beverage. The three most important ingredients used in the production of beer are malt, hops, and yeast. Malting involves controlled wetting (by steeping) and germination of seeds under conditions conducive to production of desirable physical and chemical changes associated with the germinative process while holding weight losses due to germination and respiration to a minimum (Cook,

Table 18-2. Composition in Percent of Milled Barley and Products of Barley Milling[a]

Product	Moisture	Protein	Fat	N-free extract	Crude fiber	Ash
Dehulled barley	12.5	10.6	1.7	72.1[b]	1.6	1.5
Pearls	12.5	7.8	1.0	76.2	1.4	1.1
Pearling dust	12.5	9.5	1.4	74.3	0.8	1.5
Feedmeal	12.0	12.5	2.0	64.0	5.0	3.5
Bran	10.5	14.0	3.5	57.1	10.0	4.9
Husks	10.4	3.6	1.0	49.2	28.6	7.2

[a]Adapted from Rohrlich and Bruckner, 1966; as-is basis.
[b]In hulled barley.

Table 18-3. Composition in Percent of Typical Barley Products

Barley product	Moisture	Protein[a]	Fat[a]	Fiber[a]	Ash[a]
Chester	9.0–10.0	11.0–12.0	2.0–2.5	1.0–1.5	0.7–1.3
Portage	9.0–10.0	11.0–12.0	1.0–1.5	1.0–1.5	0.7–1.3
Quick-cooking	9.0–10.0	10.0–12.0	1.0–1.5	0.5–1.0	0.7–1.3
Flakes	10.0–12.9	10.0–12.0	1.0–1.5	0.5–1.0	0.7–1.3
Flour	—	13.0–15.0	1.9–2.9	1.2 max	1.0–2.0

[a]Dry-matter basis.
Courtesy Quaker Oats Co., Chicago.

1962). Malting develops the amylolytic enzymes that modify starch to fermentable carbohydrates, the source of alcohol in beer. Malt also contains a series of proteases and cytolytic enzymes. The products of proteolytic degradation (along with other components) act as flavor precursors and as nutrients for the yeast during fermentation. The malted grain is dried to halt growth and stop enzymatic activity and to develop a storable product of desired color and flavor. Drying is followed by removal of malt sprouts. The malting process is given in the form of a flowsheet in Figure 18-1.

In the old system of malting, steeped barley was germinated on concrete floors in cool, moist rooms and turned by hand. Floor malting was replaced by pneumatic-type malting in which conditioned air was forced through the grain. Malting is performed in drums or compartments. Newer construction involves mainly compartments and greater mechanization to reduce labor costs and to obtain a more uniform product. The first commercial continuous-malting plant is the Domalt system from Canada. Pneumatic malting systems require large quantities of air adjusted to a desired temperature and saturated with water vapor.

Barley in Malting

Barley occupies a unique position in malting and brewing. That position stems from the fact that during malting barley produces many hydrolytic

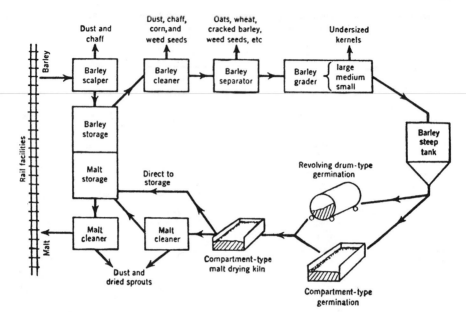

Figure 18-1. Typical malt house flowsheet. (From A.D. Dickson.)

enzymes including relatively large amounts of alpha- and beta-amylases. The combination of the two amylolytic enzymes results in a more complete and rapid degradation of starch than in malts from most other cereal grains. This degradation of starch is accompanied by the breakdown of other grain components (mainly proteins and nonstarchy polysaccharides) and yields an optimally modified malt (Pomeranz, 1975a). In barley, the husks are cemented to the kernel and remain attached after threshing. The husks protect the kernel from mechanical injury during commercial malting, strengthen the texture of the steeped barley, and contribute to a more uniform germination of the kernels. The husks are also important as a filtration aid in the separation of extract components during mashing, and they contribute to the flavor of the malt and astringency of the beer.

Composition of Barley

On a dry-matter basis, covered barley contains 63–65% starch, 1–2% sucrose, about 1% other sugars, 1–1.5% soluble gums, 8–10% hemicellulose, 2–3% lipids, 8–13% protein ($N \times 6.25$), 2–2.5% ash, and 5–6% other components. In regular barley, the linear starch component (amylose) comprises 24% of the total starch.

The proteins in barley are composed of four groups varying in solubility. The albumin fraction comprises less than 10% of the proteins, the globulins about 20%, the hordeins (soluble in 70% alcohol) 30%; the remaining 40% of the proteins are glutelins. About one-half of the amino acid residues in hordeins are either glutamine or proline; the amounts of aspartic acid, glycine, and lysine are small. The amino acid composition of the glutelins resembles that of hordeins.

Barley lipids are concentrated in the embryo and the aleurone layer. Although the whole grain has only 2% petroleum-ether–extractable components, isolated embryos contain 15% lipids. The predominant constituent fatty acids are linoleic, oleic, and palmitic, with the unsaturated components accounting for nearly 80% of the total.

Mature barley may contain over 2% of fructosans. Unlike starch, which is restricted to the starchy endosperm, the fructosans are distributed throughout the grain. Sucrose is virtually restricted to the embryo and aleurone; it represents 12–15% of the embryo but only 1–2% of the whole grain. Raffinose is also a major embryo constituent—about 5% of the dry weight. The husks contain over two-thirds of the grain's cellulose; the cell walls of the starchy endosperm lack true cellulose.

Malt Types

Malting barley in the United States was reviewed by Peterson and Foster (1973). Two general types of malts are produced commercially—brewers' and distillers' malts. Brewers' malts are made from barleys of plumper, heavier kernels with a mellow or friable starch mass. They are steeped and germinated at moisture contents ranging from 43% to 46%; the final temperature used in drying the malts is in the 71–82°C range. These malts are dried to about 4% moisture. The high final drying temperature reduces the enzymatic activities of the malt, darkens the malt and the wort made from it, and increases the malt flavor and aroma. Distillers' (or high-diastatic) malts are made from small-kerneled barleys high in protein content and enzymatic potentialities. The barleys are steeped and malted at higher moistures (45–49%) and dried at lower temperatures (49–60°C) to higher finished moisture contents (5–7%) than the brewers' type of malt.

According to MacLeod (1976), the success of malting initially depends on securing an optimum balance of oxygen and water within the embryo so that germination can proceed rapidly. Subsequent modification of the endosperm is related to the gibberellin-mediated synthesis of hydrolytic enzymes. Problems arise in enhancing endosperm modification without, at the same time, increasing excessive seedling growth. The relationship between malt and beer, the role of barley quality, and the biochemistry of malting and brewing were reviewed by Enari (1981). The most important biochemical process during malting and mashing are amylolysis, proteolysis, and cytolysis. Malt enzymes that participate in starch degradation are alpha-amylase, beta-amylase, debranching enzymes, alpha-glucosidase, beta-fructosidase, phosphorylase, and glucosyl transferases. The pH optima are 5.7 and 5.3, and the temperature optima are 55–65°C and 50–60°C for the alpha-amylase and beta-amylase, respectively.

About 90% of the proteolytic activity is due to two thiol proteinases with pH optima of 3.9 and 5.5, and 10% is due to three metalloproteinases with pH optima of 5.5, 6.9, and 8.5. In addition, there are a large number of peptidases (four carboxypeptidases, four aminopeptidases, and two alkaline dipeptidases).

Of the four beta-glucanases, beta-glucan solubilase is present in barley, green malt, and kilned malt; the others (endo-beta-1,4-, endo-beta-1,3-, and barley-endo-beta-glucanase) are present in green and kilned malt only. Endo-beta-1,4-glucanase and barley-endo-beta-glucanase have pH optima of 4.5–4.8; their optimum and inactivation temperatures are 40–45°C and 55°C, respectively. Endo-beta-1,3-glucanase and beta-glucan solubilase have higher optimum and inactivation temperatures (60–62°C and 70–73°C, respectively). Their pH optima are 4.6–5.5 and 6.6–7.0, respectively.

Some of the modern techniques for accelerated malting include hot-water steeping, spray steeping, abrasion, and the use of additives (Axcell et al, 1984). The use of enzymes in practical brewing was summarized by Woodward (1978).

Conventional malt quality parameters include the following:

Acrospire length: a guide to the level and uniformity of modification.
Diastatic power and alpha-amylase: as indices of amylolytic activity and capacity to degrade adjuncts in the brewing process.
Kolbach index, soluble nitrogen, and *free amino nitrogen*: as indices of nitrogen modification.
Viscosity (at 70°C or by the European Brewery Convention method), *water-insoluble hemicelluloses*, and *water-soluble gums*: as indices of degradation of beta-glucans by endo-beta-glucanases to dextrins.

According to Axcell et al (1984), those indices are of limited value and should be replaced or supplemented by determinations of (1) *barley variety*, as it governs alpha-amylase levels in malt, protein modification, beta-glucan content and degradation, and starch gelatinization characteristics; (2) *germinative energy* and *vigor*; (3) *gross composition* (mainly moisture and protein); and (4) *modification indices* and their uniformity, including colorimetric determinations of *beta-glucans* in individual kernels, *alcohol-soluble* (as index of undegraded hordein), *small starch granules* (which disappear faster in malting than large granules), *fermentability*, and *kilning monitors* (an enzyme marker, color, or other heat-sensitive factor).

Comparative ranges in composition of barley and malt are given in Table 18-4. When cross sections through an untreated barley kernel and kilned malt are compared, in the untreated barley kernel there is a protein matrix in which intact starch granules are imbedded, while in the kilned malt the protein matrix has basically disintegrated and is in part coagulated on the

Table 18-4. Comparative Ranges in Composition of Barley and Malt

Property	Barley	Malt (brewers' and distillers')
Kernel weight (mg)	32–36	29–33
Moisture (%)	10–14	4–6
Starch (%)	55–60	50–55
Sugars (%)	0.5–1.0	8–10
Total N (%)	1.8–2.3	1.8–2.3
Soluble N (% of total)	10–12	35–50
Diastatic power (°L)	50–60	100–250
Alpha-amylase (20° units)[a]	trace	30–60
Proteolytic activity[b]	trace	15–30

[a]20°C dextrinizing units, a unit of alpha-amylase activity.
[b]Arbitrary units.
Source: Wiebe, 1968.

surface of the starch granules. Several of the latter are degraded extensively (see also Pomeranz, 1974; Pomeranz and Sachs, 1972b).

The pattern of enzyme modification of malted grain is critical to the science and technology of malt production (Palmer and Bathgate, 1976). Brown and Morris emphasized as early as 1890 that failure to understand this basic feature of malting grain is at the very root of inefficient processing. The term modification describes the sum total of physical and chemical changes that take place during malting. According to MacLeod (1967), optimum modification is a "rather nebulous but nonetheless real condition that has resulted from the transformation of endospermic constituents" to yield a maximum of extractable solids while minimizing losses in weight of malt products and excessive degradation of high-molecular-weight components. According to Kneen and Dickson (1967), modification of cell wall material is primarily responsible, during malting, for the physical changes that convert hard and vitreous barley to friable and mellow malt. Early in malting, proteolytic enzymes are elaborated and distributed throughout the kernel. Proteolysis renders about 40% of the total protein soluble in dilute salt solutions. Kneen and Dickson (1967) postulated that the rigid kernel structure stems from the proteins in the barley endosperm and that degradation of the proteins contributes to the physical modification of the kernel.

Brown and Morris (1890) concluded that the dorsal (nonfurrowed) surface of the grain modified more rapidly than the regions of the ventral furrow. According to light-microscopic studies of Dickson and Shands (1941), structural changes in the endosperm of malting barley are mainly digestion of endosperm cell walls and dissolution of the protein matrix, in which the starch granules were embedded. There was little structural change in the starch granules except immediately adjacent to the scutellum. Dissolution of endosperm cell walls progressed from the scutellum to the distal end and was followed by dissolution of the protein matrix. Pomeranz

(1972) observed with scanning electron microscopy that partial breakdown of cell walls in the center of the starchy endosperm of malted barley was accompanied by extensive dissolution of the protein matrix and freeing of small starch granules that were previously embedded in that matrix. The effect on the appearance of the starch granules was small. Extent of matrix degradation differed widely in malts from barleys whose protein content varied. Whereas Briggs (1972) and Clutterbuck and Briggs (1973) stated that endosperm modification started over the surface of the scutellar epithelial cells, Palmer (1974) postulated that the role of epithelial cells in modification is insignificant.

More recently, Palmer et al (1985) reported on a study in which enzyme development in the scutellum of the embryo was associated with aleurone contamination. The aleurone layers of excised endosperms of barley produced alpha-amylase in the absence of growing embryos, suggesting that aleurone-free scutellar tissue does not have the potential to synthesize and secrete significant levels of endosperm-degrading enzymes into the starchy endosperm of germinated barley.

Spray steeping, in contrast to multiple steeping, reduced endosperm breakdown and increased beta-D-glucan release and extract viscosity. Spray steeping also depressed the development of alpha-amylase, endo-beta-1,3:1.4-glucanase, and pentosanase but encouraged development of endo-beta-1.3-glucanase. Gibberellic acid was essential for optimal enzyme development and release. However, even in the presence of gibberellic acid a significant amount of pentosanase was retained in the aleurone. It was postulated that steeping methods govern malt quality by promoting optimal enzyme development and secretion into, and uniform hydration of, the starchy endosperm. According to Marchylo et al (1985), the embryo is involved in the synthesis of alpha-amylase I, rather than of alpha-amylase II.

Dronzek et al (1972) found enzymatically degraded starch granules (mainly the larger A-type) near the aleurone layer in wheat sprouted for 2 days. MacGregor (1980), however, has shown that the changes are quite complex. Malted barley contains three main, heterogeneous groups of alpha-amylase. Basically, small granules were hydrolyzed faster than large granules. Starch degradation started at the endosperm edge adjacent to the embryo.

Fretzdorff et al (1982) studied modification in a kilned malt sample by a combination of histochemistry, light microscopy, and transmission and scanning electron microscopy. Hydrolysis of cell walls, proteins, and starch was most extensive in the starchy endosperm area adjacent to the scutellar epithelium. Some hydrolysis occurred in areas adjacent to the aleurone layers; hydrolysis decreased as distance increased from the embryo end to the distal end and from the aleurone layer to the center of the starchy endosperm. Although no rigid sequence of hydrolysis was observed, generally, cell-wall hydrolysis was more extensive than protein hydrolysis and starch hydrolysis seemed to take place gradually in the late stages of malting and kilning. Small granules were hydrolyzed more extensively than large granules.

The main by-product of malting is called "malt sprouts"; about 3–5% of malt sprouts are obtained in malt cleaning (Pomeranz and Robbins 1971). They are easily separated from the kilned malt by passing the malt through revolving reels of a wire screen. The sprouts include rootlets mainly 5–8 mm long and 0.3–0.4 mm thick. The sprouts have a yellow-brown color and a slightly bitter taste. Excessive amounts of rootlets are an economic loss. Consequently, the maltster is interested in optimally modified malts with smallest amounts of malt sprouts. Scanning electron micrographs of malt sprouts are shown in Figure 18-2.

Composition of malt sprouts varies with the source, malting length, and method of preparation and storage. The malting sprouts contain on an as-is basis 25.3–34.2% N-compounds, 1.6–2.2% fat, 8.6–11.9% fiber, 6.0–7.1% ash, and 35.2–43.9% N-free extract. Malt sprouts are used mainly in feed formulation.

ADJUNCTS IN BREWING

Adjuncts are used in brewing to impart certain characteristics to wort and beer, but mainly for reasons of economy. As the marketing of beers becomes more competitive and the prices of raw materials rise, greater attention is being paid to the problem of improving the utilization of raw materials. More than 90% of brewers' extract consists of carbohydrates, of which 70% is fermented in the production of beer. Obviously, carbohydrates represent one of the most important components in brewing. The use of cereal adjuncts has increased, and brewers in many countries incorporate up to 30% of these materials in their grist. In the United States and Canada, the proportion of adjunts used may in some instances be as great as 50%.

Adjuncts should be inexpensive and should contain more carbohydrates than malt. The adjuncts most commonly used in American brewing practice are corn or rice grits. In addition to an increase in the total amount of adjuncts in brewing, there has also been a change in the kinds of adjunts used. Corn and rice have a reciprocal usage relation, due to market price fluctuation of the two adjuncts. Some brewers feel that the alteration of flavor resulting from the use of corn is not great enough to justify using brewers' rice if there is any price differential. Others prefer brewers' rice to any other adjunct if the cost differential is not prohibitive. Materials used per barrel of beer (31.0 gal) total about 45 lb. According to Hartwick (1983), malt and malt products comprised about 65%, adjuncts about 35%, and hops and hops products about 0.5% of the materials used in brewing in the United States in 1980. There has been a large increase in the use of commercial corn syrups in beer production. The amount of hops is only

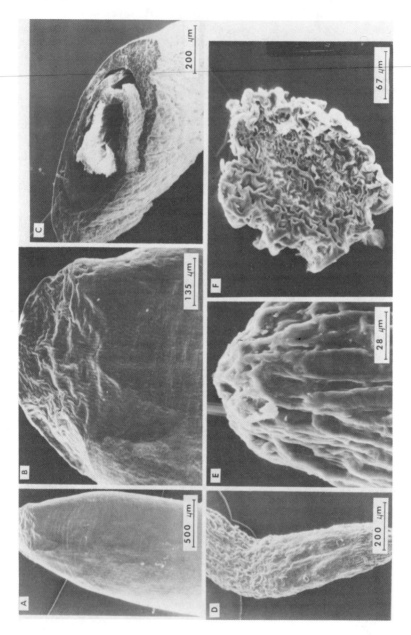

Figure 18-2. Scanning electron micrographs of rootlets and acrospires from Klages barley (malted 7 days and kilned). (A) Acrospire surface, (B) Acrospire tip, (C) Cross section through acrospire, (D) Rootlet surface, (E) Rootlet tip, (F) Cross section through rootlet.

about half that used in the United States some 50 years ago. In addition to corn and rice grits, small amounts of barley and sorghum are used. Wheat and wheat products are used in Australia. Corn and rice are preferred in the United States to obtain relatively pale, bland, and less filling beers. Adjuncts are claimed to improve the shelf life of beer.

Hops were used first in brewing by the Finns and were introduced by the Estonians to the Slavs and the Germans. The famous Hallertau hop fields of Germany were well established by 840 AD. Hops have medicinal and dietetic properties, act as a preservative, ensure the soundness and stability of beer, and impart a unique and pleasant aroma and flavor to fermented malt liquors. The hops of commerce are dried inflorescences of the female plant of the diocious, perennial species *Humulus lupulus*.

The axis of the hop cone is known as the "strig." It is bent at obtuse angles along its length and has at each angle four bracteoles enclosed by two bracts. The bracteoles are oval-shaped and slightly incurved at the base to carry the seed. The bracts are coarser, somewhat larger, and generally pointed at the tip. The rachis (strig) is covered with downy, linear, epidermal hairs. Specialized cup-shaped glandular hairs (lupulin glands) develop on the surface of the lower parts of the bracteoles and over the surface of the perianth; sometimes a few are found on the bracts. The lupulin glands contain the brewing principles, the bitter resins, and the essential oils. A membranous covering on the glands prevents escape of the lupulin.

THE BREWING PROCESS

The brewing process is given in the form of a flowsheet in Figure 18-3. The initial process in brewing of beer is mashing. The term "mashing" means extraction of malt with water, though in practice the process is accompanied by additional changes. Prior to mashing, the dried malt is milled. The milled product and water are put into a mash tub in which a series of controlled time-temperature treatments enhance dissolution of materials present in malt and enzymatic degradation of insoluble compounds.

There are two basic forms of mashing. Infusion mashing is the traditional and simplest method, in which the mixed grist and liquor are allowed to stand in a mash tub and the temperature is gradually raised but kept below the boiling point. This process somewhat resembles the making of tea. Decoction mashing is the traditional method in the preparation of a malt extract for bottom-fermented beers and is the main mashing method used in central Europe. In the decoction method, part of the mash is withdrawn, boiled, and returned to the mash tub to raise the temperature of the whole mash. The method used in the United States and Canada to incorporate relatively large amounts of adjuncts is called double mash. Two mashes are prepared: one mainly of malt with a small proportion of

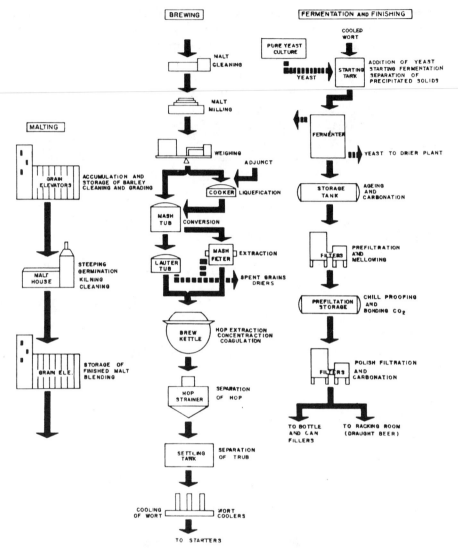

Figure 18-3 Malting and brewing flow chart. (Courtesy, Jos. Schlitz Brewing Co., Milwaukee.)

uncooked cereal; the other of a higher adjunct level with a smaller amount of malted barley (Yates, 1979). They are mashed separately at 29.5–38°C and allowed to stand for 30–60 minutes to dissolve the mash components and activate the enzymes. The main adjunct mash is heated at 70°C in a mash copper (cereal cooker) for 20 minutes. The saccharified mash is boiled for 45 minutes and mixed with the main malt mash at 68.5–70°C in a third vessel. After 15 minutes of intense saccharification, the temperature is raised to 73°C by the hot liquor or by steam. The wort is then run off in a lauter tub. Up to 60% of malted adjunct can be used (in the total malting grist) with this method.

In all processes the mash passes to a lauter tub with a false bottom. The grain husks collect on this false bottom and form a filter bed, through which the sugar-rich extract (wort) is strained. The spent grains are separated from the wort. The clear extract (sweet wort) is boiled with hops. The hot wort is strained to remove leaves and stems of hops. The hopped wort is cooled, yeast is mixed in, and the mixture is pumped to settlers where it remains for 10–12 hours. It is then transferred to fermentation tanks, where it remains until fermentation is completed and the fermented sugars are converted to carbon dioxide and ethanol. The fermented wort is allowed to age and is prefiltered, chill-proofed (by proteases to hydrolyze undegraded proteins and prevent their precipitation during chilling), filtered, carbonated, and filled into appropriate containers. Bottled or canned beer is either pasteurized or sterilized by ultrafiltration.

The following are some typical brewing figures:

Specific gravity of beer	1.012
Original extract of wort (solids before fermentation)	12.0%
Apparent extract of beer (Balling or Plato)	3.3%
Real attenuation (fermented solids)	7.0%
Alcohol (by volume)	4.4%
Extract, fermented	58.2%
Sugars in original extract (% of total)	66.5%
pH	4.6%
Amylodextrins	traces

A typical beer in the United States contains in addition to 90% water, 4% alcohol, 4% carbohydrates, 0.8% inorganic salts, 0.3% nitrogenous compounds, 0.2% organic acids, 0.5% CO_2, and 0.2% other compounds (Hartwick, 1983).

Chemical Changes during Brewing

The modifications during mashing involve simple dissolution of materials, already present in the malt in a soluble form, and enzymatic changes. The most important quantitative reaction during mashing is the conversion of starch into dextrins and maltose.

The question of changes in the nitrogenous compounds is more complex. Mashing generally favors the liberation of large peptides rather than the degradation of small ones. Boiling the wort stabilizes its composition by inactivating amylases and cytoclastic enzymes, by elimination (coagulation) of unstable colloidal protein (trub), and by extraction of wort components from hops. According to Hough et al (1971), the carbohydrates account for

91–92% of the hopped wort extract, of which 68–75% is fermentable by yeast. Of the wort carbohydrates, maltotetraose, more complex carbohydrates, and dextrins account for 23–28% of the extract, are not normally fermented, and persist into the finished beer. The nitrogenous constituents of wort account for up to 5% of the extract. They include amino acids, peptides, polypeptides, proteins, nucleic acids, and their degradation products. Most of the amino acids in the wort, except for proline, are assimilated during fermentation. The majority of the proteins are not assimilated during fermentation, and those that persist into the beer slowly react with other components to form nonbiological haze.

Beer solids consist of about 80% carbohydrates (mainly dextrins with a small amount of unfermented mono- and disaccharides). Gumlike materials range from 10% to 20% of the extract, but only about one-eighth of those are high-molecular-weight material. The nitrogenous materials show a wide range of complexity. Comparison of beer with wort indicated that the high-molecular-weight nitrogenous compounds were substantially unaltered by fermentation and persisted in the beer. As molecular weight decreased, increasing amounts disappeared during fermentation, and the simplest compounds (ie, most amino acids) were present in beer in trace amounts only. The extract usually contains 3–4% mineral matter and small amounts of miscellaneous compounds from hops and fermentation.

One of the characteristics most appreciated by beer lovers is beer foam and head retention (Lyons, 1984). Typically, good foam quality in the United States means a satisfactory high head retention value (at least 110 seconds), foam with a good appearance (small, uniform bubbles to produce a creamy rather than coarse open texture), and a reasonable cling of the beer foam. Beer foam formation involves foam or head formation, foam collapse, bubble collapse, and bubble cling, lacing, or foam adhesion.

The positive factors that contribute to foam formation and stability include proteins and glycoproteins, hop iso-alpha acids, some metallic components, viscosity, and foam stabilizers. Negative factors include lipids (in decreasing order diglycerides, free fatty acids, monoglycerides, and triglycerides), alcohol, sulfites, and surface-active agents (Lyons, 1984). Practically all steps in the brewing process affect beer foam; the raw materials improve it by contributing glycoproteins, the brewhouse process (as it affects excessive glycoprotein breakdown, lipid leaching, and alcohol formation), and postfermentation (as related to the amount of residual lipids and of yeast).

Asano and Hashimoto (1980) reported that "foaming proteins" consisted of three fractions with molecular weights of 90,000–1,000,000, 40,000, and 15,000. The three fractions were surface-active but differed in the mechanism of their contribution to foaming. The higher- and medium-molecular-weight fractions combined with isohumulones through epsilon amino groups to form more surface-active complexes and enhanced foaming. Immunological studies showed that "foaming proteins" were formed mainly during germination of barley. The amount of "foaming proteins" decreased considerably during brewing, particularly during kettle boiling. The finished beer contained 160–620 mg/L of those proteins.

Causes for gushing in beer can be divided into two types: Primary gushing is associated with the quality of barley and malt; secondary gushing is caused by faults during beer production or packaging. Primary gushing is most probably caused by mold growth on barley or malt. Kitabatake (1978) isolated a peptide associated with carbohydrates and polyphenols that induced gushing in beer when added at levels of 0.5 ppm to nonhopped wort. The peptide was acidic in nature; resistant to several proteases, carbohydrases, and a polyphenoloxidase; and degraded by pronase.

Haze formation in beer has been the subject of numerous investigations. Three kinds can be distinguished: biological haze due to microorganisms, chemical haze from the action of various chemical agents (starch, metallic aggregates), and colloidal haze due to aggregation of colloids. Most beer components in beer are in a colloidal state and carry an electrical charge. Colloidal stability of the beer depends to a large extent on the size and complexity of those aggregates. There is also interadsorption of colloids; this accounts for the heterogenous nature of the colloidal deposits from beer.

Both proteins and polyphenols are essential for haze formation; the haze also contains carbohydrates and metallic elements. Most if not all barley proteins are found in beer, though some predominate more than others. The mineral content and composition of hazes differ appreciably from the parent beer, indicating selective precipitation. The importance of metallic components in haze formation is not clear. They may be active initiators of haze formation or may be involved (as cations or chelates) indirectly through association with proteins and polyphenols that form the haze. There have been reports indicating that electronegative proteins are associated with haze formation; no significant correlation could be established, however, between protein components of low isolectric point and shelf life of beer.

In a recent review, Dadic (1984) summarized the factors that affect nonbiological beer stability. Stability has become the most important parameter of beer quality because of the trend toward lighter and paler beers. Physical (colloidal) instability of beer results in formation of chill and permanent hazes through polyphenol-protein interactions involving the participation of oxygen, sugars, and metallic catalysts. There is an interrelationship between colloidal (physical, physicochemical, haze, chill) and flavor stabilities; it involves to a large extent the presence of polyphenols (tannins).

Measures to improve shelf life can be divided into prebrewing, brewing, and postbrewing. The most important measure taken prior to brewing is selection of barleys and malt with the optimal contents of anthocyanogens

Table 18-5. Brewing Measures

Treated compounds	Reaction mechanism	Stabilizer or method
Protein	Adsorption	Silica gel (xero/hydro), bentonite
	Precipitation	Tannic acid, "Thermo-shock"
	Hydrolysis	Papain, pepsin, ficin, bromelain
Polyphenol	Adsorption	Nylon 66, PVPP, SBF[a]
	Chemical reaction	Formaldehyde
	Precipitation	
Polysaccharide	Hydrolysis	Amyloglucosidase, alpha-amylase, endo-beta-glucanase
Oxygen	Chemical reduction	Ascorbic acid, CySH, SO_2
	Enzymatic reduction	Glucose oxidase
Mineral	Chelation	EDTA

[a]Matsuda et al U.S. Pat. 4,008,339; Feb. 25, 1977.

and catechins (tanninogens) and nitrogenous substances. Brewing measures fall into four categories; accelerating or delaying haze formation, use of antioxidants, and "no-additives" or "processing agents." The various treatments are listed in Table 18-5. Acceleration can be achieved by adding haze-forming agents (tannins, anthocyanogens, gelatin-rutin complexes, or other polyphenols) and filtration. To delay haze formation, proteolytic enzymes, silica gels (including bentonite), or adsorbents of polyphenol-protein complexes (including polyvinylpyrrolidone, perlon, nylon) can be added. Antioxidants can be added to slow the rate of oxidation, reduce the undesirable effects, or act as oxygen scavengers. Some antioxidants act through breaking a chain reaction (ie, by removing a chain-propagation step). Others prevent or retard the introduction of chain reaction initiators or active radicals into the system. The most promising antioxidants are of the phenolic type akin to some beer components. The "no-additives" or "processing agents" include several brewing measures and refinements. They include controlled lautering and sparging, adequate mashing (including controlled oxidation), minimal pasteurization, and reduced hop rates. Postbrewing measures are mainly directed to a reduction of excess air and removal of yeast; temperature control during filtration, storage, and transportation; and use of incandescent light in warehouses.

Types of Beer

Ancient tablets found in Mesopotamia describe 16 types of beer made as early as 4000 BC. An Assyrian tablet of 2000 BC indicates that beer was among the provisions on Noah's Ark. The Chinese had in 2300 BC a beerlike beverage called "kun." The ancient Incas knew a brew called "sora" which was replaced by a brew called "chica."

A beverage resembling English beer was brewed in America at least 33 years before the Pilgrims landed. The Mayflower was bound for Virginia, but put in at Plymouth Rock because the ship drifted off course as the ship's log recorded: "We could not take time for further search or consideration, our victuals being much spent, especially our beer." Thomas Jefferson persuaded Bohemian master brewers to emigrate to America and teach their advanced skills to American brewers. Basically, there are four major types of beer: ales, stout, porter, and lager. The eight different types of ale are: mild, bitter, light, pale, India pale, Scotch, Burton, and brown. Stout is a strong, heavy beer made from well-roasted barley and caramelized sugar. Porter is a weak form of stout; it is produced in very small quantities. Lager beers (as described in detail later) differ from ales and stouts, as they are brewed by different fermentation processes.

A major difference between British and Continental European practices of beer making, often confusing to consumers, arises from the behavior of the yeast in fermentation. In Britain, brewers still employ types of yeast (eg, *Saccharomyces cerevisiae*) that rise to the surface during fermentation. "Top-fermentation" beers come in many varieties, from relatively pale ales to porter (a dark heavy ale that is popular in Ireland) and stout (a dark heavy beer with a relatively high alcohol content). Top-fermented beers are still preferred in New England. Elsewhere in the United States, the brewing industry derives its traditions from the brewmasters who came from Germany in the 19th century to set up large commercial breweries. They brought with them the new technique of bottom fermentation, which employs a yeast *(Saccharomyces carlsbergensis)* that settles to the bottom during fermentation. The German "lager" beers (the term indicates that the beer has been aged (lagered) after fermentation) are usually paler than ales and are more mellow in flavor. There are three main types: Muenchener, Dortmunder, and Pilsener, named after the towns in which they were first brewed and listed here in the order of their colors, from dark to pale. Pilsener beer has a high hop content and a stable head. Muenchener has a more pronounced malt flavor and darker malt, and less hops are used in its manufacture. Under the pressure of standardization arising from mass distribution, U.S. beers have tended to become increasingly pale and resemble Pilsener beer except that they have less hop flavor and are more highly carbonated. A seasonal exception to the trend is traditional bock beer; a dark, full-flavored beer that is brewed in winter and marketed in spring. Today caramel and sometimes black malts are used. Originally, bock beer was brewed in Bavaria for Easter celebrations; it derives its name from the German word for goat, the zodiacal sign for the month of March. The various types of beer produced in Britain (bitters, mild ones, bitter and sweet stouts), the United States, Australia, and Continental Europe (including the bocks and doublebocks, cambic beers, etc) were described by Yates (1979).

The two main approaches that have been taken in the production of low-calorie beer are (1) reduction of the residual carbohydrates by using

Table 18-6. Comparison of Low- and Regular-Carbohydrate Beers

	Regular	Low
Apparent extract (%)	2.20	0.25
Real extract (%)	3.80	1.61
Original extract (Plato°)	10.75	7.63
Real degree of attenuation (%)	65.9	79.2
Alcohol by weight (%)	3.56	3.03
Reducing sugars (as maltose, %)	0.96	0.55
Protein ($N \times 6.25$, %)	0.32	0.25
Acidity (as lactic acid, %)	0.12	0.10
pH	4.22	4.24
Color (°SRM)	2.85	2.50
Bitterness units (ASBC)	13.39	13.33
Air contents (cc/10–12 fl oz)	0.97	2.68
Gas volumes (air corrected)	2.58	1.00

Adapted from *Brewers Digest 1984, 59*(3), 38.

limit dextrinase or amyloglucosidase and as complete removal of resulting sugars as possible by fermentation, and (2) separation of a protein-rich fraction to provide adequate body, foam formation and retention, overall acceptance, and combination of the protein fraction with a small amount of carbohydrate fraction (Pomeranz, 1975b).

Regular and low-carbohydrate beers are compared in Table 18-6. Classical beer types brewed in the world are listed in Table 18-7.

The German beer law, which dates back to the 1516 Bavarian Purity law, specifies that for the production of bottom-fermented beers only barley malt, hops, yeast, and water are allowed. Top fermented beers follow the same regulations, but wheat malt may be included (Narziss, 1984). For special beers, pure beet-cane-invert sugar is allowed. Beer production in 1980 (in 1,000 hL) totaled about 940,000, of which 352,000 was produced in America and 444,000 in Europe.

By-Products

The main by-products of the brewing process are spent grains, trub (break), spent hops, and yeast. Proximate-gross composition of some of the products of the brewing industry are compared in Table 18-8; most by-products are incorporated into feed formulations.

SOME NEW DEVELOPMENTS

Numerous reports have been published on the feasibility of beer brewing based on the adding of an enzyme complex to barley and other unmalted raw materials. Even though feasible, widespread use of unmalted barley plus enzymes as substitute for malt in U.S. brewing practices has not been

Table 18-7. Classical Beer Types Brewed in the World

Type	Character	Origin	Alcohol (Vol%)	Flavor features
Bottom-fermented				
Münchener	Lager/ale	Munich	4–4.8	Malty, dry, mod. bitter
Vienna (Märzen)	Lager	Vienna	5.5	Full bodied, hoppy
Pilsner	Lager	Pilsen	4.5–5	Full bodied, hoppy
Dortmunder	Lager	Dortmund	5+	Light hops, dry, estery
Bock	Lager	Bav., U.S., Can.	6	Full bodied
Doppelbock	Lager/ale	Bavaria	7–13	Full bod., estery, winey
Light beers	Lager	U.S.	4.2–5	Light bodied, light hops
Top-fermented				
Saisons	Ale	Belgium, France	5	Light, hoppy, estery
Trappiste	Ale	Bel. Dutch Abbeys	6–8	Full bodied, estery
Kölsch	Ale	Cologne	4.4	Light, estery, hoppy
Alt	Ale	Düsseldorf	4	Estery, bitter
Provisie	Ale	Belgium	6	Sweet, ale-like
Ales	Ale	UK, U.S., Can., Aus.	2.5–5	Hoppy, estery, bitter
Strong/old ale	Ale	United Kingdom	6–8.4	Estery, heavy, hoppy
Barley wine	Ale/wine	United Kingdom	8–12	Rich, full, estery
Stout (Bitter)	Stout	Ireland	4–7	Dry, bitter
Stout (Mackeson)	Stout	United Kingdom	3.7–4	Sweet, mild, lact. sour
Porter	Stout	London, U.S., Can.	5.7–5	Very malty, rich
Wheat-malt beers (So. Ger.)				
Weizenbeer	Lager/ale	Bavaria	5–6	Full bodied, low hops
(Berliner) Weisse	Lager	Berlin	2.5–3	Light flavored
Gueuze-Lambic	Acid ale	Brussels	5+	Acidic
Hoegards wit	Ale	E. of Brussels	5	Full bodied, bitter

This table was taken from a broader chart provided by A.A. Leach of the Brewers' Society, U.K.

Table 18-8. Average Percent Composition[a] of Raw Materials and By-products of the Brewing Industry

	Moisture	Protein	Fat	Crude fiber	Ash	N-free extract
Malt	7.7	12.4	2.1	6.0	2.9	68.9
Malt sprouts	7.6	27.2	1.6	13.1	5.9	44.6
Brewers' dried grains	7.2–7.7	21.1–27.5	6.4–6.9	15.3–17.6	3.9–4.2	39.4–42.9
Hops[b]	12.5	17.5	18.7[c]	13.2	7.5	27.5
Spent hops	6.2	23.0	3.6	24.5	5.3	37.4
Yeast	4.3	50.0	0.5	0.5	10.0	34.7

[a]From Leavell, 1942; as-is basis.
[b]From Luers, 1950.
[c]Total ether extract; includes nonlipid components; additional component, 3.0% tannins.

adopted. The most attractive methods from the standpoint of economy and quality control involve the use of high-enzyme malts (ie, abraded, treated with gibberellic acid) and mixtures of amylases from microbial sources. In recent years, there has been a steady increase in the United States in the use of liquid adjuncts (syrups). The mashing process may exert a great effect on wort composition and is perhaps the most important single stage in malting and brewing. A great deal of attention has been paid to accelerated mashing and separation of wort. Many patents have been granted for continuous mashing processes, but none has gained wide acceptance. The continuous processes have high energy requirements and are very complicated mechanically.

The trend in the area of hops has been toward the utilization of alpha acids added after fermentation. The addition of hop extracts affords both economic advantages and improved control of bitterness. The tendency has been, however, to add a small amount of hops at the boiling stage to impart the odor of hop oils to the brew.

Proteins precipitated during the boiling of wort can be removed by centrifuges or filters. In the Whirlpool separator, the wort is injected tangentially into a vertical cylindrical vessel and the protein precipitates and is deposited in a vertical cone. The vertical vessels are used in conjunction with a hop strainer which removes the coarse spent hops. Energy requirements for the Whirlpool separator are smaller than the requirements of conventional centrifuges or filters.

Considerable advances have been made in the technology of beer fermentations from both the production and developmental points of view. The use of large batches can reduce the cost. The capital costs of the equipment can be further reduced by fermenting concentrated worts and diluting the beer finally to the desired concentration. In other developments, the concept of separately fermenting malt components and adjunct carbohydrates has been used to still further economies of capital cost.

The time required to separate yeast from beer may sometimes exceed the time required for actual fermentation. Introduction of centrifugal yeast separators has reduced the total fermentation time and introduced much flexibility in selection of yeast strains and fermentation vessels.

The cylindroconical fermentation vessel is a highly efficient fermenter in which a deep cylindrical vessel has a conical base for yeast sedimentation. Cylindroconical fermentation vessels can be used for the production of both top-fermentation and bottom-fermentation beers. When large fermenters are used, the carbon dioxide concentration in the beer increases and it is necessary to take special precautions to preserve beer sterility. With due care, however, excellent results can be obtained if fermentation and storage are carried out in large cylindrical-conical vessels.

Continuous fermentation can result in considerable savings. One of the continuous fermentations employs a multivessel system which allows free escape of yeast, with the stream of beer from the main fermentation zone.

In a second type (plug-flow fermentation), wort is forced through a yeast-Kieselguhr mixture in a plate filter. In the latter type, the concentration of yeast relative to wort is very high, the flow of the wort is rapid, and little or no growth of yeast cells takes place. Beer produced in such plug-flow fermentations may have an undesirable flavor. That flavor can be eliminated by a second fermentation. Since there is no growth of yeast, this rapid fermentation requires an extraneous supply of yeast.

Because pasteurization can alter the flavor of beer, there is great interest in sterile filtration through membranes and in chemical preservatives. Whereas filtration is used quite extensively, chemical preservation has yet to be accepted by the brewers.

BIBLIOGRAPHY

Asano, K.; Hashimoto, N. *Rep. Res. Lab. Kirin Brewery* **1980**, *23*, 1–13.
Axcell, B.; Morrall, P.; Tulej, R.; Murray, J. *MBAA Technol. Q.* **1984**, *21*(3), 101–106.
Briggs, D.E. *Planta (Berl.)* **1972**, *108*, 351.
Brown, H.T.; Morris, G.H. *J. Chem Soc.* **1890**, *57*, 458.
Clutterbuck, V.J.; Briggs, D.E. *Phytochemistry* **1973**, *12*, 537.
Cook, A.H. (Ed.). "Barley and Malt; Biology, Biochemistry, Technology," Academic Press; New York; **1920.**
Dadic, M. *MBAA Techn. Q.* **1984**, *21*(1), 9–26.
Dickson, J.G.; Shands, H.L. *Am. Soc. Brewing Chem.* **1941**, 1.
Dronzek, B.L.; Hwang, P.; Bushuk, W. *Cereal Chem.* **1972**, *49*, 232.
Enari, T.-M. *Proc. EBC Cong.* **1981**, 69–80.
Fretzdorff, B.; Pomeranz, Y.; Bechtel, D.B. *J. Food Sci.* **1982**, 47, 786–791.
Hartwick, W.A. In "Biotechnology," Vol. 5. Rehm, H.J.; Reed, G., (Eds.). Verlag Chemie; Weinheim; **1983**, pp. 165–228.
Hough, J.S.; Briggs, D.E.; Stevens, R. "Malting and Brewing Science." Chapman and Hall; London; **1971.**
Kneen, E.; Dickson, A.D. *Kirk-Othmer Encycl. Chem. Technol.* **1967**, *21*, 861.
Kitabatake, K. *Bull. Brewing Sci.* **1978**, *24*, 21–32.
Leavell, G. U.S. Dept. Agric. Bureau Anim. Ind. Bull. No. 58: Washington, D.C.; **1942.**
Luers, H. "Die Wissenschaftlichen Grundlagen von Malzerei und Brauerei." Verlag H. Huber; Nuremberg, West Germany; **1950.**
Lyons, T.P. *Brewers Digest* **1984**, *59*(3), 22–24.
Matsuda, L. U.S. Patent 4,008,339, **1977.**
MacGregor, A.W. *MBAA Technol. Q.* **1980**, *17*(4), 215.
MacLeod, A.M. *J. Inst. Brewing* **1967**, *73*, 146.
MacLeod, A.M. *MBAA Technol. Q.* **1976**, *13*(3), 193–198.
Marchylo, B.A.; MacGregor, A.W.; Kruger, J.E. *J. Inst. Brewing* **1985**, *91*, 161–165.
Narziss, L.J. *Inst. Brewing* **1984**, *90*, 351–358.
Palmer, G.H. *J. Inst. Brewing* **1974**, *80*, 13.
Palmer, G.H.; Bathgate, G.N. *Adv. Cereal Sci. Technol.* **1976**, *1*, 237.
Palmer, G.H.; Gernah, D.I.; McKernan, G.; Nimmo, D.H.; Laycock, G. *Am. Soc. Brewing Chem.* **1985**, 18–28.
Peterson, G.A.; Foster, A.E. *Adv. Agronomy* **1973**, *25*, 328–378.
Phillips, C.L.; Boerner, E.G. "Present Uses of Barley and Barley Products." U.S. Dept. Agric., Bureau Agric. Econ.: Washington, D.C.; **1935.**
Pomeranz, Y. *Cereal Chem.* **1972**, *49*, 5–19.
Pomeranz, Y. *Cereal Chem.* **1974a**, *51*, 545–552.
Pomeranz, Y. *CRC Crit. Rev. Food Technol.* **1974b**, 377–394.
Pomeranz, Y. *Bull. Assoc. Operative Millers* **1975a**, 3503–3512.
Pomeranz, Y. *Brewers Digest* **1975b**, *50*(10), 38, 40, 42.
Pomeranz, Y.; Robbins, G.S. *Brewers Digest* **1971**, *46*(5), 58.

Pomeranz, Y.; Sachs, I.B. *Cereal Chem.* **1972a**, *49*, 1–4.

Pomeranz, Y.; Sachs, I.B. *Proc. Am. Soc. Brewing Chem.* **1972b**, 24–29.

Rohrlich, M.; Bruckner, G. "Das Getreide. I. Das Getreide und Seine Verarbeitung," 2nd ed. Paul Parey; Berlin; **1966**.

Wiebe, G.A. "Barley; Origin, Botany, Culture, Winterhardiness, Genetics, Utilization, Pests." Agriculture Handbook No. 338, Agric. Res. Service, U.S. Dept. Agric.; Washington, D.C.; **1968**.

Woodward, J.D. *Brewers Digest* **1978**, *53*(5), 38–44.

Yates, S. *Chem. Ind. (Lond.)* **1979**, 887–893.

Chapter 19

Oats, Sorghums and Millets, and Rye

OATS

Oat products are milled to provide oatmeal for porridge and oatcake, rolled oats for porridge, oat flour for baby foods, and ready-to-eat breakfast cereals. Oat milling is a major cereal industry for the production of breakfast cereals. Rolled oats and oatmeal are high in protein, fat, and energy value and are rich sources of calcium, phosphorus, iron, and thiamine. They have a high nutritive value because they are made from oat groats, which are obtained by removing the fibrous hull and adhering portions from the oat grain. Groats correspond to the caryopsis of wheat; the bulk of the bran, the aleurone layer, and germ remain with the portion used as human food. Hence, rolled oats and oatmeal are essentially whole grain products (Doggett, 1970).

Oat hulls are an important by-product of oat milling because they yield furfural by conversion of the pentosans in the hulls. Furfural is used extensively in the manufacture of phenolic resins and as a solvent. A diagram of an oat milling process is given in Figure 19-1. Typical composition of oats and oat products are also given in Table 19-1. (See also Caldwell and Pomeranz, 1973).

For milling purposes, plumpness, soundness, and freedom from heat damage, foreign odors, wild seed, smut, must, and mold are important. Only high-grade oats are employed in milling. The initial step in oat milling involves cleaning, drying, or slow roasting to reduce the moisture to about 6% to increase the brittleness of the hulls and facilitate their removal. The milling process involves the following steps: grading, hulling, separating the hulls and unhulled oats from the groats, rolling the groats into flakes, and parching. The parched oats are passed through graders, which sort them according to length into five or six groups. Each stream is sent to separate hulling stones. The groats from the largest oats yield the best grade of rolled oats. The hullers consist of two "stones." The lower one is flat and stationary;

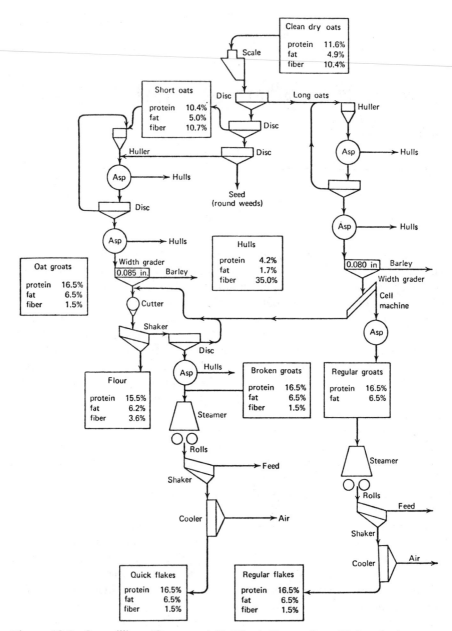

Figure 19-1. Oat milling. (Courtesy A.B. Ward, Kansas State University.)

Table 19-1. Typical Composition of Oats and Oat Products from Same Milling, Including Certain Vitamin and Mineral Constituents

Analysis	Unit	Moisture basis	Dry-milling oats	Finished groats	Hulls	Oat shorts	Oat flour, chips, and meal
Test weight	lb/bu	as is	38.5	52.0			
Kernel weight	mg	as is	23.0	14.9			
Moisture	%	—	7.0	7.0	6.5	7.5	7.5
Protein	%N × 6.25	as is	12.1	15.8	4.2	9.5	15.5
Crude fat	%	as is	5.1	7.2	1.7	3.2	6.2
Crude fiber	%	as is	11.0	1.5	32.9	22.0	3.6
Thiamine	mg%	dry	0.65	0.77	0.15	0.44	0.78
Riboflavin	mg%	dry	0.14	0.14	0.16	0.35	0.17
Niacin	mg%	dry	1.15	0.97	1.04	1.62	1.25
Ash	%	7%	3.4	1.9	6.0	6.7	2.1
Iron	%	7%	0.0051	0.0044	0.0112	0.0120	0.0047
Calcium	%	7%	0.065	0.053	0.092	0.178	0.051
Phosphorus	%	7%	0.341	0.466	0.100	0.329	0.414

Courtesy Quaker Oats Co., Chicago.

the upper one is slightly conical and rotates rapidly. The space between the stones is adjusted to less than the length of the grain but slightly more than the length of the groat. As the oats are carried from the center to the periphery of the stones, they are turned endwise so that pressure is exerted on the ends. This shatters the hull and releases the groat.

From the hulling stones, the mixture of unhulled oats, hulls, whole and broken groats, and flour is passed through a horizontal sifting reel to remove oat flour and through several air separators to remove the hulls. The remaining mixture of unhulled oats and groats is graded to take out broken groats. The mixture is then passed through a series of apron machines, provided with pockets to receive the groats and carry them over the machine while the unhulled oats slide off the apron at the bottom.

In the manufacture of the standard type of slow-cooking choice rolled oats, whole groats are steamed with live steam at atmospheric pressure. This sterilizes and partially cooks the groats and increases the moisture content to a level that is essential to prevent excessive breakage and production of flour. For quick-cooking rolled oats, the groats are cut by steel rotary cutters. The medium sizes are used to produce quick-cooking rolled oats; they are about one-third of the groat size. The cut groats are heated, similar to the whole groats.

The final step, flaking, involves feeding steamed groats between two large steel rolls. In producing quick-cooking rolled oats, the rolls are adjusted to produce a thinner flake than that of the standard slower-cooking type. Heat treatments of groats for the two types of rolled oats are similar; the quicker cooking properties are attained by decreasing the size and thickness of the flake.

Composition of commercial oat products is listed in Table 19-2. The steel oat product is an oat that has been hulled and cut to a smaller size. The steam table product is a dehulled oat that has been steamed and rolled into a thick flake. The regular Buckeye rolled product is a dehulled oat that has been steamed and rolled into a large flake. The Quick Buckeye rolled product is a dehulled oat that has been steamed and rolled into a thin flake for quick cooking. The flour is a dehulled oat that has been steamed and ground to produce a stable flour in which enzyme activity has been minimized.

Medium-quality grade No. 2 white oats yield about 42% good-quality rolled oats, 30% hulls, and 28% other products (including oat shorts, oat middlings, cereal grains, weed seeds, and other material removed in the cleaning process).

SORGHUMS AND MILLETS

Component parts of mature sorghum (milo) kernels and their chemical composition are given in Table 19-3. Proximate compositions of major

Table 19-2. Composition in Percent of Typical Oat Products

Oat product	Moisture	Protein[a]	Fat[a]	Fiber[a]	Ash[a]
Steel oat	9.0–12.0	16.5–18.5	7.0–8.0	1.4–1.8	2.0–2.5
Steam table	9.0–12.0	16.5–18.5	6.0–9.0	1.4–1.8	2.0–2.5
Regular buckeye rolled	9.0–12.0	16.0–18.5	6.0–9.0	1.2–1.8	2.0–2.5
Quick buckeye rolled	9.0–12.0	16.0–18.5	7.0–8.0	1.2–1.8	2.0–2.5
Flour	8.0–12.0	17.0–19.0	7.0–8.0	0.7–1.6	1.8–2.2

[a]Dry-matter basis.
Courtesy Quaker Oats Co., Chicago.

Table 19-3. Component Parts of Mature Grain Sorghum and Their
Average Proximate Composition

Kernel part	Proportion of kernel (%)	Composition of kernel parts (%, db)			
		Starch	Protein	Fat	Ash
Whole kernel	100	71.2	12.6	3.5	1.8
Endosperm	84	79.1	12.0	0.4	0.4
Bran	7	18.1	6.4	4.6	2.5
Germ	9	7.5	19.1	24.0	10.0

Calculated and averaged from data of Hubbard et al, 1950; Miller, 1958; Watson, 1967; and Rooney, 1973.

cereal grains are compared with sorghum and millets in Table 19-4. Sorghum is similar in composition to corn, but contains slightly more starch and protein and less lipids than corn. Corn contains yellow pigments, of which most sorghum grains contain only small quantities. The vitamin contents of the two grains are similar. Sorghum is unique among cereal grains in that the coating contains relatively large amounts (0.2–0.5% of kernel weight) of wax with properties much like those of carnauba wax. Amino acid contents of major cereal grains and sorghums and millets are compared in Table 19-5.

Dry Milling

The conventional roller-milling process can be employed for manufacture of whole or refined sorghum products. Sorghums are harder to grind than wheat, barley, or oats, and slightly easier than corn. During cracking, grits and a break flour are obtained. The grits are used for feeding purposes and are comparable in composition to the whole grain. Part of the fibrous material can be removed by sieving and aspiration. The break flour (10–15% yield) contains little protein (4.0–4.5%, as compared to 8%–9% in whole grain). Roller-milled flours are sieved to yield products varying in extraction and composition. A product obtained in 70% yield contains 0.5% ash and 0.8% fiber and is reasonably free from ob-

Table 19-4. Proximate Composition of Cereals and Millet Grains[a]

	Protein (%)	Fat (%)	Carbo-hydrates (%)	Minerals (%)	Fiber (%)	Calcium (mg/100 g)	Phos-phorus (mg/100 g)	Iron (mg/100 g)	Thiamine (mg/100 g)	Ribo-flavin (mg/100 g)
Wheat	12.1	1.7	69.4	2.7	1.9	48	423	11.5	0.49	0.29
Rice	7.2	2.3	75.1	1.3	0.7	10	230	4.5	0.29	0.21
Barley	11.5	1.3	69.3	1.2	3.9	26	215	3.0	0.47	0.20
Oatmeal	13.6	7.6	62.8	1.8	3.5	50	380	3.8	0.54	0.12
Maize	11.1	3.6	66.2	1.5	2.7	10	348	2.0	0.42	0.17
Sorghum	10.4	1.9	72.6	1.6	1.6	25	222	5.8	0.37	0.27
Bajra (*Pennisetum typhoidenum*)	11.6	5.0	67.5	2.3	1.2	42	269	14.3	0.33	0.16
Ragi (*Eleucine coracana*)	7.3	1.3	72.0	2.7	3.6	344	283	17.4	0.42	0.10
Italian millet (*Setaria italica*)	12.3	4.3	60.9	3.3	8.0	31	290	12.9	0.59	0.08
Pani Varagu (*Panicum miliaceum*)	12.5	1.1	70.4	1.9	2.2	14	206	11.5	0.20	0.18

[a]*Nutritive value of Indian foods and the planning of satisfactory diets* (Indian Council of Medical Research), 1966.

Table 19-5. Amino Acid Content (per 16 g of Total Nitrogen) of Cereals and Millet Grains

	Trypto-phan	Threo-nine	Isoleu-cine	Leucine	Lysine	Methio-nine	Cystine	Total	Phenyl alanine	Tyrosine	Valine	Arginine	Histi-dine
							Sulphur-containing amino acids						
Wheat grains (whole flour)	1.15	2.69	4.05	6.26	2.56	1.42	2.05	3.47	4.61	3.49	4.32	4.46	1.90
Wheat (white flour)	1.12	2.62	4.19	7.02	2.08	1.20	1.82	3.02	5.01	3.12	3.94	4.05	1.82
Rice (brown, converted, white)	1.02	3.73	4.46	8.21	3.76	1.71	1.30	3.01	4.78	4.35	6.66	5.49	1.60
Barley	1.17	3.15	3.97	6.48	3.15	1.34	1.87	3.22	4.82	3.39	4.69	4.80	1.74
Corn (corn meal)	0.61	3.98	4.62	12.96	2.88	1.86	1.30	3.15	4.54	6.11	5.10	3.52	2.06
Sorghum	1.12	3.58	5.44	16.06	2.72	1.73	1.66	3.39	4.98	2.75	5.71	3.79	1.92
Pearl millet (*Pennisetum glaucum*)	2.03	3.75	5.20	14.29	3.14	2.21	1.25	3.46	4.14	—	5.58	4.29	1.97
Ragi millet	1.28	4.06	5.98	9.33	3.04	4.06	2.82	6.88	3.95	—	7.12	1.50	1.18
Foxtail millet (*Setaria italica*)	0.99	3.10	7.60	16.06	2.10	2.80	—	—	6.70	—	6.90	3.60	2.10
Little millet (*Panicum miliare*)	0.61	3.39	6.70	10.90	1.79	2.30	—	—	4.80	—	6.10	4.70	1.90

Source: Carr and Watt, 1957.

Table 19-6. Composition in Percent of Dry-Milled Grain Sorghum Products

Product	Protein	Oil	Fiber	Ash
Whole grain	9.6	3.4	2.2	1.5
Pearled	9.5	3.0	1.3	1.2
Flour (crude)	9.5	2.5	1.2	1.0
Flour (refined)	9.5	1.0	1.0	0.8
Brewers' grits	9.5	0.7	0.8	0.4
Bran	8.9	5.5	8.6	2.4
Germ	15.1	20.0	2.6	8.2
Hominy feed	11.2	6.5	3.8	2.7

Source: Hahn, 1969; as-is basis.

jectionable specks. For the production of high-extraction flours (ie, 90%), impact grinding is preferred as it requires less space and equipment than the roller-mill system. Because the sorghum kernel is round, the bran can be removed mechanically by abrasion. Without tempering, a 75–80% yield of almost completely dehulled grain (except for the portion left on the germ) can be obtained. Proper tempering to assist in loosening the hull can increase the yield to 85%. The germ can be removed by passing the dehulled kernel through cracking rolls or impact machines. Germ separation can be accomplished by flattening and sieving, air classification, or gravity separation. The final product is milled to a relatively white flour with low ash and fiber contents. The yield is 58–65%. In the United States all dry-milled sorghum grain products are for industrial uses. Some dehulled, degermed, and ground grits are used by the fermentation industry (Anon., 1971; Hoseney et al, 1981; Jackson et al, 1980; Wall and Ross, 1970).

By-products of sorghum dry milling (bran, germ, and shorts) are used in the production of hominy feed. Decreasing flour extraction in roller milling from 90% to 70% decreased oil content of the groats from 2.8% to 2.0%. The oil content decreased from 3.4% in whole sorghum grain to 0.6% in decorticated grain, in which 39.0% of the kernel was abraded; that decrease was accompanied by decreases in fiber (2.2% to 0.7%), mineral components (1.5% to 0.4%), and protein (9.6% to 6.9%) and by an increase in brewers' extract (from 85.7% to 97.0%) (Hahn, 1969).

Impact or attrition degerminators (after grain tempering and dehulling) are effective in germ separation and production (after sieving) of low-oil products. Specific-gravity separators (after dehulling, impaction, and size classification) can be used to separate the germ and endosperm during dry milling. Both waxy and nonwaxy grain sorghum can be milled and fractionated. The waxy germ, however, is more difficult to separate, since waxy grits contain more oil than nonwaxy grits. Composition of dry milled grain sorghum is summarized in Table 19-6.

Wet Milling

The sorghum wet-milling process is similar to that of corn, although finer-mesh screens are necessary for efficient operation. Chemically and microscopically, sorghum starch is similar to corn starch, which it also resembles is gelatinization characteristics. Milo starch is blander in flavor and does not develop rancidity. Other products of wet milling are oil and gluten feeds.

Millets

Pearl millet (*Pennisetum americanum*) is the largest-seeded and most widely grown millet in India and Africa (Abdelrahman et al, 1984). In whole kernels ranging in weight from 10.4 to 18.9 mg, the bran comprises 7.2–10.6%, the endosperm 73.9–76.2%, and the germ 15.5–17.4%. Generally, there was an inverse relation between bran and endosperm content. The whole grain, the endosperm, the germ, and the bran contained 13.3, 6.3, and 1.7%; 10.9, 0.5, and 0.3%; 24.5, 32.2, and 7.2%; and 17.1, 5.0, and 3.2% (dry basis) protein, fat, and ash, respectively. Functional properties and potential use as food of millets are described by Casey and Lorenz (1977).

Sweet Sorghum

Some sweet sorghum varieties produce large quantities of sugar in the stalks. The pressed juice is boiled to produce a syrup with a distinctive flavor. There is interest in using the juice for sugar and alcohol production.

RYE

Consumption of rye breads in West Germany, Poland, and the Soviet Union continues to be much higher than in other European countries and the United States. But even in those countries with a long tradition of rye bread, its consumption decreases.

The origin of rye is described by Stutz (1972). It appears that cultivated rye (*Secale cereale* L.) originated from weedy products derived from introgression of *S. montanum* and *S. vavilovii*. The rye kernel is composed of the pericarp (about 17%), endosperm (80%), and germ (3%) (Seibel, 1981). Their composition (dry-matter basis; %) is as follows:

	Protein	Ash	Fat	Carbohydrates
Whole grain	9–11	1.7–2.0	1.6–1.8	83–86
Pericarp	6–11	6–9	1.5	80–88
Aleurone	24–26	10–12	5–6	56–61
Starchy endosperm	6–9	0.5–0.8	1.8–2.2	88–92

The significance of kernel development in governing milling yields was discussed by Drews (1973). Quality characteristics of rye with regard to bread production (including alpha-amylase and pentosan levels) have been reviewed by Weipert (1972) and by Weipert and Zwingelberg (1980). Rheological properties and bread quality potentials of hard red winter wheat, rye, and triticale processed under U.S. conditions are compared by Haber et al (1976). Pedersen and Eggum (1983) report that the lysine content (g/16 g N) was 4.23 in whole rye but only 3.76 in 65% extraction flour and that a corresponding reduction in biological value was found. A reduction of 50% or more was observed for several minerals, with zinc and phosphorus being most affected.

Rye and triticale are sometimes attacked by a fungus, ergot (*Claviceps purpurea*). Other cereal grains are also attacked but comparatively rarely. Ergot is a poisonous contaminant and should be removed almost completely prior to milling. Grain that contains above 0.3% attacked kernels also presents problems in wheat, rye, and triticale cleaning and in feed where ergot tends to concentrate. It is impossible to wash out gluten from a dough made entirely of rye flour, and rye flour is thus inferior to wheat flour in yeast-leavened bread. In the United States, most of the so-called rye bread is baked from mixtures of rye and wheat flour (usually a first clear). Most of the major mills market rye blends made from a mixture of strong spring or hard winter wheat and rye. They roughly follow the pattern of 80% clear and 20% dark rye, 70% clear and 30% medium rye, and 50% clear and 50% white rye.

In the manufacture of specialty breads made largely from rye flour, a proper level of alpha-amylase activity is the major factor in determining the quality of bread. When there is insufficient alpha-amylase, the bread has a dry crumb and the crust becomes torn and cracked on cooling. Excess alpha-amylase produces bread with a wet and soggy crumb which frequently pulls away from the crust and leaves large hollow spaces in the bread.

Rye is milled into flour by a process similar to that described for wheat. As the bran in rye adheres tenaciously to the endosperm, however, it is not practical to produce clean "middlings" or to purify them by aspiration. Moreover, rye is a "tough" grain. The middlings are not so friable as those of wheat and, if ground between smooth rolls, tend to flake or flatten rather than pulverize. Consequently, it is very difficult to damage rye starch mechanically. The reduction rolls in rye milling are finely corrugated (40–50 corrugations per inch). The objective of rye milling is to produce flour during the breaking process. The break rolls are set relatively close and have finer corrugations than those used for corresponding breaks in milling wheat. Rye grain requires either little or no tempering. The highest grade of flour is produced by the first-break rolls. As the purity of the flour decreases, it becomes darker and has a more pronounced rye flavor. In American milling three main grades of rye flour are produced: white or light, medium, and dark. The light (white) rye flour represents 50–65% of the grain. Most of the rye flour used in bread making in the United States is of the white type. Medium rye flour corresponds to straight-grade wheat flour, and dark rye flour corresponds to the "clear." In addition to those three main grades, "cut" or "stuffed" straights are sometimes produced. The cut straight is a medium rye flour from which a small percentage of white rye flour has been removed, whereas a stuffed straight is a medium rye flour to which a small percentage of dark rye flour has been added. The milling of rye normally yields 65% light or patent rye, 15–20% dark rye, and 15–20% offal (Bushuk, 1976; CIGI, 1975).

According to Shaw (1970), the two problems encountered in rye selection for milling are ergot contamination and field sprouting. Basically, the cleaning flow is similar to the flow used in cleaning wheat for milling. Pocket sizes on the disc machinery, however, are slightly different, because the shape of the rye kernel is different from the wheat kernel. The average rye kernel is thinner and slightly longer than the average wheat kernel.

The moisture content of the tempered rye should be 14.5–15.0% and the temper time should be not less than 6 hours. In milling rye we do almost the opposite from what we do in milling wheat. No effort is made in milling rye to remove the bran particles from the middlings as they come from the sifters, and no purifiers are used. Very high extraction percentages are carried throughout the breaking system. All middlings and tailings rolls are corrugated. It is desirable to have ample roll surface on the first break. Grinding surface for the break rolls should be about 45% of the total surface. In contrast to wheat milling, which is a process of gradual reduction with purification and classification, rye milling does not employ gradual reduction. Unlike in wheat, all reduction rolls are corrugated. Smooth rolls would flake the stocks so that they would scalp off either to tailings or to the next reduction system and on out the tail of the mill to feed. About 60% of the total flour is produced by the first two breaks, the sizings, and the first two reductions. In wheat milling, most of the flour is produced in the first three or four reductions. Some rye mills use dusters between each head end break scalp and the next break rolls. This increases the percentage of white rye or patent rye flour.

BIBLIOGRAPHY

Abdelrahman, A.; Hoseney, R.C.; Varriano-Marston, E. *J. Cereal Sci.* **1984**, *2*, 127–133.

Anon. "Grain Sorghum Research in Texas—1970." Texas Agric. Exp. Station; College Station, TX: Consolidated Report 2938–2949; **1971**.

Bushuk, W. "Rye: Production, Chemistry, Technology." Am. Assoc. Cereal Chem.; St. Paul, MN; **1976**.

Caldwell, E.F.; Pomeranz, Y. In "Industrial Uses of Cereals." Pomeranz, Y., Ed. Am. Assoc. Cereal Chem.: St. Paul, MN; **1973**, pp. 393–411.

Carr, M.L.; Watt, B.K. In "Home Economics Report No. 4." U.S. Dept. Agric.; Washington, D.C.; **1957**.

Casey, P.; Lorenz, K. *Bakers Digest* **1977** *51*(1), 45–51.

CIGI (Canadian International Grains Institute)."Grains and Oilseeds. Handling, Marketing, Processing" 2nd ed. CIGI; Winnipeg, Canada; **1975**.

Desikachar, H.S.R. *J. Sci. Ind. Res. (India)* **1975**, *34*, 231–237.

Doggett, H. "Sorghum." Longmans, Green & Co.; London; **1970**.

Drews, E. *Getreide Mehl Brot* **1973**, *27*, 305–311.

Haber, T.; Seyam, A.A.; Banasik, O.J. *Bakers Digest* **1976**, *50*(3), 24–27, 53.

Hahn, R.R. *Cereal Sci. Today* **1969**, *14*, 234–237.

Hoseney, R.C.; Varriano-Marston, E.; Dendy, D.A. *Adv. Cereal Sci. Technol.* **1981**, *6*, 71–144.

Hubbard, J.E.; Hall, H.H.; Earle, F.R. *Cereal Chem.* **1950**, *27*, 415–420.

Jackson, D.M.; Grant, W.R.; Shafer, C.E. "U.S. Sorghum Industry," ESCS-USDA, Agric. Econ. Rep. 457; Washington, D.C.; **1980**.

Miller, D.F. (Ed.). "Composition of Cereal Grains and Forages." Natl. Acad. Sci., Natl. Res. Council Publ. 585: Washington, D.C.; **1958**.

Pedersen, B.; Eggum, B.O. *Qual. Plant. Plant Foods Human Nutr.* **1983**, *32*, 185–196.

Rooney, L.W. In "Industrial Uses of Cereals." Pomeranz, Y., Ed. Am. Assoc. Cereal Chem.; St. Paul, MN; **1973**, pp. 316–342.

Salisbury, D.K.; Wichser, W.R. *Bull. Assoc. Operative Millers* **1971**, 3242–3247.

Seibel, W. *AID Verbraucherdienst* **1981**, *26*, 230–233.

Shaw, M. *Bull. Assoc. Operative Millers* **1970**, 3203–3208.

Stutz, H.C. *Am. J. Botany* **1972**, *59*, 59–70.

Wall, J.S.; Ross, W.M. (Eds.). "Sorghum Production and Utilization." Avi Publ. Co.; Westport, CT; **1970**.

Watson, S.A. In "Starch Chemistry and Technology," Vol. II: "Industrial Aspects." Whistler, R.L.; and Paschal, E.P. Eds. Academic Press; New York; **1967**.

Weipert, D. *Z. Acker Pflanzbau* **1972**, *135*, 269–278.

Weipert, D.; Zwingelberg, H. *Getreide Mehl Brot* **1980**, *34*, 97–100.

Youngs, V.L.; Peterson, D.M.; Brown, C.M. *Adv. Cereal Sci. Technol.* **1982**, *5*, 49–105.

Chapter 20

Extrusion Products

Cereals can be processed into foods by extrusion. Regular extrusion is used primarily for the production of alimentary pastes. More recently, high-temperature, short-time (HTST) extrusion is used extensively to produce instant and infant foods, expanded pet foods, feeds, and cereals for industrial uses.

ALIMENTARY PASTES (PASTA)

According to a legend, pasta products were first introduced to the royal courts of Italy by Marco Polo upon his return from China in the 13th century. In the 15th century, the Italians learned production of pasta from the Germans. The first mechanical process to extrude pasta products, rather than cut them from a sheeted dough, dates back to the earliest 19th century. Although mixers, kneaders, hydraulic presses, and drying cabinets have been used since the turn of this century, continuous extrusion is only 50 years old, and fully automated processing, weighing, and packaging is only 30 years old. Pasta was introduced to the United States by immigrants before the Civil War. The basic raw material for the production of high-quality pasta products is semolina from durum wheat.

Durum wheat, which is generally hard and has a tough, horny endosperm, is the preferred raw material in the production of semolina for the production of pasta products. Semolina from durum wheat requires less water to form a dough and produces a translucent pasta product of acceptable cooking and eating properties. A variety of pasta products can, however, be manufactured from a wide range of wheats milled to various granulations. Figure 20-1 depicts diagrammatically how durum wheat is milled into semolina or flour.

Commercial semolina should pass through a U.S. No. 20 sieve and should contain a maximum of 3% flour (passing through a No. 100 sieve). A uniform fine particle size is specified. According to FDA standards of identity, egg noodles and egg spaghetti must contain 5.5% egg solids, by weight, in the final product. Optional ingredients (in specified maximum

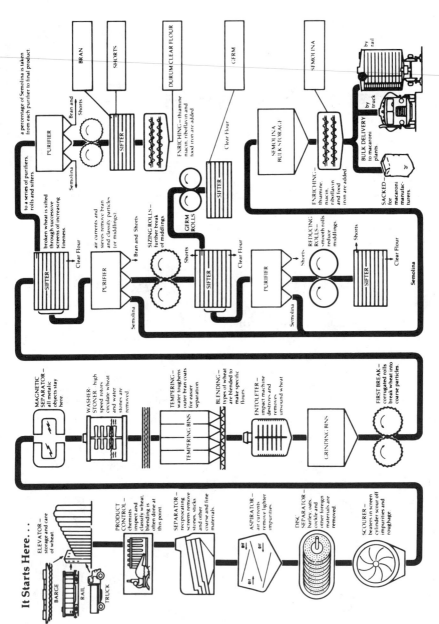

Figure 20-1. How durum wheat is milled into semolina or flour. (Courtesy Durum Wheat Institute.)

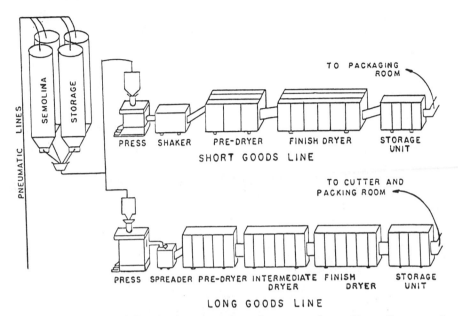

Figure 20-2 Material flow in the processing of pasta products. (From Fraase et al, 1974.)

amounts) include seasonings, enrichment (minerals and vitamins), soy flour and soy protein, vegetables, and gluten. In commercial practice, alimentary pastes are formed by extrusion on large automatic machines (capacities up to 1,500 lb per hour) that perform several operations. Material flow in the processing of pasta products is depicted in Figure 20-2. Water is added (along with other ingredients) to make a stiff dough with about 31% water. The dough is forced under pressure through dies of an extrusion auger. Processing elements of a single-screw extruder are shown in Figure 20-3.

The die for long goods, (eg, spaghetti), consists of a metal plate with drilled holes. The extruded spaghetti is folded on a rack and cut to length. Long goods are dried on hanging rods under carefully controlled conditions. Short goods—eg, elbow macaroni—are dried on endless belts or screens. Introduction of Teflon-lined brass dies and accelerated drying has enhanced both production efficiency and product quality. Modern presses are equipped with a vacuum chamber to remove air bubbles from the pasta before extrusion; otherwise, the finished product has reduced mechanical strength and a white-chalky appearance, and oxidation of carotenoid pigments imparts an unattractive gray color. Drying is the most difficult step to control; conventional, stepwise drying takes 8–18 hours. Microwave drying (about 1½ hours) has yet to be applied successfully to drying of long goods Banasik, 1981; Hummel, 1966).

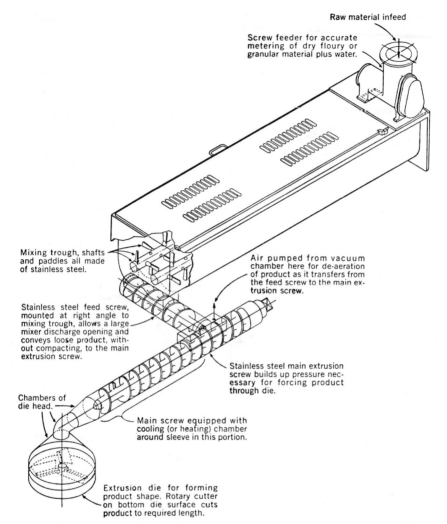

Figure 20-3. Processing elements of a single-screw extruder. (Courtesy Buhler Corp., Minneapolis.)

Pasta products marketed in Europe and the western hemisphere include spaghetti—small diameter, solid rods; macaroni—hollow tubes; noodles—flat strips or extruded oval strips; and miscellaneous products—cut by revolving or blade cutters. The various pasta shapes are illustrated in Figure 20-4.

Banasik et al (1976) studied structural changes that take place in processing durum wheats into spaghetti. A comparison of two cultivars (Rolette and Leeds) showed differences in protein distribution, especially in the subaleurone layers. Uneven structure of the aleurone layer may be responsible for excessive amounts of bran particles in semolina.

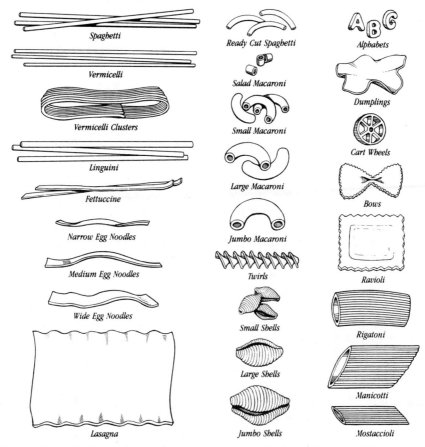

Figure 20-4. Pasta products. (Courtesy Conagra Co. Ltd., Omaha, NE.)

A high ratio of adhering to wedge proteins was postulated essential to the production of the best spaghetti. Production of alimentary pastes is accompanied by starch gelatinization. During cooking, the coagulated proteins form a fibrillar structure that is essential to consumer acceptable end products.

Matsuo et al (1980) studied by scanning electron microscopy changes in pasta dough during spaghetti processing in a laboratory-scale continuous press. Adding water in the preliminary mixing stage, before dough formation, changed the tight, compact structure of semolina to a more open structure. During dough formation in the extrusion auger, a discontinuous protein matrix became predominant. Starch granules aligned along the direction of flow as the dough reached the end of the extrusion auger. As processing continued, the protein matrix became more ordered. No continuous network of protein sheets of fibrils was attained. This indicated that full gluten development did not occur. Such development was demonstrated, however, if the water absorption was increased from

27% to 60%. It was suggested (Dexter and Matsuo, 1977; Dexter et al, 1978) that the absence of complete gluten development during pasta processing may explain why no significant differences were detected in solubility changes of semolina processing for wheats of widely differing spaghetti-making quality.

Protein quality and quantity are significant factors affecting cooking quality of spaghetti, particularly with respect to the maintenance of firmness and cooking quality (Grzybowski and Donnelly, 1979). Dexter and Matsuo (1978) reported on a relationship between high-quality spaghetti cooking quality and a long pasta dough mixing time. Still, though desirable mixing characteristics may be a prerequisite, they do not guarantee superior spaghetti cooking quality. Protein solubility fractionations revealed that insoluble residue protein was the protein fraction most responsible for variations in gluten strength, dough mixing, and overall spaghetti cooking quality (Dexter and Matsuo, 1980). Spaghetti stickiness was influenced by raw material granulation and protein content but was not related to sprout damage. Stickiness was significantly correlated to cooking loss, cooked weight, degree of swelling, compressibility, recovery, and firmness. However, even when all the above factors were included in a stepwise regression equation, less than 50% of the variance in stickiness could be predicted (Dexter et al, 1983).

Various proteins can be used to produce high-protein durum wheat pasta products with greatly improved nutritional value. Generally, the traditional taste, appearance, and mouth feel are altered (Banasik and Dick, 1982). Durum wheat semolina made into an expanded product has an appealing appearance and good taste.

The use of extruded *Triticum aestivum* wheat semolina was studied by Kim and Rottier (1980). The extruded products, in which the starch is partly gelatinized, can be used as binders for meat products. In eclairs, custard fillings, and meat croquettes, the extruded products drastically shorten the cooking time and simplify the cooking process of these products. Extruded semolina products are highly suitable for the preparation of sponge and high-ratio cakes. According to Mercier (1980), extrusion temperatures of cereal starches govern their structures that may range from a spaghetti-like to a strongly expanded product. Extrusion at high temperatures may decrease in vitro susceptibility of starch-lipid complexes to alpha-amylase degradation. Complex formation, on the other hand, improves taste and texture. Complexing (which involves interaction with amylose) reduces starch solubility and excessively tougher products and enhances consumer acceptance.

According to the U.S. Federal Standards of Identity (U.S. FDA, 1981), enriched pastas must contain 13–16.5 mg of iron, 4–5 mg of thiamine, 1.7–2.2 mg of riboflavin, and 27–34 mg of niacin per pound. Enrichment with calcium (500–625 mg/lb) or vitamin D (250–1,000 USP units/lb) is optional and not common (Douglass and Matthews, 1982). U.S. Federal

Table 20-1. Nutrient Content of Pasta Products

Nutrient (per 100 g)	U.S. products			Asian products		
	Macaroni	Noodle	Protein-enriched	Saimin	Chow funn	Soba
Moisture (g)	9.8	9.2	10.9	11.6	7.2	11.4
Protein ($N \times 5.7$, g)	12.8	14.0	20.5	10.1	12.0	11.5
Fat (g)	1.6	4.2	2.5	0.4	1.4	2.2
Carbohydrate (g)	75.1	71.7	64.5	73.7	75.8	71.9
Ash (g)	0.7	0.9	1.2	4.2	3.6	3.0
Fe (mg)	4.1	4.9	4.3	ND[a]	4.9	4.0
Mg (mg)	48	71	ND	ND	46	83
K (mg)	152	191	ND	895[b]	1,161[b]	ND
Na (mg)	8	21	5	931[b]	929[b]	ND

[a]*No data.*
[b]Source added.
Adapted from Douglass and Matthews, 1982.

Standards of Identity specify that macaroni products with fortified protein must contain 20% protein ($N \times 6.25$, 13% moisture basis) and have a protein efficiency ratio of not less than 95% that of casein.

In addition to products made from durum semolina, other pastas are available. Asian pasta products include udon, somen, and saimin made from wheat sources other than durum wheat; chow funn made from rice or wheat; and soba made from buckwheat. Their composition is compared in Table 20-1.

EXTRUSION COOKING

An extrusion cooker may be considered as a continuous reactor capable of simultaneous transporting, mixing, shearing, and forming of food materials under elevated temperature and pressure at short resistance times (Linko et al, 1984). Changes that take place in HTST extrusion are described by El-Dash (1981), Harper (1981), Hauck (1981), Matson (1982), and Faubion et al (1982).

According to Linko et al (1981), the basic components of an HTST extrusion cooking system include (1) continuous, uniform, and controlled feeding of processing materials to the extruder; (2) availability of equipment to precondition the materials with steam at controlled temperatures; (3) uniform application of steam and/or water; (4) an extrusion assembly for process materials; (5) temperature control during the whole process; (6) control of residence time in the extruder to optimize temperature, shear, and agitation; (7) control of exudate shape and size; and (8) availability of equipment to dry, cool, size and treat the product through the addition of flavors, vitamins, fats, etc. A typical arrangement is shown in Figure 20-5;

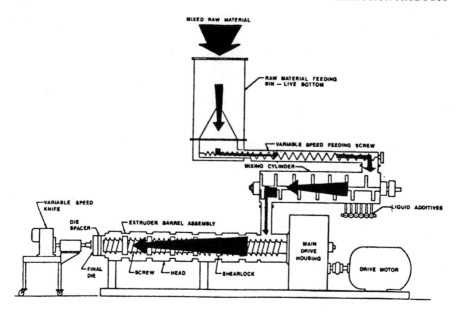

Figure 20-5. Typical arrangement of line bin feeder, preconditioner, HTST extrusion cooker. (Source: Wenger Mfg. Co., Inc., Sabetha, KS; from Smith and Ben-Gera, 1980.)

a multiple extrusion process is shown in Figure 20-6. Extrusion can be used for production of foods, feeds (eg, pet foods, fish feed, and gelatinized cereals for ruminants), and products for industrial purposes (eg, pregelatinized or modified starches and flours). The extruded foods include breakfast cereals and snacks; fortified cereal extrudates; instant or quick-cooking noodles or pasta; alimentary pastes from nonwheat flours; crackers, wafers, crisp-bread products; extruded flour for baking; dietetic foods; confectionery products; linear oligosaccharides for baby foods; engineered foods (ie, 60% gluten, 12.5% wheat flour, 25% soy grits, and a small amount of additives); pregelatinized and extrusion-modified starches, flours, and grits; and miscellaneous foods (instant soup mixes, precooked instant beverage powders, milk substitutes, croutons, breadings, high-fiber products).

According to Linko et al (1984), starch gelatinization at low moisture levels allows both the pretreatment of cereal starches and flours as intermediates to be used in bread and cake baking, and direct processing of cereal-based materials into crispy flat breads, biscuits, crumbs, etc. Extrusion processing conditions also may be adjusted for inactivation of enzymes and use of high alpha-amylase activity flours in bread making. Extrusion cooking can be used in the production of high-fiber breads, specialty breads, breads from sprouted grain, products from wheats of inferior bread-making quality, and high-ratio cakes (without flour chlorination) (as

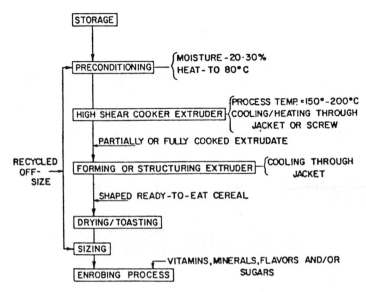

Figure 20-6. Typical ready-to-eat cereal process utilizing extrusion. (From Tribelhorn and Harper, 1980.)

described by Jowitt, 1984; Linko et al, 1984; Zeuthen et al, 1984). The bread-making and extrusion cooking processes are compared in Figure 20-7.

The effects of food components on performance in HTST extrusion were described by Linko et al (1981) and El-Dash (1981). The composition and particle size of the ingredients to be extruded have a marked effect on the characteristics of the extrudate (El-Dash, 1981). Water-solubility index (WST) and water absorption index (WAI) decrease as percent of protein increases. Expansion was highest when gluten protein was 16%; soy concentrate reduced expansion. A high fat content had a deleterious effect on expansion. Extrusion temperature can modify expansion, WAI, and WSI as a function of the starch or flour amylose content.

According to Linko et al (1981), moisture content is of greater concern in single-barrel than in double-barrel extruders; in the latter, the effect is pronounced in materials below 20% moisture. Similarly, particle size may be critical in single-barrel extruders; if the material is very coarse, the starch may not be well gelatinized; if the material is very fine, it may cause choking. Fine particles may overheat; the hazard is reduced if the moisture is low. Coarse particles (above 1.2 mm) may be undertreated, but less so if the moisture is high. Emulsifiers and surface-active agents act at the 0.1% level as lubricants and starch-complexing agents; they influence stickiness, expansion, and carbohydrate solubility and digestibility. Fats and oils up to 3% lubricate but may weaken dough and affect viscosity. Sugars and syrups may affect texture, and NaCl improves expansion and cohesion. The effects are not always predictable, and they are affected by the overall

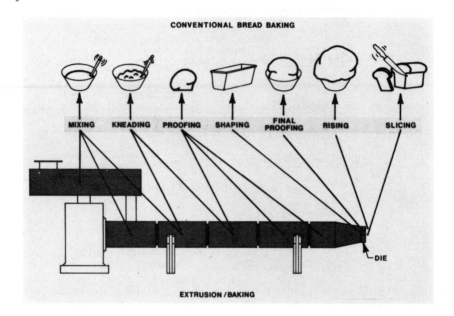

Figure 20-7. Comparison of bread-baking process and extrusion cooking process. (Courtesy Wenger Mfg. Co., Inc., Sabetha, KS.)

composition, moisture, and conditions of extrusion, equipment, etc. In general, expansion is governed by changes following starch gelatinization. Expansion depends on starch content, type of starch, temperature, and moisture content; it decreases as moisture increases and is maximum at 15% moisture. High fat (above 5%) reduces expansion. It was suggested (Linko et al, 1981) that hard red winter wheat flour, wheat middlings, and bran produced a dense, fully cooked breakfast cereal and that a low-protein soft red winter flour in combination with corn starch, white corn flour, and/or rice flour produced a light, frothy, and crisp product. The mixture should be low in gluten to encourage expansion and low in fat to reduce hardness. For cohesiveness and mechanical strength, use of material that was not denatured extensively before extrusion was recommended.

HTST extrusion products are available in various forms as snacks and in breakfast cereals as part of a great array of convenience foods that are available to the consumer. Practically all of today's ready-to-eat cereals are fortified. They provide in a single-ounce serving 25–100% of the daily recommended requirements of several vitamins and minerals. Several breakfast foods are also protein-fortified. When combined with 4 oz of milk, they also provide 15–20% of the daily recommended requirements of high-quality protein. An example of nutrition information on a package of cereals is given in Figure 20-8.

**NUTRITION INFORMATION
PER SERVING**

SERVING SIZE: One ounce (1⅓ cup) Corn
flakes alone and in combination with
½ cup vitamin D fortified whole milk.

SERVINGS PER CONTAINER: 12

	CORN FLAKES	
	1 oz.	with ½ cup whole milk
CALORIES	110	190
PROTEIN	2 gm	6 gm
CARBOHYDRATES	24 gm	30 gm
FAT	0 gm	4 gm

**PERCENTAGE OF U.S. RECOMMENDED
DAILY ALLOWANCE (U.S. RDA)**

	CORN FLAKES	
	1 oz.	with ½ cup whole milk
PROTEIN	2	10
VITAMIN A	25	25
VITAMIN C	25	25
THIAMINE	25	25
RIBOFLAVIN	25	35
NIACIN	25	25
CALCIUM	*	15
IRON	10	10
VITAMIN D	10	25
VITAMIN B$_6$	25	25
FOLIC ACID	25	25
PHOSPHORUS	*	10
MAGNESIUM	*	4

*Contains less than 2 percent of the U.S. RDA
of these nutrients.

Figure 20-8. Example of nutrition information on cereal package.

One of the oldest food forms of wheat, bulgur, has been reintroduced to the modern market. Bulgur, which originated in the Middle East, is wheat that has been parboiled, dried, and partially debranned. Bulgur tastes like wild rice.

BIBLIOGRAPHY

Banasik, C.J. *Cereal Foods World* **1981**, *26*, 166–169.
Banasik, C.J.; Dick, J.W. *Cereal Foods World* **1982**, *27*, 553–557.
Banasik, O.J.; Haber, T.A., Seyam, A.M. *Macaroni J.* **1976**, *58*, 18.
Dexter, J.E.; Matsuo, R.R. *Cereal Chem.* **1977**, *54*, 882.
Dexter, J.E.; Matsuo, R.R. *Cereal Chem.* **1978**, *55*, 44–57.
Dexter, J.E.; **1978** Matsuo, R.R. *J. Agric. Food Chem.* **1980** *28*, 899–902.
Dexter, J.E.; Dronzek, B.L.; Matsuo, R.R. *Cereal Chem.* **1978**, *55*, 23.
Dexter, J.E.; Matsuo, R.R.; Morgan, B.C. *J. Food Sci.* **1983**, *48*, 1545–1551.
Douglass, J.S.; Matthews, R.H. *Cereal Foods World* **1982**, *27*, 558–561.
El-Dash, A.A. In "Cereals—Renewable Resource, Theory and Practice." Pomeranz, Y.; Munck, L., Eds. Am. Assoc. Cereal Chem.; St. Paul, MN; **1981**, pp. 165–217.
Faubion, J.M.; Hoseney, R.C.; Seib, P.A. *Cereal Foods World* **1982**, *27*, 212–214, 216.
Fraase, R.G.; Walsh, D.E.; Anderson, D.E. "An Analysis of the Economic Feasibility of Processing Pasta Products in North Dakota." ND Agric. Exp. Station Bull. No. 496; **1974**.
Harper, J.M. "Extrusion of Foods," Vols. I, II. CRC Press; Boca Raton, FL; **1981**.
Hauck, B.W. *Cereal Foods World* **1981**, *26*, 170–173.
Grzybowski, R.A.; Donnelly, B.J. *J. Agric. Food Chem.* **1979**, *27*, 380–384.
Hummel, C. "Macaroni Products—Manufacture, Processing, and Packaging," 2nd ed. Food Trade Press; London; **1966**.

Jowitt, R. (Ed.) "Extrusion Cooking Technology." Elsevier Applied Science Publishers; London; **1984**.

Kim, J.C.; Rottier, W. *Cereal Foods World* **1980**, *24*, 62.

Linko, P.; Colonna, P.; Mercier, C. *Adv. Cereal Sci. Technol.* **1981**, *4*, 145–235.

Linko, P.; Antila, J.; Linko, Y.-Y.; Mattson, C. *Proc. Symp. Adv. Baking Sci. Technol.* Kansas State University; **1984**, pp. Q1–Q12.

Matson, K. *Cereal Foods World,* **1982**, *27*, 207–210.

Matsuo, R.R.; Dexter, J.E.; Dronzek, B.L. *Cereal Chem.* **1980**, *55*, 744.

Mercier, C. *Getreide Mehl Brot* **1980**, *34*, 52.

Smith, O.B.; Ben-Gera, I. In "Food Process Engineering," Vol. I: "Food Processing Systems." Linko, P.K.; Malkki, Y.; Olkku, J.; Larinkari, J., Eds. Applied Science Publ.; London; **1980**.

Tribelhorn, R.E.; Harper, J.M. *Cereal Foods World* **1980**, *25*, 134–136.

U.S. Food and Drug Administration. Code of Federal Reg. Title 21, Part 139, Macaroni and Noodle Products. U.S. Government Printing Office: Washington, D.C.; **1981**.

Zeuthen, P., et al (Eds.). "Thermal Processing Quality of Food." Elsevier Applied Science Publ.; London; **1984**.

Chapter 21

Industrial Uses of Cereals

CEREALS AS A SOURCE OF CHEMICALS

Traditionally, cereals are considered food and feed grains. However, large and widespread production beyond the demands of the principal markets have encouraged exploration of industrial uses of cereals (Pomeranz, 1973; Pomeranz and Munck, 1981). Cereal grains, milled products, and by-products of milling are finding increasing use in a variety of nonfood applications.

About 1 ton of wheat straw is produced along with 1 ton of wheat grain. Total production of straw pulp on a worldwide basis is about 14 million metric tons. By using better collection methods, the straw potential could be increased to 1 billion metric tons. Only 10% of the world's straw used for pulp production would be about 30 million tons of pulp. But for such an increase there must be economic incentives and more economical methods for collecting, handling, transporting, and storing straw so that the price at the mill could be more competitive with prices of such other raw materials as hardwoods.

Residues and by-products of cereals are excellent sources of furfural, a basic raw material in many industrial technologies. The potential furfural content ranges from 12% in rice hulls to 22% in oat hulls and 23% in corn cobs.

Cereal biomass can be processed several ways. It can be pyrolyzed to produce sugar, olefinic compounds, charcoal, and gaseous fuel. Carbohydrates can be hydrolyzed to fermentable sugars to make ethanol and then converted into ethylene and butadiene. Biomass can be digested by anaerobic bacteria to produce methane. Biomass can be reacted with carbon monoxide, using heat, pressure, and a catalyst to produce an oil. Lignin can be used to make phenol in benzene production.

Cereal polymers (carbohydrates and proteins) can be converted into monomers, which can provide the raw material and flexibility as basic raw materials. Equally promising are novel uses of undegraded polymers. Examples are modification of starch and graft polymerization to give novel polymers; utilization of the unique structure of lignin to produce special, thermosetting-like types of resins; utilization of cereal proteins (ie, zein, gluten) in fiber production; and production of new plasticizers, lubricants, and surface-active agents.

Most industrial uses of cereals depend on the properties of the main component, starch, which ranks in many developed countries as a major industrial chemical. To meet specific requirements the properties of starch (and in many cases of flours) have been modified by physical treatments or through chemical and/or enzymatic reactions.

Starch can be used as a source of polyols, a basic raw material in the production of rigid urethane foam for insulation, paints, and surfactant cleaning agents; in plastics to achieve various degrees of water solubility and biodegradability; and in rubber to process crude latex to finished products. Among recent developments are the microbial polysaccharides, which entail the transformation of starch into new polymers. Less complete chemical transformation of the starch molecule but marked change in properties have been achieved by graft polymerization of petroleum-derived monomers onto the starch molecule. Interesting discoveries significant for industrial application of cereal products lie in the field of rubber technology. Xanthide derivatives of starch are excellent rubber reinforcers, functioning similarly to carbon black, but giving translucent rather than opaque rubbers. Rubbers in powdered form can be prepared by coprecipitation of rubber latices and starch xanthide. The xanthide encases the rubber particles and prevents their sticking together to form large aggregates.

Raw materials and technology exist for basing a portion of the chemical industry on four fermentation products: ethanol, isopropanol, n-butanol, and 2,3-butanediol. Comparing a biochemical industry and one based on coal favors fermentation in several ways. Fermentation would put smaller pressure on fossil reserves of raw materials; would use simpler technologies and relatively larger numbers of semiskilled workers; would be most economical in small dispersed units, would not require large installations, and would not create major environmental problems. And its by-products can be used as fertilizers or feeds whereas by-products of synthetic fuels may be toxic and present health and environmental hazards. To make the fermentation route attractive, however, the current price of fermentable biomass must be substantially reduced.

Industrial uses constitute an economic market for grain ethanol, one in which the product is competitive with ethanol derived from petroleum and natural-gas liquids. The industrial market for ethanol could grow significantly, mainly by use of grain ethanol as an intermediate in the production of chemicals now derived from petroleum and natural gas.

TOTAL UTILIZATION OF CEREALS

Several by-products from processing cereals are widely used in commercial feeds. Some of the materials are dry and can be stored and used inexpensively. Examples are screenings from wheat cleaning, by-products of flour milling, by-products of cornmeal or cornflakes production, by-products of wet corn milling, brewers' and distillers' by-products, surplus yeast from brewing, and by-products of spent grains from brewing.

Industrial (nonfood) uses comprise a small percentage (about 5% in the U.S.) of dry-milled corn products. The main uses (in decreasing order) are as animal feeds, as adjuncts in brewing (as a source of fermentable sugars), in breakfast cereals, and in other foods.

The main nonfood uses are in decreasing order: as pregelled corn flours in foundry cores and molds; in ore refining; in building materials (including gypsum boards and fiber boards); as binders for charcoal briquettes; in fermentations (including processes that include vitamins, antibiotics, etc.); in oil well drilling muds; as additives, adhesives or binders in paper, paperboard, and corrugated products; and in a great variety of products that range from carriers of vitamins or of pesticides to binders in explosives.

Wheat or low-grade wheat flours can be separated into many products that find wide applications.

Wheat flours: in paper sizing and coating (cationic, carbamoylethylated, acid-modified, or enzyme-modified); as adhesive or laminating mixtures in corrugated box boards, paper bonding, plywood industry, decorative woods and veneers, detergent formulations.

Starch: in paper sizing and coating (cationic, xanthates and xanthides, acid-modified, mechanically modified, enzyme-modified), fiber or textile finishing, printing mixtures, paper bonding, adhesives, plywood industry, alcohol production.

Gluten: in paper manufacture (epoxidized or carbamoylated), surface-active agents, adhesives, monosodium glutamate and glutamic acid, edible and/or soluble packaging fabrics, coatings, gums, sausage casings.

Wheat germ (as such or as oil): in production of antibiotics, vitamins, pharmaceuticals, skin conditioners.

Wheat bran: in production of furfural, in the production and/or carriers of enzymes, antibiotics, and vitamins.

After separation in the wet-milling industry, 1 bu (56 lb) of corn yields 1.7 lb corn oil, 3.0 lb 60% protein corn gluten meal, 14.5 lb 21% protein corn gluten feed, 2.5–2.6 gal of 200-proof ethanol, and 15 lb CO_2. Thus, all parts of the corn can be used except the waste heat from the corn milling/alcohol plant, and that is being looked into as a heat source for "greenhouses" with CO_2 fed in to accelerate growth and maturity of vegetables grown hydrophonically. Such integrated approaches are a must and have the greatest potential to make cereals and other biomass economically feasible, renewable resources.

CEREALS AS A SOURCE OF ENERGY

Residues after the harvest of crops have been proposed as an energy source. Wheat cropping typically produces 2.5 tons of residue per acre, corn about 2.9 tons, and sorghum 4 tons. The benefits are limited by high energy costs to collect, transport, and process the residues. The energy to collect 1 ton of corn residues or wheat straw is 43,100 kcal and 50,500 kcal, respectively.

The conversion of glucose to ethanol is represented by the formula

$$C_6H_{12}O_6 \quad \rightarrow \quad 2C_2H_5OH + 2CO_2$$

180 g	92 g
673 kcal	655 kcal
12.88 lb	1 gal (84,356 BTU)

Thus, the conversion of solid sugar to a liquid form involves a loss or reduction of weight by one-half with the concomitant loss of only 10% of the energy. The alcohol produced can be used as fuel.

The production of alcohol as a source of fuel has been the subject of many investigations. In the United States, Canada, South Africa, Australia, and Europe several companies engage in production of alcohol from cereals. It has been estimated that converting the entire U.S. production of corn, wheat, barley, oats, grain sorghum, and the indigenous sugar crop would yield ethanol equivalent to 15% of the U.S. gasoline consumption. Some surveys concluded that ethanol production uses more energy than it produces. Differences of opinion on the energy balance derive mainly from variations in interpretation.

The result depends strongly on assumptions about use of crop residues for fuel and the rating of gasohol (a 90:10 mixture of fuel and alcohol). In terms of total nonrenewable energy, "gasohol" is close to the energy break-even point. On the other hand, in terms of petroleum or petroleum-suitable energy, "gasohol" is an energy producer, as most energy inputs into the process can be supplied by nonpetroleum sources like coal.

BIBLIOGRAPHY

Pomeranz, Y. (Ed.). "Industrial Uses of Cereals." Am. Assoc. Cereal Chem.; St. Paul, MN; **1973**.

Pomeranz, Y.; Munck, L. (Eds.). "Cereals—A Renewable Resource; Theory and Practice." Am. Assoc. Cereal Chem.; St. Paul. MN; **1981**.

Appendix A

WHEAT—HARD RED SPRING*

Analytical results for wheat are reported on a 13.5% moisture basis, and for flour on a 14.0% moisture basis. The AACC methods cited are those of the American Association of Cereal Chemists given in *Cereal Laboratory Methods*, Eighth Edition, 1983. The ICC methods are those of the International Association of Cereal Chemistry.

Sampling Cargoes

As the grain is loaded onto a vessel at a terminal elevator, a continuous series of samples is taken by a mechanical grain sampler at the point of discharge of the grain from the elevator. These small samples are bulked together, and then thoroughly mixed and subdivided using a Boerner or a Dean Gamet Precision Sample Divider to obtain a representative portion of the whole parcel discharged from the elevator to the vessel. The Grain Inspection Division of the Canadian Grain Commission checks the sample for grade, test weight and moisture content. A representative portion of the sample is retained by the Inspection Division for the official loading sample for the cargo, and another representative portion of the sample is sent to the Grain Research Laboratory for the determination of protein content and for compositing of weighted grade average samples on which milling, baking and analytical tests are made. The grades No. 1 and No. 2 C.W. red spring wheat are exceptions to this procedure. For these two grades, which are segregated by protein content to provide shipments of guaranteed minimum protein levels, the individual samples representing the grain loaded onto a vessel during prescribed time intervals, are thoroughly mixed and checked for protein content at the seaport using near infrared reflectance spectroscopy. The protein results for these individual cargo-loading portions are subsequently verified by the Kjeldahl

*Courtesy of Grain Research Laboratory, Canadian Grain Commission.

procedure. These loading portions are used by the laboratory at the end of each quarter of the crop year in preparing the weighted grade average sample on which the quality data are obtained.

Test Weight

The weight per hl is determined with the Schopper Chondrometer using the 1 l container.

Weight per thousand kernels is determined with an electronic seed counter using 20 g of grain from which all broken kernels and foreign material have been previously removed by hand-picking. The calculated weight of 1000 kernels is reported on a 13.5% moisture basis.

Moisture

For wheat: the determination is made by the Inspection Division at regular intervals during loading of individual cargoes, and by the laboratory on both individual and grade composite samples, using the Model 919 moisture meter calibrated against the AACC method (two-stage 130°C air-oven method). For flour: a 10g sample is heated for 1 hour in a semi-automatic Brabender oven at 130°C. Results are reported as percent.

Protein

(N × 5.7) is determined on a 1g sample by the AACC method. Results are reported as percent.

Alpha-amylase

Activity of wheat and of flour is determined by the method of Kruger and Tipples. *Cereal Chem.*, 58:271–274, 1981.

Falling Number

This is determined on a 7g sample of ground wheat by the method of Hagberg (*Cereal Chem.*, 38:202–203, 1961). Wheat (300g) is ground in a Falling Number Laboratory Mill 3100 (ICC Standard Method No. 107). Results are reported in seconds.

Milling

This is carried out in an Allis-Chalmers Laboratory mill using the GRL sifter flow as described by Black et. al., *Cereal Foods World*, 25:757–760, 1980.

Wet Gluten

Ten g of flour and 6 ml of distilled water are mixed by hand for about 2 minutes. The dough is then washed for 12 minutes in a Theby Gluten Washer using a salt-phosphate buffer of pH 6.7; this is followed by 2 minutes hand-washing. The resulting gluten is worked between the fingers until it becomes tacky, and is then weighed. Results are reported as percent.

Ash

This is determined on a 4g sample in a silica dish incinerated overnight at 585°C. After cooling, the dish and ash are weighed, the ash brushed out, the dish reweighed, and the weight of ash determined by difference. Results are reported as percent.

Flour Color

A color index is obtained with the Kent-Jones and Martin Flour Color Grader which gives the relative reflectance (with filter No. 58) of a flour-water slurry. Results are reported as arbitrary scale units; the lower the number the brighter the flour.

Starch Damage

This is determined on a 5g sample of flour by the method of Farrand (*Cereal Chem.*, 41:98–111, 1964). Results are reported in Farrand units.

Amylogram

Sixty-five g of flour (14% moisture basis) and 450 ml distilled water are used with the Brabender Amylograph and the pin stirrer; other details are as in the AACC method. Peak viscosity is reported in Brabender Units.

Gassing Power

This is determined on a 10g sample by the AACC Pressuremeter method using a modified pressuremeter (*Trans. Am. Assoc. Cereal Chem.*, 13: 147–151, 1955). Results are reported as ml pressure for total gas evolved after 6 h fermentation.

Farinogram

Fifty g of flour (14% moisture basis) are mixed in a small stainless steel farinograph bowl (63 rpm drive) for 15 minutes with sufficient distilled water to give a maximum dough consistency centered about the 500 Brabender Unit line. Farinograph absorption is the amount of water which must be added to a flour of 14.0% moisture to give the required consistency, and is reported as percent. Dough development time is the time required for the curve to reach its maximum height.

Extensigram

Doughs are made from 300g flour (14.0% moisture basis), 6g salt, and distilled water equal to the farinograph absorption less 2.0 percentage units to compensate both for the salt and for the substitution of the large stainless steel farinograph bowl. Doughs are mixed for 1 minute, rested for 5 minutes, and mixing is then continued until the curve is centered about the 500 Brabender Unit line. Curves are drawn for duplicate doughs at 45 and at 135 minutes though doughs are also rounded and shaped at 90 minutes. Average curves for 45 and 135 minutes are reproduced, but measurements (length in centimeters, height in Brabender Units, and area in square centimeters) are reported only for the 135 minutes curve (solid line). The Extensigraph is set so that the 100 Brabender Units equal a 100 g load.

Alveogram

The ICC Standard Method No. 121 is followed.

Baking

This is carried out by the Remix baking test procedure of Irvine and McMullan. *Cereal Chem.*, 37:603–613, 1960, as described in detail by Kilborn and Tipples. *Cereal Foods World*, 26:624–628, 1981.

Appendix B

WHEAT—DURUM*

Analytical results for wheat are reported on a 13.5% moisture basis, and for semolina on a 14.0% moisture basis. The AACC methods cited are those given in *Cereal Laboratory Methods,* Seventh Edition, 1962.

Sampling Cargoes

The grain is sampled continuously by an automatic sampler to obtain a representative portion of the whole parcel as it is transferred from the elevator to the vessel. Each individual cargo sample is thoroughly mixed and subdivided with a Dean Gamet Precision Sample Divider. The Grain Inspection Division of the Canadian Grain Commission checks the sample for grade, test weight, and moisture content. Part of the sample is retained by the Inspection Division for the official loading sample for the cargo, and part of the sample is forwarded to the laboratory. Subsequent subsampling, and thorough mixing of the composite sample for each grade on which milling, pasta processing, and analytical tests are made is done with a Boerner Sample Divider.

Test Weight

The weight per hl is determined with the Schopper Chondrometer using the 1 l container.

Weight per thousand kernels is determined with an electronic seed counter using 20g of grain from which all broken kernels and foreign material have been previously removed by hand-picking. The calculated weight of 1000 kernels is reported on a 13.5% moisture basis.

*Courtesy of Grain Research Laboratory, Canadian Grain Commission.

Moisture

For wheat: the determination is made by the Grain Inspection Division on individual cargoes and by the laboratory on average samples using the Model 919 Moisture Meter calibrated against the AACC method (two-stage 130°C air-oven method). For semolina: a 10g sample is heated for 1 h in a semi-automatic Brabender oven at 130°C. Results are reported as percent.

Vitreous and "Vulgare" Kernels

These determinations are made by the Grain Inspection Division on a 50g sample of clean wheat. The vitreous kernels are hand-picked and weighed. These kernels are returned to the original sample, and the "vulgare" kernels are then separated and weighed. Results are reported as percent, "as is" moisture basis.

Milling

Wheat is tempered overnight and milled at 16.5% moisture in an Alis-Chalmers laboratory mill. Semolina, corresponding in granulation and dress to a commercial third sizing, is obtained with the aid of a small-scale purifier. Results are reported as extraction percent, wheat basis. The mill room is controlled for temperature (22°C) and humidity (60%).

Protein

Protein (N $\times$ 5.7) is determined on a 1g sample by the AACC method. Results are reported as percent.

Wet Gluten

Ten g of semolina and 5 ml of distilled water are mixed by hand for about 2 minutes. The dough is then washed for 10 minutes in a Theby Gluten Washer using a salt-phosphate buffer of pH 6.7; this is followed by hand-washing for 2 minutes. The resulting gluten is worked between the fingers until it becomes tacky; it is then weighed. Results are reported as percent, 14% moisture basis.

Ash

This is determined on a 4g sample in a silica dish incinerated overnight at 585°C. After cooling, the dish and ash are weighed, the ash brushed out, the dish reweighed, and the weight of ash determined by difference. Results are reported as percent.

Yellow Pigment

This is determined on an 8g sample of semolina or of ground spaghetti, extracted overnight with 40 ml water saturated *n*-butyl alcohol. After filtering the extract (No. 1 Whatman paper), light transmission is determined in a spectrophotometer at a wavelength of 435.9 nm. Concentration is calculated on the basis of beta-carotene, and results are reported as parts per million.

Spaghetti

Fifty g of semolina with sufficient distilled water to bring to 33.0% absorption is mixed in a 50g farinograph bowl, as described in *Cereal Chemistry*, Vol. 49, No. 6, 707–711, 1972. Spaghetti is dried with a controlled decrease in relative humidity for 28 hours at 39°C and at the high 70°C temperature used by many processors.

Spaghetti Color

Whole strands of spaghetti are mounted on white cardboard for color measurements. Dominant wavelength, purity, and brightness are determined, using the Ten Selected Ordinates method, in a Beckman, Color DB-G Spectrophotometer (*Handbook of Colorimetry*, A.C. Hardy, The Technology Press, Massachusetts Institute of Technology, Cambridge, 1936).

Farinogram

Fifty grams of semolina (14% moisture basis) is mixed with distilled water (31.5% absorption) in a small stainless steel farinograph bowl (59 rpm drive), using the rear sensitivity setting. Mixing time, reported in minutes, is the time required to reach the peak of the curve.

Appendix C

BARLEY*

Samples

Samples of new-crop barley are submitted to the research laboratory by the elevator managers of all firms operating licensed elevators in the Prairie Provinces. Only samples qualifying for the Six-row, the Two-row, and the No. 1 feed grades are selected for inclusion in the survey; the number of samples selected is proportional to Statistics Canada's August estimates of production in each crop district. Selection starts when harvesting begins and continues until the production-based quota of samples has been met for each crop district.

Test Weight

The weight per hl is determined with the Schopper Chondrometer using the 1 l container.

Moisture

For whole barley: determined with a Tag-Heppenstall moisture meter using the Grain Research Laboratory's conversion chart. For ground material: determined by the ASBC air-oven method (135°C). Results are reported in percent.

Plump Kernels

These are determined by passing a sample over a 6/64″ by 3/4″ slotted sieve. Barley remaining on top of the sieve is reported as percent plump barley.

*Courtesy of Grain Research Laboratory, Canadian Grain Commission.

Protein

Protein (N × 6.25) is determined on a 1g sample by the procedure of the American Association of Cereal Chemists, *Cereal Laboratory Methods*, Seventh Edition, 1962 with the modifications described in *Laboratory Practice*, 22:38–39, 1973 and in the *Journal of the Science of Food and Agriculture*, 24:343–348, 1973. Results for barley are reported as percent, dry basis. To determine the amount of soluble protein of malt extract, 20 ml of extract is evaporated down to a thick syrup; protein analysis is the same as for barley and results are reported as percent, dry basis.

Saccharifying Activity

The saccharifying activity of barley is determined on a 2.5g sample by the method in *Canadian Journal of Agricultural Science*, 35:252–258, 1955.

Diastatic Power

The diastatic power of malt is determined by the ASBC method for diastatic activity. Results for both barley and malt are reported as degrees Lintner, dry basis.

Malt Extract

The malt extract is determined on a 50g sample using an Anton-Parr DMA 50 density meter as described in *Journal of American Society of Brewing Chemists*, 37:105–106, 1979. Results are reported as percent, dry basis.

Subject Index